MW01378233

Cloud-Based Design and Manufacturing (CBDM)

Dirk Schaefer
Editor

Cloud-Based Design and Manufacturing (CBDM)

A Service-Oriented Product Development Paradigm for the 21st Century

Editor
Dirk Schaefer
George W. Woodruff School of Mechanical
Engineering
Georgia Institute of Technology
Atlanta, GA
USA

ISBN 978-3-319-07397-2 ISBN 978-3-319-07398-9 (eBook)
DOI 10.1007/978-3-319-07398-9
Springer Cham Heidelberg New York Dordrecht London

Library of Congress Control Number: 2014942267

Printed on acid-free paper

Springer is part of Springer Science+Business Media (www.springer.com)

Preface

Cloud-Based Design and Manufacturing (CBDM) refers to a new service-oriented product realization paradigm for the twenty-first century in the broader context of distributed and collaborative product development. CBDM fosters knowledge and resource sharing as well as highly efficient rapid product development with reduced cost through social networking and negotiation platforms that exist between service providers and consumers. A design and manufacturing cloud is a collaborative and distributed system consisting of a collection of interconnected physical and virtualized service pools of design and manufacturing resources, as well as associated search and retrieval capabilities. CBDM systems are anticipated to become the backbone of future intelligent and semantics-based Web 3.0 applications for design and manufacturing in the broader context of Social Product Development.

The purpose of this book is to provide an introductory overview of one of the most topical developments in the context of advanced design and manufacturing.

As the title suggests, "Cloud-Based Design and Manufacturing: Status and Promise" gives an overview of the current status and promise of CBDM. First, Wu et al. introduce their definition and vision of CBDM. This is followed by a discussion of the characteristics of CBDM systems as well as the similarities and differences between CBDM and more traditional paradigms, such as web- and agent-based approaches. The chapter continues with the presentation of a CBDM prototype system developed at Georgia Tech and concludes with an outline of current and future research directions in the context of CBDM.

"Multi-User Computer-Aided Design and Engineering Software Applications" discusses multi-user Computer-Aided Design and Engineering software applications as a new paradigm for product development, considering past collaborative research and the emerging wave of cloud-based social and gaming tools. Red et al. consider how multi-user architectures will change the single-user paradigm from serial to simultaneously collaborative, promote new on-demand access methods like cloud serving, and bring long hoped for efficiencies to product development.

In "Distributed Resource Environment: A Cloud-Based Design Knowledge Service Paradigm", a cloud-based design knowledge service paradigm is introduced. Zhang et al. propose a distributed resource environment, which enables companies to utilize collective open innovation and rapid product development with reduced cost. Definition, functionality, structure, and characteristics of their distributed resource environment are presented, followed by a cloud-based knowledge service framework for managing knowledge sources in distributed environments.

"Research and Applications of Cloud Manufacturing in China" sheds some light on the impact of cloud manufacturing on the manufacturing industry as a whole. Zhang et al. consider potential impacts of cloud manufacturing in the context of advanced manufacturing, intelligent manufacturing, sustainable manufacturing, agile manufacturing, and personalized social production modes.

In "Future Manufacturing Industry with Cloud Manufacturing", Li et al. provide a comprehensive overview of cloud manufacturing-related research and development activities in China and provide a snapshot of the state of the art.

"Enabling Product Customisation in Manufacturing Clouds" proposes a concept and architecture to enable the dynamic customization of products based on the availabilities of the production network from the cloud manufacturing concept of Manufacturing-as-a-Service (MaasS). Yip et al. provide an overview of MaaS and a related architecture, which includes core components for product configuration, manufacturing service management, and the integration of factory IT-systems.

In "A Manufacturing Ontology Model to Enable Data Integration Services in Cloud Manufacturing using Axiomatic Design Theory", Valilai and Houshmand propose and discuss a manufacturing ontology model aimed at enabling data integration services in cloud manufacturing environments, based on axiomatic design theory.

"Distributed, Collaborative and Automated Cybersecurity Infrastructures for Cloud-Based Design and Manufacturing Systems" is dedicated to Cybersecurity in the context of CBDM. Thames provides an overview of emerging global-scale cyber information exchange frameworks that will enable cybersecurity in future CBDM environments. In addition, a reference architecture utilizing information obtained from global cyber exchange for dynamic cyber protection of CBDM systems is proposed.

The book concludes with "Teaching Creativity in Design Through Project-Based Learning in a Collaborative Distributed Educational Setting", in which Ito et al. present a case study on teaching creativity in a distributed cloud-based project-based learning environment.

CBDM is a new and exciting paradigm anticipated to significantly impact and reshape product development in distributed collaborative settings. The utilization of CAD and CAE software as a service through the cloud was only the beginning. Cloud-based design used in concert with cloud-based 3D printing services quickly led to the formation of the so-called "makers movement," which is referred to as a New Industrial Revolution. One can anticipate that over the next five years additional types of manufacturing systems and services on a large scale will be provided and utilized through cloud-based environments. Since this exciting field is just at its infancy, a lot is yet to be discovered both in terms of fundamental research and potential application fields.

This book is the first collection of works related to various aspects of CBDM. I hope you find it informative, and perhaps it will spark new ideas and visions that you, the reader, will be sharing with the world in a future publication.

Atlanta, GA, USA
Winter 2013/14

Dirk Schaefer

Contents

Contributors

Xudong Chai Beijing Simulation Center, Beijing, China

Jonathan Corney Department of Design, Manufacture and Engineering Management, University of Strathclyde, Glasgow, UK

David French Brigham Young University, Provo, UT, USA

Nevan C. Hanumara Department of Mechanical Engineering, Massachusetts Institute of Technology, Cambridge, MA, USA

Ammon Hepworth Brigham Young University, Provo, UT, USA

Baocun Hou Beijing Simulation Center, Beijing, China

Mahmoud Houshmand Advanced Manufacturing Laboratory, Department of Industrial Engineering, Sharif University of Technology, Tehran, Iran

Anrui Hu School of Automation Science and Electrical Engineering, Beihang University, Beijing, China

Pei Huang DG-HUST Manufacturing Engineering Institute, Beijing, China

Ralph C. Huntsinger The California State University—Chico Campus, Chico, CA, USA

Tetsuo Ichikawa Institute of Health Bioscience, The University of Tokushima, Tokushima, Japan

Teruaki Ito Institute of Technology and Science, The University of Tokushima, Tokushima, Japan

Ananda Jagadeesan Department of Design, Manufacture and Engineering Management, University of Strathclyde, Glasgow, UK

Greg Jensen Brigham Young University, Provo, UT, USA

J. Lane Thames Research and Development Department, Tripwire Inc., Alpharetta, GA, USA

Tan Li Beijing Simulation Center, Beijing, China

Bo Hu Li Beihang University (BUAA), Beijing, People's Republic of China; Beijing Simulation Center, Beijing, People's Republic of China; School of Automation Science and Electrical Engineering, Beihang University, Beijing, China; Beijing Simulation Center, Beijing, China

Xiang Li School of Mechanical Engineering, Shanghai Jiao Tong University, Shanghai, China

Tingyu Lin Beijing Simulation Center, Beijing, China

Yonghong Liu SANY Group Co., Ltd, Changsha, China

Jingeng Mai School of Automation Science and Electrical Engineering, Beihang University, Beijing, China; Beihang University (BUAA), Beijing, People's Republic of China

Yi Qin Department of Design, Manufacture and Engineering Management, University of Strathclyde, Glasgow, UK

Ursula Rauschecker Fraunhofer IPA, Stuttgart, Germany

Edward Red Brigham Young University, Provo, UT, USA

Lei Ren School of Automation Science and Electrical Engineering, Beihang University, Beijing, China; Beihang University (BUAA), Beijing, People's Republic of China

David W. Rosen The George W. Woodruff School of Mechanical Engineering, Georgia Institute of Technology, Atlanta, GA, USA

Dirk Schaefer The George W. Woodruff School of Mechanical Engineering, Georgia Institute of Technology, Atlanta, GA, USA

Alexander H. Slocum Department of Mechanical Engineering, Massachusetts Institute of Technology, Cambridge, MA, USA

Brett Stone Brigham Young University, Provo, UT, USA

Fei Tao School of Automation Science and Electrical Engineering, Beihang University, Beijing, China; Beihang University (BUAA), Beijing, People's Republic of China

Omid Fatahi Valilai Advanced Manufacturing Laboratory, Department of Industrial Engineering, Sharif University of Technology, Tehran, Iran

Yongzhi Wang China CNR Corporation Limited, Beijing, China

Dazhong Wu The George W. Woodruff School of Mechanical Engineering, Georgia Institute of Technology, Atlanta, GA, USA

Youbai Xie School of Mechanical Engineering, Shanghai Jiao Tong University, Shanghai, China

Chen Yang Beijing Simulation Center, Beijing, China

Chao Yin Institute of Manufacture Engineering, Chongqing University, Chongqing, China

Arthur L. K. Yip Department of Design, Manufacture and Engineering Management, University of Strathclyde, Glasgow, UK

Xinpei Zhao Beijing ND Tech Corporation Limited, Beijing, China

Lin Zhang Beihang University (BUAA), Beijing, People's Republic of China; School of Automation Science and Electrical Engineering, Beihang University, Beijing, China

Zhinan Zhang School of Mechanical Engineering, Shanghai Jiao Tong University, Shanghai, China

Chun Zhao Beihang University (BUAA), Beijing, People's Republic of China

Zude Zhou Wuhan University of Technology, Wuhan, China

Longfei Zhou School of Automation Science and Electrical Engineering, Beihang University, Beijing, China

Acronyms

AM	Additive manufacturing
CAD	Computer-aided design
CAE	Computer-aided engineering
CAM	Computer-aided manufacturing
CBD/CBM/CBDM	Cloud-based design/Cloud-based manufacturing/ Cloud-based design and manufacturing
DMCloud	Design and manufacturing cloud
FEA	Finite element analysis
HaaS	Hardware-as-a-service
IaaS	Infrastructure-as-a-service
IoT	Internet of things
PaaS	Platform-as-a-service
SaaS	Software-as-a-service
SMEs	Small- and medium-sized enterprises
SNA	Social network analysis
SOA	Service-oriented architecture

Cloud-Based Design and Manufacturing: Status and Promise

Dazhong Wu, David W. Rosen and Dirk Schaefer

Abstract The information technology industry has benefited considerably from cloud computing, which allows organizations to shed some of their expensive information technology infrastructure and shifts computing costs to more manageable operational expenses. In light of these benefits, we propose a new paradigm for product design and manufacturing, referred to as cloud-based design and manufacturing (CBDM). This chapter introduces a definition and vision for CBDM, articulates the differences and similarities between CBDM and traditional paradigms such as web- and agent-based technologies, highlights the fundamentals of CBDM, and presents a prototype system, developed at Georgia Tech, called the Design and Manufacturing Cloud (DMCloud). Finally, we conclude this chapter with an outline of future research directions.

Keywords Cloud-based design and manufacturing · Cloud-based design · Cloud-based manufacturing · Design and manufacturing cloud

1 Introduction

Today's highly competitive global marketplace is refining the way many design and manufacturing companies do business. The force of globalization has connected individuals and organizations from all across the globe, enabling them to

---

D. Wu · D. W. Rosen · D. Schaefer (✉)
The George W. Woodruff School of Mechanical Engineering, Georgia Institute of Technology, 813 Ferst Drive, NW, Atlanta, GA 30332-0405, USA
e-mail: dirk.schaefer@me.gatech.edu

D. Wu
e-mail: dwu42@gatech.edu

D. W. Rosen
e-mail: david.rosen@me.gatech.edu

D. Schaefer (ed.), *Cloud-Based Design and Manufacturing (CBDM)*,
DOI: 10.1007/978-3-319-07398-9_1,

share data, information, and knowledge in a collective manner. Friedman first introduced three major phases of globalization (Friedman 2005): Globalization 1.0, driven by countries and empires; Globalization 2.0, driven by global corporations; and Globalization 3.0, driven by individuals aided by technology. Among the many new technologies driven by globalization, cloud computing is one of the major advances in the field of computing. In recent years, the information technology (IT) sector has significantly benefited from cloud computing through (1) on-demand self-services, (2) ubiquitous network access, (3) rapid elasticity, (4) pay-per-use, and (5) location-independent resource pooling (Linthicum 2009). One particular benefit is that cloud computing allows for faster and more flexible development and implementation of IT solutions compared to traditional infrastructures and service models. Although cloud computing has been adopted in the IT field, it has just emerged on the horizon of other application domains such as product design and manufacturing. As a result, few studies have investigated the potential of cloud computing for the field of product design and manufacturing. Of particular need is a well-accepted definition of cloud-based design and manufacturing (CBDM) along with theoretical frameworks and prototypes that can guide the development of CBDM systems.

To address resource sharing and optimal allocation, Li et al. (2010) introduced cloud-based manufacturing (CBM), also referred to as *cloud manufacturing*, based on cloud computing, SOA, and networked manufacturing. Afterwards, by introducing cloud-based design (CBD) into the context of CBM, Wu et al. (2012, 2014a, b) generalized and extended the original concept of CBM to CBDM. The following definition of CBDM sets the stage for the discussions that follow.

> CBDM refers to a service-oriented networked product development model in which service consumers are able to configure, select and utilize customized product realization resources and services and reconfigure manufacturing systems through Infrastructure-as-a-Service (IaaS), Platform-as-a-Service (PaaS), Hardware-as-a-Service (HaaS), and Software-as-a-Service (SaaS) in response to rapidly changing customer needs. CBDM is characterized by on-demand self-service, ubiquitous access to networked data, rapid scalability, resource pooling, and virtualization. The types of deployment models include private, public, and hybrid clouds.

While research pertaining to CBDM is still in its infancy, several companies are already developing components for commercial CBDM systems. For example, General Electric (GE) and the Massachusetts Institute of Technology (MIT) are jointly developing a crowdsourcing platform to support the ongoing adaptive vehicle make portfolio of the Defense Advanced Research Projects Agency (DARPA) (2013). The new crowdsourcing platform is expected to enable a global community of experts to design and rapidly manufacture complex industrial systems such as aviation systems and medical devices by connecting data, design tools, and simulations in a distributed and collaborative setting. Another frequently quoted example is MFG.com, which connects service consumers that request design and manufacturing services to service providers. Consumers provide technical product specifications and select qualified service providers based on geographic locations, certifications, manufacturing capacity, or a combination of these factors. The above examples are intended to provide an impression of the

Table 1 Service providers and their services

Provider		Service
IaaS	Google drive, dropbox	Online storage, file syncing
	Amazon elastic compute cloud	Virtual machines
PaaS	Microsoft windows azure	Developing and hosting web applications
	Amazon relational database service	Database query system for analysis of large datasets
	Salesforce.com, netsuite	Developing user interfaces and social network sites
HaaS	Ponoko, shapeways	Additive manufacturing
	MFG.com, Quickparts.com	Supplier search engine, cloud-based e-Sourcing
SaaS	Autodesk 360 platform	CAD file editing, mobile viewing, cloud rendering
	Dassault Systemes	3D modeling
	Sabalcore	High performance computing for FEA/CFD

types of CBDM services currently being offered by some of the major players in this field. Additional service providers who offer select services related to IaaS, PaaS, HaaS, and SaaS are listed in Table 1. We further elaborate the discussion of CBDM-related services and business models in Sect. 3.1.

Many emerging research issues related to CBDM need to be addressed, including computing architectures, economic impact analyses, resource allocation and scheduling, service recommender algorithms, distributed process planning, semantic web, cyber security, open standards, negotiation protocols, and so on (Wu et al. 2012). This chapter provides a review of the current state of academic research and industry implementation in the field of CBDM and discusses potential areas of research to help bridge the gap between cloud computing and product design and manufacturing. Section 2 provides a brief history and several well-accepted definitions of cloud computing, and Sect. 3 reviews current and recent research projects in the field of CBDM and examples from industry. Section 4 articulates the fundamentals of CBDM, including a vision of CBDM, CBD, and CBM, two aspects of CBDM, and differences between CBDM and traditional design and manufacturing methods. Section 5 describes a prototype system for CBDM referred to as the Design and Manufacturing Cloud (DMCloud), developed at Georgia Tech. Finally, Sect. 6 concludes with closing remarks and directions for future research.

2 Cloud Computing

To set the stage for CBDM, this section provides an overview of existing definitions for cloud computing as follows:

- "Cloud computing is a model for enabling ubiquitous, convenient, on-demand network access to a shared pool of configurable computing resources (e.g., networks, servers, storage, applications, and services) that can be rapidly provisioned and released with minimal management effort or service provider interaction" (NIST 2011).

- "Cloud computing refers to both the applications delivered as services over the internet and the hardware and systems software in the datacenters that provide those services. The services themselves have long been referred to as (SaaS).... The datacenter hardware and software is what we will call a Cloud" (Armbrust et al. 2010).
- "Clouds are a large pool of easily usable and accessible virtualized resources (e.g., hardware, development platforms, and/or services). These resources can be dynamically reconfigured to adjust to a variable load (scale), allowing also for an optimum resource utilization. This pool of resources is typically exploited by a pay-per-use model in which guarantees are offered by the infrastructure provider by means of customized SLAs" (Vaquero et al. 2009).
- "A cloud is a type of parallel and distributed system consisting of a collection of interconnected and virtualized computers that are dynamically provisioned and presented as one or more unified computing resources based on service-level agreements established through negotiation between the service provider and consumers" (Buyya et al. 2008).
- "Cloud computing is both a user experience (UX) and a business model. It is an emerging style of computing in which applications, data and IT resources are provided to users as services delivered over the network. It enables self-service, economies of scale and flexible sourcing options...an infrastructure management methodology—a way of managing large numbers of highly virtualized resources, which can reside in multiple locations...." (IBM 2010).

While the term *cloud computing* was coined in 2007, the concept behind cloud computing—delivering computing resources through a global network—was rooted during the 1960s (Licklider 1963). The term *cloud* is often used as a metaphor for the Internet and refers to both hardware and software that deliver applications as services over the Internet (Armbrust et al. 2010). When looking backward, one realizes that cloud computing derives from pre-existing and well-established concepts such as the client-server architecture, utility computing, grid computing, virtualization, service-oriented architecture, and SaaS (Bohm et al. 2010). One milestone in the history of computing is utility computing, proposed by John McCarthy in 1966. The idea behind utility computing is that "computation may someday be organized as a public utility." Because of a wide range of computing-related services and networked organizations, utility computing facilitates the integration of IT infrastructures and services within and across virtual companies (Parkhill 1966). Another milestone was that Ian Foster and Carl Kesselman proposed the concept of grid computing in 1999. A computational grid refers to a hardware and software infrastructure that provides dependable, consistent, pervasive, and inexpensive access to high-end computational capabilities (Foster and Kesselman 1999). Since cloud and grid computing share a similar vision, Foster et al. identified the main differences between grid computing and cloud computing (Foster et al. 2008). The most significant difference is that cloud computing addresses Internet-scale computing problems by utilizing a large pool of computing and storage resources whereas grid computing is aimed at large-scale computing problems by harnessing a network of resource-sharing commodity computers and dedicating resources to a single computing problem (Vaquero et al. 2009; Foster et al. 2008).

The innovative nature of cloud computing can be viewed from technical and business perspectives. From a technical point of view, cloud computing is an advancement in computing history that evolved from calculating machines with binary digit systems, and mainframe computers with floating-point arithmetic, to personal computers with graphical user interfaces and mobility, and from the Internet, which offers computing resources via distributed and decentralized client-server architectures, and eventually to utility, grid, and cloud computing (Bohm et al. 2010). From a business point of view, cloud computing is a breakthrough that has changed the mode of IT deployment and service pricing strategies.

3 Cloud-Based Design and Manufacturing

3.1 Recent and Current Research Projects

This section reviews current and recent research initiatives pertaining to CBDM. To the best of our knowledge, the first cloud manufacturing project was funded by China's National High-Tech Research and Development program and National Basic Research Program. The goal of the project was to "realize the general sharing of global manufacturing resources, reduce time-to-market, improve quality of service, as well as reduce manufacturing costs." The cloud manufacturing concept proposed by Li et al. (2010) refers to a service-oriented, knowledge-based smart manufacturing system which encompasses the entire product development lifecycle from market analysis to design, manufacturing, production, testing, and maintenance. Meanwhile, the goal of the ManuCloud project (2010), launched by the European Commission's Seventh Framework Programme (EC FP7) with € 5 million ($\approx$ $6,700,000), is to "develop a service-oriented IT environment as basis for the next level of manufacturing networks by enabling production-related inter-enterprise integration down to shop floor level." Recently, the Engineering and Physical Science Research Council (EPSRC) in the United Kingdom funded a project, titled "Cloud Manufacturing—Towards a Resilient and Scalable High Value Manufacturing" with £2.4 million ($\approx$ $4,040,000). The objective of this research is to "develop a holistic framework and understand its role within global manufacturing networks through: seeking the appropriate products, sectors, scales and volumes; identifying the impacted lifecycle stages from design to manufacture, maintenance and re-cycling; understanding how new product design and manufacturing will be influenced by product lifecycle data; and finally analyzing how future products will be influenced by cloud manufacturing enabling local on-demand supply of components and services."

Another successful project on CBDM conducted in the U.S. is part of the Manufacturing Experimentation and Outreach (MENTOR) program of the DARPA. The MENTOR effort is part of the Adaptive Vehicle Make (AVM) program portfolio. Three teams were awarded contracts for the MENTOR program, including Georgia Tech. The vision for the MENTOR program is to "develop an integrated, distributed design and manufacturing infrastructure that can support a progressive set of prize challenge competitions through integrated CAD, CAE, and CAM tools." The goal of this project is to "engage students from participating high schools in a series of collaborative design and distributed manufacturing experiments" (Rosen et al. 2012). The developed prototype system, design and manufacturing Cloud (DMCloud), builds upon an integrated distributed manufacturing infrastructure with tools such as CNC machine tools and additive manufacturing machines (i.e., 3D printers) through a network of high schools dispersed across the U.S. This prototype system enables students to learn and participate in product development as a continuum of design, analysis, simulation, prototyping, and manufacturing activities (Patel et al. 2013). In the DMCloud, IaaS provides students with a platform virtualization environment along with high-performance computing servers and storage space. PaaS provides students with a ubiquitous computing and development environment. Specifically, the DMCloud is constructed from existing technologies such as Sakai, Moodle, Drupal, Wiggio, and Google Docs. HaaS provides students with a heterogeneous hardware environment including 3D printers, milling machines, lathes, laser cutters, and other CNC machines. Providing students with access to web-based software applications over the Internet, SaaS eliminates the need to install and run software on their own computers. The software includes engineering design, analysis, and simulation tools from Dassault Systems. The DMCloud is currently implemented as a private DMCloud, but it can easily be extended to be a public or hybrid CBDM system. The prototype system, DMCloud, will be detailed in Sect. 5.

3.2 Select Examples of CBDM in Industry

In addition to research projects being conducted in academia, several companies are developing and testing similar commercial systems, most notably in the consumer product industry with rapid prototyping manufacturing resources. These companies utilize cloud-based services as a technology enabling their ventures and connecting designers with manufacturing resources over the Internet. According to *The Economist*, Quirky (2012) offers users access to a complete product creation enterprise. The business model of Quirky incorporates the originating designers into the wealth-sharing model and provides them with a portion of the profits that their products yield. *The Economist* also discusses Shapeways, a company offering 3D printing services over the Internet. In contrast to the vetting process used in the Quirky business model, Shapeways provides users immediate access to 3D printers to build any object that they want. Tables 2, 3, 4 and 5 list some of the example

Table 2 Examples in IaaS

Provider	Service	Price scheme
Rackspace	Internet hosting	Starting at $17/month
Amazon elastic compute cloud (EC2)	Virtual machines	$0.48/h for 4 cores, 15 GB memory
Google compute engine		$0.163/h for 2 cores, 1.8 GB memory
Amazon simple storage service (S3)	Online storage, file syncing	$0.055/GB/month over 5000 TB/month
Google drive		Free for 25 GB; $4.99/month for 100 GB $9.99/month for 200 GB $49.99/month for 1 TB $99.99/month for 2 TB
Dropbox		Free for 2 GB $19.99/month for 100 GB

Table 3 Examples in PaaS

Provider	Service	Price scheme
Google app engine	Developing and hosting web applications	$9/app/month
Microsoft windows azure		Free for up to 60 min of CPU/day, 10 sites, 1 GB storage, 20 MB of MySQL (first 12 months) $0.02/h, up to 240 min of CPU/day, 100 sites, 1 GB storage, 20 MB of MySQL (first 12 months)
Google BigQuery	Database query system for analysis of massively large datasets	$0.12/GB/month, limit: 2 TB $0.035/GB, limit: 20,000 queries/day, 20 TB of data processed/day $0.02/GB, limit: 20,000 queries/day
Amazon Relational Database Service (RDS)		$0.025/h for Micro DB Instance $0.090/h for Small DB Instance $0.180/h for Medium DB Instance $0.365/h for Large DB Instance $0.730/h for Extra Large DB Instance
Salesforce	Workflow automation, sales teams, enterprise analytics, custom websites	$125/user/month for Enterprise $250/user/month for unlimited

service providers, services they deliver, and price schemes in the IaaS, PaaS, HaaS, and SaaS arenas, respectively. These service providers include established companies such as Amazon, Google, and Salesforce as well as emerging startup companies such as Sabalcore and TeamPlatform. These companies may shape the CBDM arena over the next few years.

Table 4 Examples in HaaS

Provider	Service	Price scheme
Shapeways	3D printing	Starting from $0.75/cm3 for sandstone Starting from $1.40/cm3 for strong plastic Starting from $8.00/cm3 for stainless steel Starting from $20.00/cm3 for sterling silver
Cubify.com		$1299 for 140 × 140 × 140 mm, 16 colors, plastic $2499-3999 for 275 × 265 × 240 mm, 18 colors, plastic

Table 5 Examples in SaaS

Provider	Service	Price scheme
Autodesk 360 platform	Storage, DWG editing, mobile viewing, rendering, design optimization, structure analysis	Free for 5 GB storage
TeamPlatform	Sharing and viewing CAD files, synchronize CAD files, track changes, visual search, CAD Meta-Data search, 3D printing quoting, project management	Free for up to 10 workspaces, up to 5 guests, up to 5 shared pages and forms, 1 GB $25 for unlimited workspaces, guests, shared pages and forms, storage
CadFaster MyCadbox	View and share CAD models	Free for sharing 10 models; $9.99/month for up to 100 models
Sabalcore	High-performance computing for FEA/CFD	$0.20–$0.29/core-hour for premium service $0.20/core-hour for high-volume service
Penguin computing	High-performance computing for CAE	$0.10/core-hour/GB/day $0.27/core-hour/50 GB/day

4 Fundamentals of Cloud-Based Design and Manufacturing

4.1 A Vision for CBDM

This section describes a vision for CBDM from the perspectives of service offerings, operational processes, and reference models. Figure 1 depicts future service offerings in an idealized CBDM environment. As shown in Fig. 1, the core of a CBDM system is a knowledge management system (KMS) that creates knowledge repositories based on data and knowledge bases. The KMS can foster knowledge and resource sharing between service providers and consumers along with intelligent search engines and negotiation mechanisms. An intelligent search engine helps identify who key service providers and consumers are, which service providers know what, which service providers know which providers/consumers, and more importantly, what resources and information they possess. A negotiation

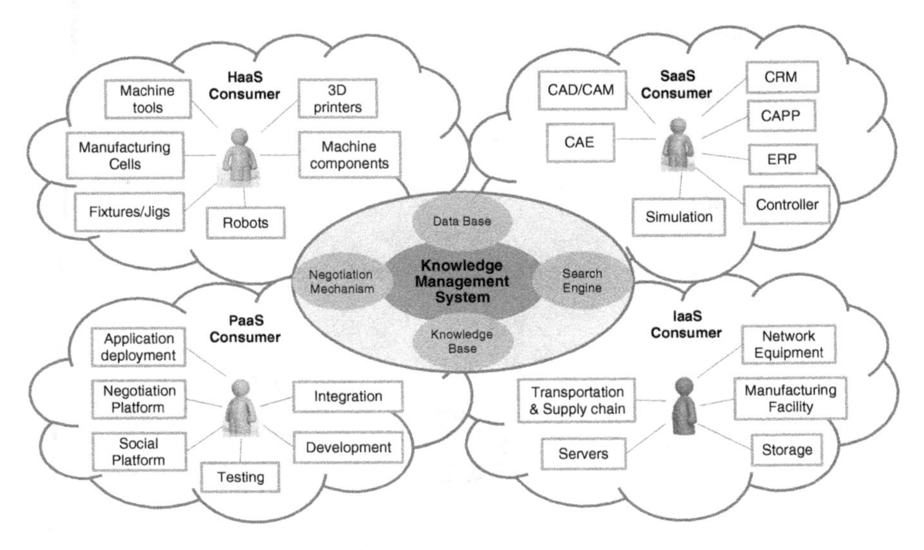

Fig. 1 Service offerings in CBDM (Wu et al. 2013a, b, c)

platform requires a series of negotiation mechanisms so that both service providers and consumers can find optimal design and manufacturing solutions (e.g., minimal costs and lead times). Furthermore, CBDM consists of four main service models:

- IaaS provides consumers with access to an IT infrastructure (e.g., network equipment) and computing resources (e.g., high-performance servers).
- PaaS provides consumers with access to computing platforms that allow the creation of web applications without buying and maintaining the software and infrastructure associated with it.
- HaaS provides consumers with access to manufacturing hardware (e.g., machine tools, 3D printers, and hard tooling). HaaS allows service consumers to rent hardware from providers as needed without upfront investments.
- SaaS provides consumers access to application software. SaaS allows service consumers to run engineering software through a web-based or thin client interface without purchasing full software licenses.

More specifically, PaaS provides a web portal serving as the front end of a CBDM system. IaaS, HaaS, and SaaS provide IT infrastructures and computing resources, manufacturing hardware, and application software, respectively, serving as the back end of a CBDM system. Moreover, Fig. 2 illustrates how CBDM systems may work from an operational perspective. For example, if consumer A submits a request for quotation (RFQ) for designing and machining a turbine blisk prototype the search engine of the CBDM system returns a list of alternative design and machining service providers (e.g., providers B and C) based on specifications in the RFQ. Then these design service providers estimate prices for their designs and respond to the RFQ. Based on these quotations and other

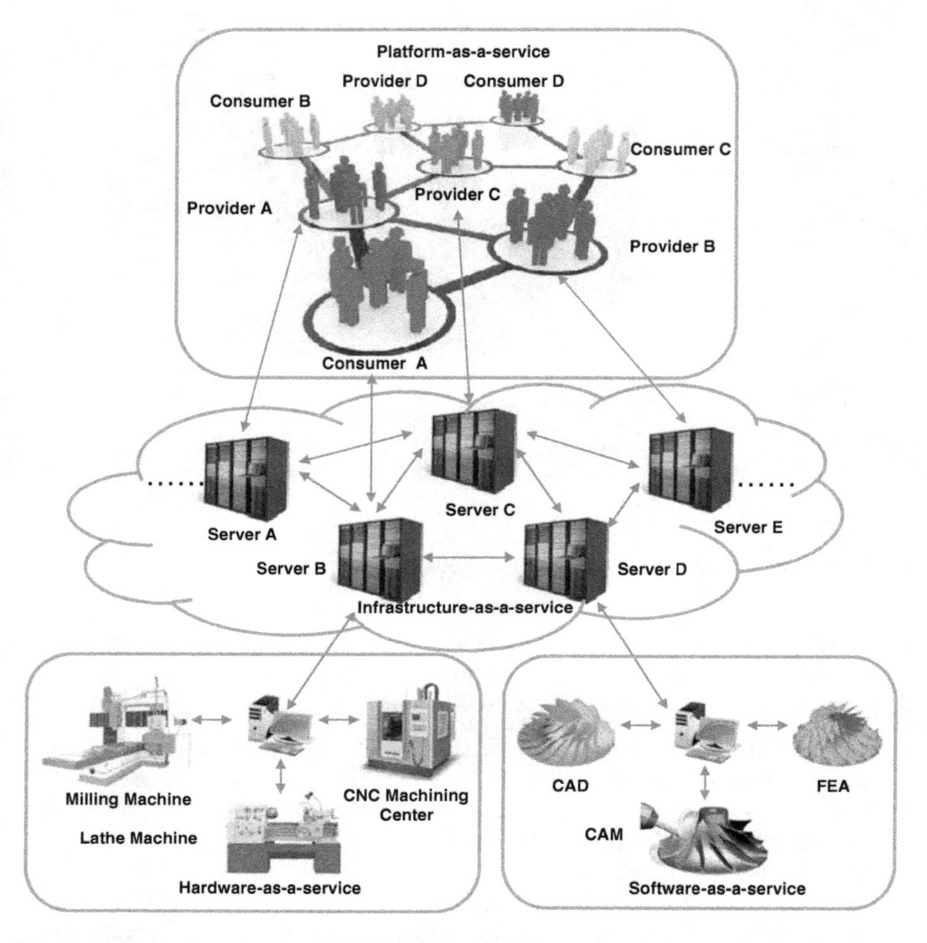

Fig. 2 Operational perspective in CBDM (Wu et al. 2013a, b, c)

qualifications (e.g., service providers' experience), the service consumer selects one of these design service providers, which conducts geometric modeling and structure and thermal analyses via CAD and FEA software in the cloud. Once a detailed design, including 3D digital models and CAD drawings, is finished, manufacturing service providers estimate machining times using tool path planning and simulation software in the cloud. Based on the estimated machining times, hourly labor rates, and material volumes, manufacturing service providers estimate the manufacturing costs and lead times and then respond to the RFQ from consumer A.

In addition to the above design and manufacturing process, Fig. 3 defines a high-level CBDM conceptual reference model including a set of actors and functions. Major actors in CBDM include cloud consumers, cloud providers, cloud

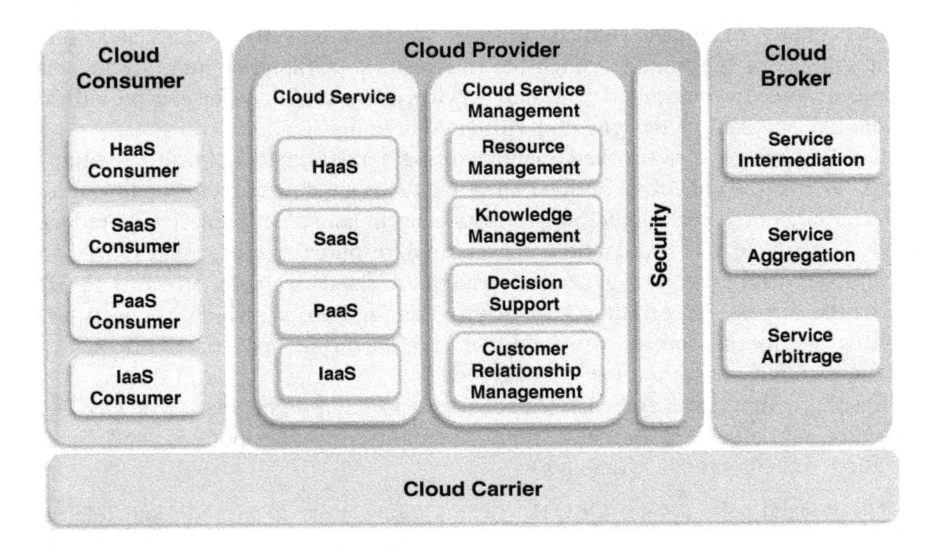

Fig. 3 CBDM conceptual reference model (Wu et al. 2013a, b, c)

brokers, and cloud carriers (NIST 2011). CBDM consumers may request cloud services (i.e., IaaS, PaaS, HaaS and SaaS) from service providers directly. Cloud providers manage manufacturing resources, knowledge, decision support, customer relationships, and physical and virtual securities. As the complexity of the integration of cloud services increases, cloud consumers may request cloud services from cloud brokers instead of cloud providers. Cloud brokers are third-party businesses that perform service intermediation, service aggregation, and service arbitrage; they act as an intermediary between service consumers and providers during negotiations. Meanwhile, cloud brokers assist service consumers in the deployment and the integration of applications from multiple service providers. In addition, cloud carriers provide network equipment to cloud consumers and providers. Cloud carriers also define service level agreements (SLAs) between service consumers and providers.

4.2 CBDM = CBD + CBM

According to the definition in the introduction of this chapter, CBDM has two strands: CBD and CBM. CBD refers to a crowdsourcing-based design model that leverages cloud computing, service-oriented architecture (SOA), Web 2.0 (e.g., social network sites), and semantic web technologies to support cloud-based engineering design services in distributed and collaborative environments. For example, Autodesk developed the Autodesk 123D, a platform that allows users to convert photos of objects into 3D models, create or edit 3D models, and build prototypes on a 3D printer through the Internet. 100kGrarages.com allows a service consumer to find qualified design service providers based on the providers' specialties and sample designs in a virtual community.

CBM refers to a networked manufacturing model that exploits on-demand access to a shared collection of diversified and distributed manufacturing resources to form temporary, reconfigurable production lines that enhance efficiency, reduce product lifecycle costs and allow for optimal resource loading in response to variable-demand customer generated tasking. A good example of CBM is 3D Hubs, the largest 3D printing service provider in Europe, which connects 3D printing service consumers with additive manufacturing service providers in a local community using the pay-per-use model.

Based on the vision for CBDM, some of the key characteristics of CBDM, including on-demand self-service, rapid scalability, resource pooling, and virtualization, are summarized as follows:

- On-demand self-service: CBDM systems allow users to provide and release engineering resources (e.g., design software, manufacturing hardware) as needed on demand. Communication between users takes place on a self-service basis.
- Rapid scalability: CBDM systems allow users to share manufacturing cells, general purpose machine tools, machine components (e.g., standardized parts and assembly), material handling units, and personnel (e.g., designers, managers, and manufacturers) in response to rapidly changing market demands.
- Resource pooling: CBDM systems allow for design and manufacturing resource pooling. Design and manufacturing resources include engineering hardware (e.g., fixtures, molds, and material handling equipment) and software (e.g., computer-aided design and FEA program packages). Wireless sensor networks, capturing the status and availability of manufacturing resources, facilitate effective and efficient resource allocation and planning.
- Virtualization: CBDM systems enable users to separate engineering software packages, computing and data storage resources from physical hardware through virtualization.

4.3 Differences Between CBDM and Web- and Agent-Based Systems

While CBDM overlaps with web- and agent-based design and manufacturing, it differs from these paradigms in several ways:

- From a computing perspective, the difference between CBDM and web- and agent-based design and manufacturing lies in two essential characteristics of cloud computing: multi-tenancy and virtualization. Figure 4 illustrates a computing architecture for CBDM systems. In the CBDM computing architecture, multi-tenancy enables a single instance of application software to serve multiple

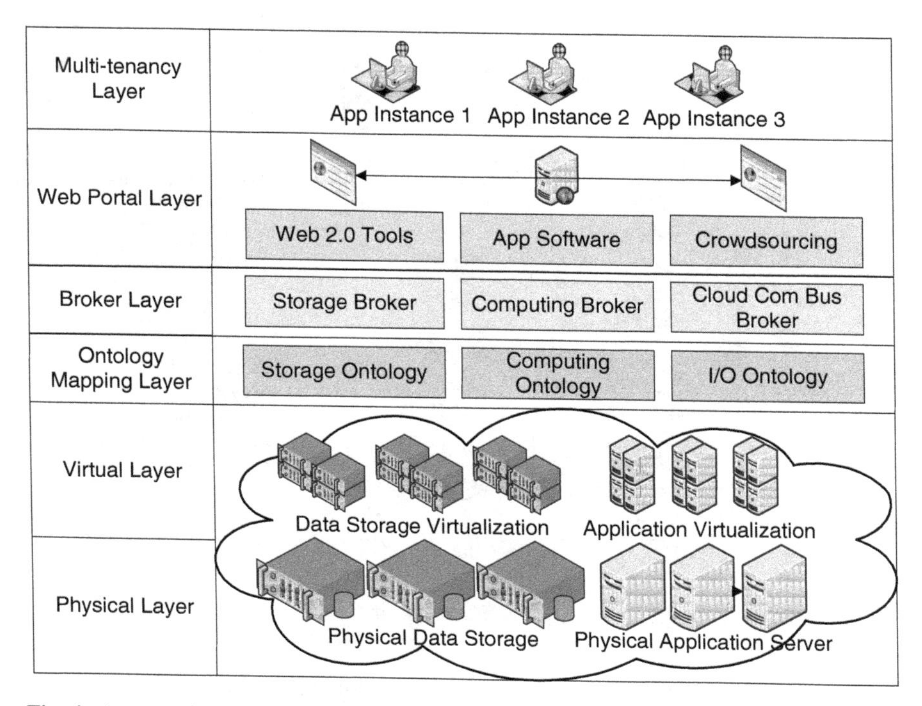

Fig. 4 A computing architecture for CBDM systems

tenants. Similar to cloud computing, CBDM employs the multi-tenancy, which allows users to sharing software applications. CBDM also utilizes virtualization for sharing computing resources. In addition, CBDM requires ontology mapping that assists in defining correspondence between ontologies in various domains.

- From a data storage perspective, in web- and agent-based design and manufacturing, product-related data are stored at designated servers, and users know where they are as well as who is providing them. However, in CBDM, networked data are stored not only in user's computers but also in virtualized data centers that are generally hosted by third parties (see the virtual and physical layers in Fig. 4). Physically, these networked data may span across multiple data centers. In other words, users may neither know who service providers are nor where data are stored. However, these data may be accessed through a web service application programming interface (API) or a web browser. The advantages of cloud-based data storage are the following: (1) it provides users with ubiquitous access to a broad range of data stored in the networked servers via a web service interface; (2) it can easily be scaled up and down as needed on a self-service basis; (3) companies are only charged for the storage they actually use in the cloud.
- From an operational process perspective, CBDM can leverage the power of the crowd through crowdsourcing. For instance, CBDM enables service consumers

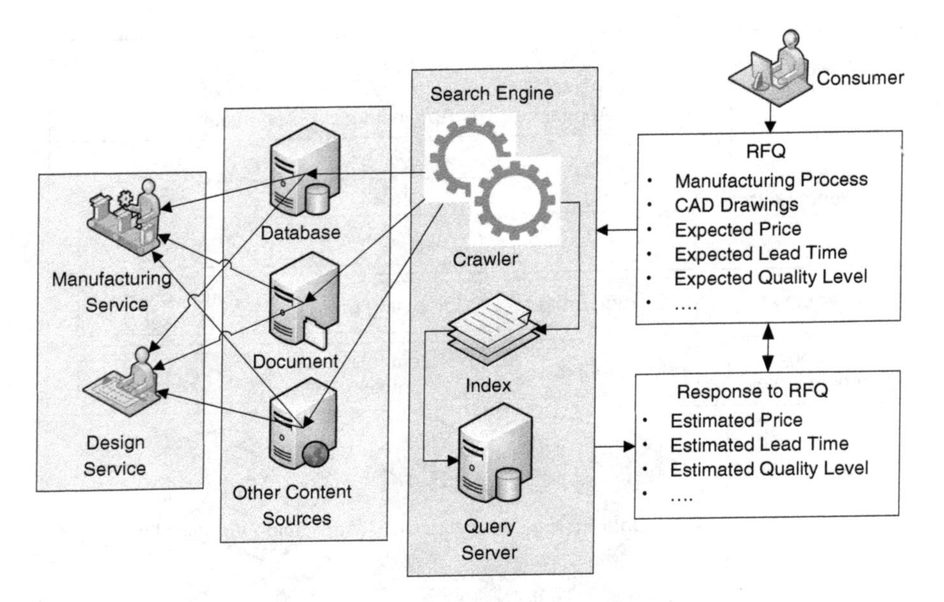

Fig. 5 Crowdsourcing process for RFQs in CBDM systems

to quickly and easily locate qualified service providers who provide design and manufacturing services such as CNC machining, injection molding, casting, or 3D printing through a crowdsourcing process. Figure 5 illustrates the crowdsourcing process that enables users to submit RFQs to a search engine and receive a list of qualified service providers. A search engine consists of a crawler, index, and query servers. The crawler gathers data related to design and manufacturing such as CAD drawings, design parameters, process variables, machine specifications from databases, document servers, and other content sources, and the crawler stores these data in the index. The index ranks these data based on metrics (e.g., price, quality, geographic location) specified by the users. The front end of a search engine is a query server that delivers results of a search query to consumers based on specifications such as expected prices, lead times, and quality levels. However, with regard to web- and agent-based systems, it is not feasible to implement such a computationally expensive crowdsourcing platform that connects service consumers and providers worldwide. Moreover, in comparison with commercial quoting systems such as Quickparts.com and MFG.com, the proposed crowdsourcing platform in CBDM not only conducts quoting for manufacturing services but also engineering design, manufacturing and computing resource allocation, and scheduling activities. Furthermore, in contrast to existing 3D printing service providers such as Shapeways, in which users upload design files and Shapeways prints objects from a single site, CBDM allows users to print their designs using any 3D printer in the cloud rather than using one at a particular site.

- From a communication perspective, CBDM uses web 2.0 technologies, as shown in Fig. 4, that allow users to leverage social media. However, users in web- and agent-based systems cannot easily gather valuable feedback about their products from customers and maintain interactive communication channels during the design and manufacturing processes.
- From a management perspective, a significant difference between CBDM and web- and agent-based manufacturing is that CBDM involves new business models, but web- and agent-based manufacturing paradigms do not. That is, CBDM also involves how design and manufacturing services can be delivered (e.g., IaaS, PaaS, HaaS, and SaaS), how they can be deployed (e.g., private cloud, public cloud and hybrid cloud), and how they can be paid for (i.e., pay-per-use).

5 Development of a Prototype System

This section presents a prototype CBDM system, DMCloud, developed at Georgia Tech. The objective of developing the DMCloud is to help students (1) learn how to use CAD and engineering analysis software, and (2) to practice product design and manufacturing in a distributed and collaborative setting (Rosen et al. 2012; Patel et al. 2013).

5.1 System Overview

The DMCloud is currently implemented as a private cloud, but it can easily be extended to be a public or hybrid DMCloud if necessary. The DMCloud builds upon an integrated collaborative design and distributed manufacturing infrastructure along with engineering design software (e.g., CAD and CAM) and manufacturing hardware (e.g., CNC machines and 3D printers). Specifically, to provide a distributed and collaborative learning environment, the DMCloud is developed using Sakai, an open-source, Java-based service-oriented software platform. To help students learn basic design and manufacturing skills, the DMCloud integrates geometric modeling and analysis software such as CATIA, developed by Dassault Systemes. To help students learn about material properties and design for manufacturing principles, the DMCloud provides an information repository that stores knowledge related to basic material properties. To help students learn about various manufacturing technologies, the DMCloud provides students with access to 3D printers, milling machines, lathes, laser cutters, and other CNC machines.

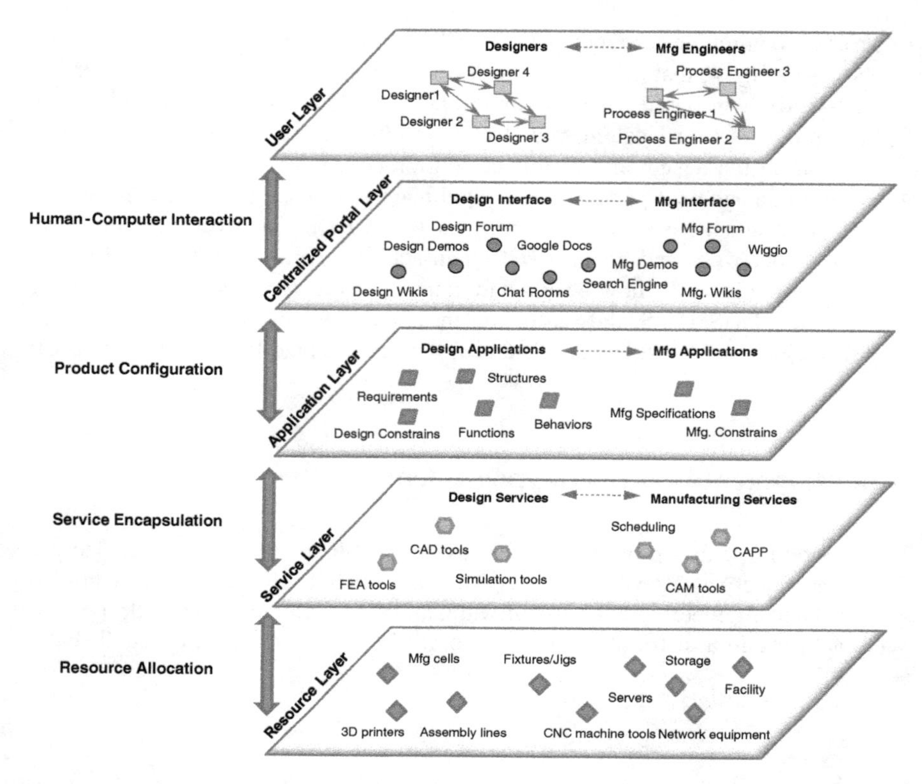

Fig. 6 System architecture of the DMCloud system (Wu et al. 2013a, b, c)

5.2 System Architecture

As shown in Fig. 6, the system architecture of the DMCloud can be captured by a five-layer conceptual model that defines the overall structure, including users, web portals, applications, services, and resources. The representation of the system architecture is a mapping mechanism between product design and manufacturing processes that links product designs to associated manufacturing processes (as shown by dotted red arrows). The centralized portal enables cloud-based human–computer interaction, facilitates effective data collection, and provides seamless integration of resources and services into the DMCloud. A product configuration process transforms data collected from the centralized portal layer into conceptual designs and high-level manufacturing specifications and constraints. Service encapsulation transforms conceptual designs into embodiment and detail designs and consolidates services based on the conceptual designs from the application layer. Resources are allocated according to the detailed designs from the service layer. The specific functions of each layer are illustrated as follows:

- User layer: The key users of the DMCloud include product designers and manufacturing engineers. In the context of the MENTOR program, major users are engineering students.
- Centralized portal layer: The key function of the centralized portal layer is to provide cloud providers and consumers with a centralized interface which facilitates communications between them. Specifically, the centralized portal provides forums, Wikis, and chat rooms. The portal also provides social networking tools such as Wiggio and document sharing tools such as Google Docs.
- Application layer: The key function of the application layer is to transform information acquired via the centralized portal to requirements, structures, functions, behaviors, design constraints, and manufacturing specifications.
- Service layer: The key function of the service layer is to provide various engineering services (e.g., CAD/CAM/CAE/CAPP). The service layer delivers detailed designs and manufacturing processes based on information from the application layer.
- Resource layer: The resource layer encompasses design- and manufacturing-related resources such as fixtures/jigs, 3D printers, CNC machine tools, manufacturing cells, assembly lines, facility, servers, network equipment, and so on.

5.3 Workflow and Services Provided by the DMCloud

This section presents the overall workflow of the DMCloud system (see Fig. 7) and services provided by the DMCloud (see Figs. 8 and 9). The DMCloud consists of a centralized interfacing server (CIS) incorporated with the Moodle learning management system, shown in Fig. 7. The CIS includes the following modules: web-based user interfaces, connectivity and virtual routing databases, content databases, authentication modules and account management, resource scheduling and synchronization, resource discovery, registration, and configuration. The CIS provides users with access to design (e.g., CAD) and manufacturing services (e.g., 3D printing and milling) through the Internet. The moodle platform provides users with a virtual learning environment which includes a discussion forum, instant messages, online announcement, and Wikis.

In addition to the workflow, the specific services provided by the DMCloud include cloud-based design, CBM, and social media.

- Cloud-based design: To serve multiple tenants using a single instance of CAD software, the DMCloud integrates CATIA, developed by the Dassault Systems using the multi-tenant software architecture. Multi-tenancy allows users to share the CATIA software tool cost-effectively and securely among multiple tenants.
- Cloud-based manufacturing: To assist in the transition from CAD models to fabrication with AM machines, the DMCloud provides several modules for AM, shown in Fig. 8. The AM-Select module allows students to identify qualified AM machines and materials that can build a specific part. The

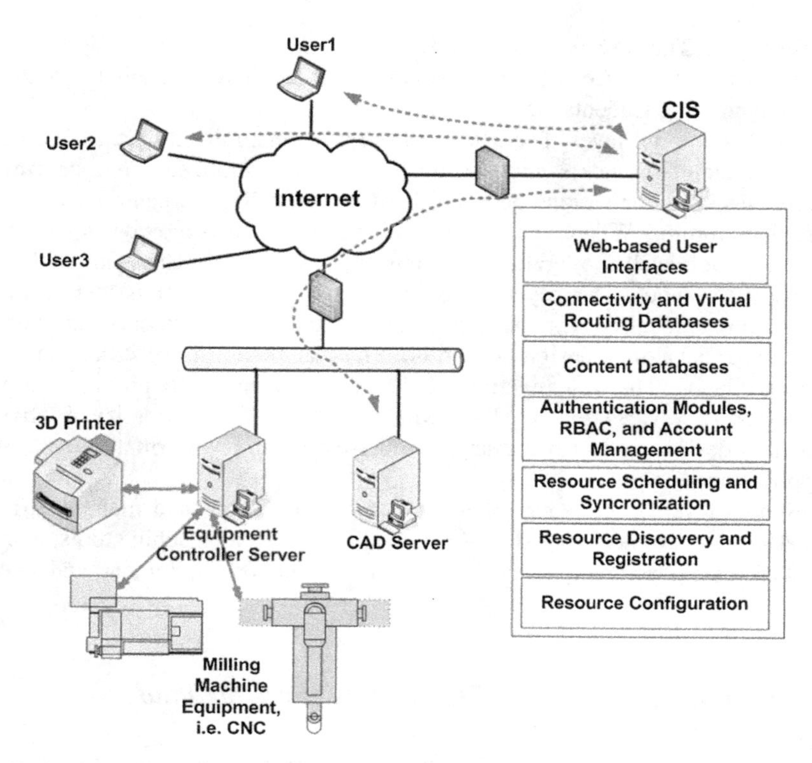

Fig. 7 Workflow of the DMCloud (Schaefer et al. 2012; Wu et al. 2013a, b, c)

AM-Advertise module allows independent manufacturing sub-systems to advertise service availability and associated service usage parameters. The AM-Request module allows service consumers to request AM services from service providers. The AM-Manufacturable module enables manufacturability analysis such as whether a specific part is manufacturable on a specific 3D printer. If a part cannot be fabricated, the AM-Manufacturable module will provide information about what properties of the part prevent the manufacturing of a part. The AM-DFAM module provides data and knowledge bases for design for additive manufacturing. The AM-Teacher module provides students with tutorials, service wizards, videos, and other learning materials.

- Social media: To facilitate communication and collaboration among students, the DMCloud supports several social networking tools (e.g., chat rooms, forum, and Wiggio), shown in Fig. 9. For example, students can easily obtain real-time feedback from others via chat rooms. A design forum allows students to experience interactive learning environments. In addition, the DMCloud allows users to host virtual meetings and video conference calls, create to-do lists, assign tasks, and share documents in cloud-based storage systems.

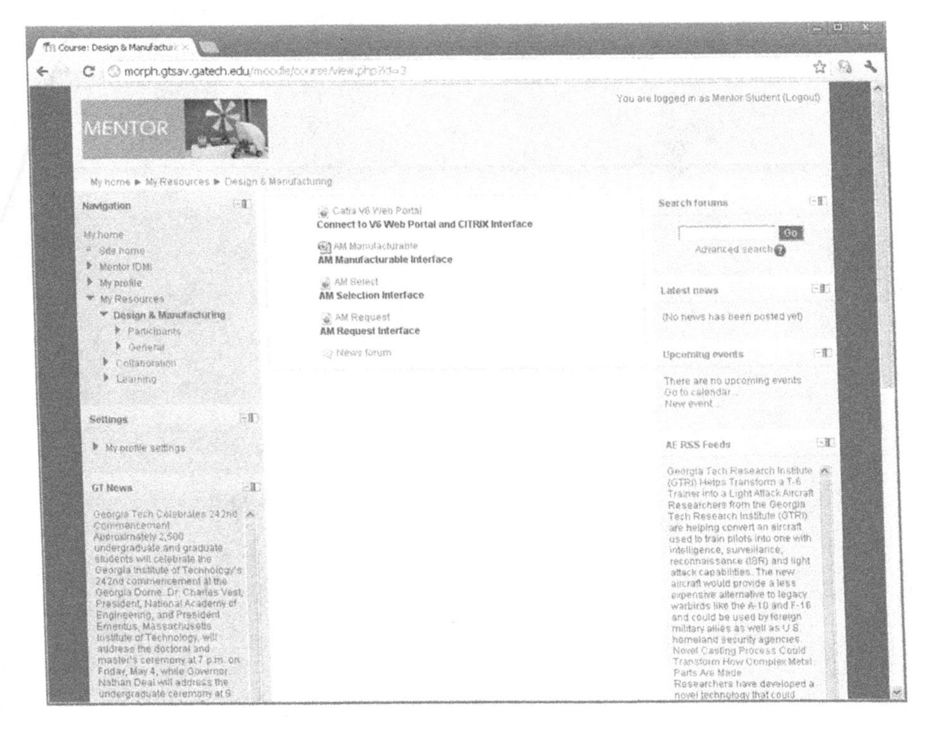

Fig. 8 Design and manufacturing module of the DMCloud portal (Wu et al. 2013a, b, c)

In view of the aforementioned services, the DMCloud has the potential to (1) allow users to access CAD/CAM software using multi-tenancy and virtualization technologies; (2) enable users to crowdsource design and manufacturing tasks; (3) facilitate coordination, communication, and collaboration between service consumers and providers; and (4) assist in design and manufacturing resource allocation and optimization.

6 Future Research Directions

Based on the review of CBDM-related research in Sect. 3, the vision for CBDM in Sect. 4, and our prototype system in Sect. 5, a significant amount of research effort in the field of CBDM has focused on articulating strategic visions, refining definitions and concepts, developing high-level system architectures, modeling manufacturing resource, scheduling, and optimizing resource allocation. To help bridge the gap between the current research progress in CBDM and the vision for CBDM as described in Sect. 4, we identify several directions for future research as follows.

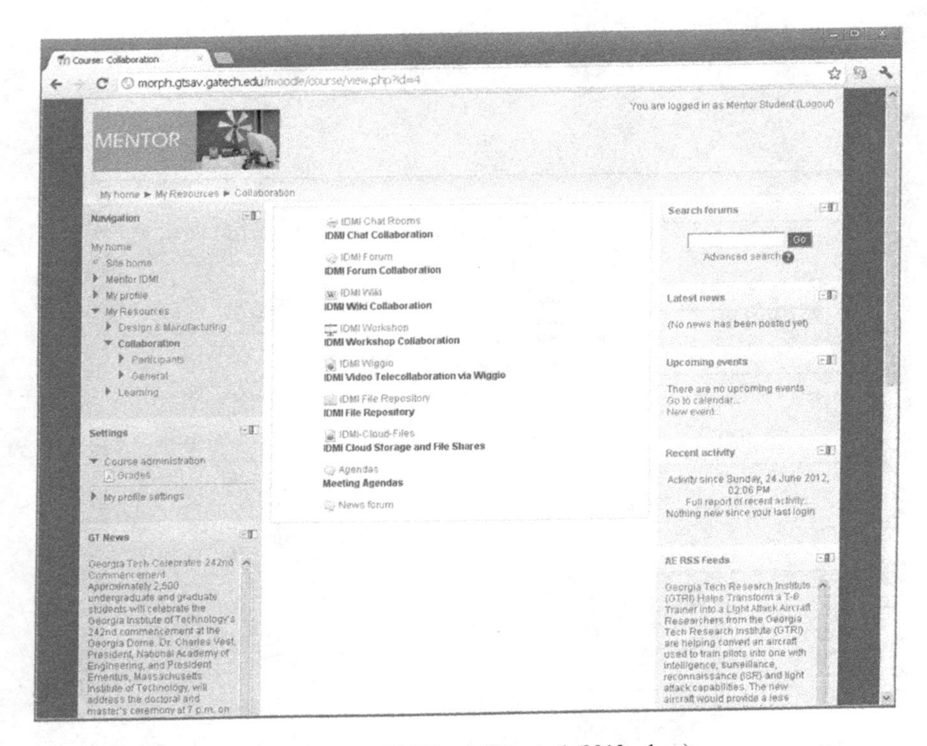

Fig. 9 Social networking tools in the DMCloud (Wu et al. 2013a, b, c)

6.1 Cloud-Based Manufacturing

While research studies related to CBM has focused on resource allocation and scheduling (Laili et al. 2013), more attentions need to be paid on modeling and simulation of material flow. Because efficient material flow helps reduce manufacturing lead time and ensure rapid scalability, one of the future research directions is to analyze concurrency and synchronization in CBM systems using analytical models or simulation. In the context of manufacturing, concurrency refers to the extent to which several manufacturing processes can be executed simultaneously. Synchronization refers to the adjustment of manufacturing paces so that multiple processes can be finished simultaneously. The purpose of concurrency and synchronization is to reduce manufacturing lead times. The modeling and simulations of CBM systems help one understand the system dynamics of CBM.

6.2 Cloud-Based Design

Unlike CBM, CBD has not yet been well investigated. As discussed in Sect. 4, incorporating social media in CBD environments is critical for implementing CBD. Because social media help users communicate and collaborate in CBD environments, social network analysis (SNA) provides both visual and mathematical analyses of communication and collaboration patterns between users. Although some researchers have investigated the social nature of engineering design processes, few studies have been devoted to identifying effective metrics for measuring the existence of connections between users and detecting users with similar interests in the context of CBD. Furthermore, because social media and social computing allow for collecting and analyzing massive amounts of social data, it is important to examine how social media in CBD influence design performance indicators such as product time-to-market, customer satisfaction, and design costs. In addition to social media, efficient design collaboration requires current CAx (e.g., CAD/CAE/CAM) applications to be converted into multi-threaded ones, in which multiple threads can run in parallel. According to Red et al. (2013), some of the important architectural issues related to current CAx applications (e.g., Solidworks, CATIA, and Unigraphics) include "multi-threading of CAx API's and GUI's, access to CAx event handlers and interrupts, client session undo's/redo's, and API's that can provide feature parameter copies rather than address handles."

6.3 Information, Communication, and Cyber Security

Because CBDM involves a complex network in which digital and physical entities are interconnected, the Internet of Things (IoT) will potentially help collect real-time design- and manufacturing-related data. According to Miorandi et al. (2012), the IoT refers to the global network interconnecting smart objects by means of extended Internet technologies including RFIDs, sensor or actuators, and machine-to-machine communication devices. The IoT can help CBDM service providers communicate and collaborate with geographically dispersed service consumers by automatically identifying and tracking smart physical objects. In addition to information and communication technology, according to Rabai et al. (2012), another challenge for adopting CBDM is how to address cyber security threats including "malicious activity, made possible by the provision of shared computing resources as well as inadvertent loss of confidentiality or integrity resulting from negligence or mismanagement." To enhance cyber security, Kaufman (2009) suggests that service providers must offer capabilities including "(1) a test encryption schema to ensure the shared storage environment safeguards all data; (2) stringent access controls to prevent unauthorized access to the data; and (3) scheduled data backup and safe storage of the backup media."

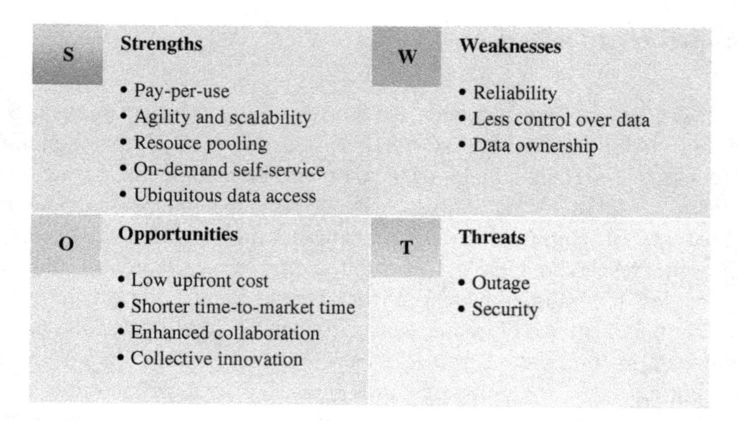

Fig. 10 SWOT analysis for CBDM

6.4 Business Model

To represent core aspects of CBDM, a clear understanding of business-related issues in CBDM is extremely important. In particular, generating creative business models for CBDM enables service providers and consumers to capture value and gain economic benefits. To understand what potential business model may fit for CBDM, we performed a SWOT analysis, shown in Fig. 10.

Identified strengths of CBDM include the pay-per-use model, agility and rapid scalability, resource pooling, on-demand self-service, and ubiquitous data access. Potential weaknesses of CBDM are reliability, less control over data, and data ownership. Opportunities of CBDM lie in low upfront costs, shorter time-to-market time, enhanced collaboration, and collective innovation. Major potential threats of CBDM include security and privacy.

Acknowledgments This work was partially funded by DARPA. Any opinions, findings, and conclusions or recommendations presented in this chapter are those of the authors and do not necessarily reflect the views of DARPA. The authors would like to thank the professors and research engineers at Georgia Tech for their assistance in developing the prototype system.

References

Armbrust M, Fox A, Griffith R, Joseph AD, Katz R, Konwinski A, Lee G, Patterson D, Rabkin A, Stoica I, Zaharia M (2010) Above the clouds: a view of cloud computing. Commun ACM 53(4):50–58

Böhm M, Leimeister S, Riedl C, Krcmar H (2010) Cloud computing and computing evolution. http://scholar.google.com/scholar?q=cloud+computing+and+computing+evolution&hl=en&btnG=Search&as_sdt=1%2C11&as_sdtp=on

Buyya R, Yeo CS, Venugopal S (2008) Market-oriented cloud computing: vision, hype, and reality for delivering it services as computing utilities. CoRR

DARPA (2013) Adaptive vehicle make. http://www.darpa.mil/Our_Work/TTO/Programs/AVM/Adaptive_Vehicle_Make_Program_Overview.aspx

Friedman TL (2005) It's a flat world, after all. http://www.nytimes.com/2005/04/03/magazine/03DOMINANCE.html

Foster I, Kesselman C (1999) The grid: blueprint for a new computing infrastructure. Morgan Kaufmann, San Francisco

Foster I, Zhao Y, Raicu I, Lu S (2008) Cloud computing and grid computing 360-degree compared. Grid Computing Environments Workshop, Austin

IBM (2010) Dispelling the vapor around cloud computing. http://www-935.ibm.com/services/us/igs/cloudforum/Mills_pres.pdf

Kaufman LM (2009) Data security in the world of cloud computing. IEEE Secur Priv 7(4):61–64

Laili Y, Tao F, Zhang L, Cheng Y, Luo Y, Sarker BR (2013) A Ranking Chaos Algorithm for dual scheduling of cloud service and computing resource in private cloud. Comput Ind 64(4):448–463

Linthicum D (2009) Cloud computing and SOA convergence in your enterprise: a step-by-step guide. Addison-Wesley Professional, Indianapolis

Li BH, Zhang L, Wang SL, Tao F, Cao JW, Jiang XD, Song X, Chai XD (2010) Cloud manufacturing: a new service-oriented networked manufacturing model. Comput Integr Manufac Syst 16(1):1–7

Licklider JCR (1963) Topics for discussion at the forthcoming meeting, memorandum for: members and affiliates of the intergalactic computer network. Advanced Research Projects Agency. http://www.kurzweilai.net/memorandum-for-members-and-affiliates-of-the-intergalactic-computer-network

Miorandi D, Sicari S, De Pellegrini F, Chlamtac I (2012) Internet of things: vision, applications and research challenges. Ad Hoc Netw 10(7):1497–1516

NIST (2011) NIST cloud computing reference architecture. Special Publication 500–292

Parkhill DF (1966) The challenge of the computer utility. Addison-Wesley Publication, Reading

Patel S, Schaefer D, Schrage D (2013) A pedagogical model to educate tomorrow's engineers through a cloud-based design and manufacturing infrastructure—motivation, infrastructure, pedagogy, and applications. In: ASEE 2013 Annual Conference and Exposition, Atlanta, Georgia, 23–26 June 2013

Quirky (2012) http://www.quirky.com/

Rabai LBA, Jouini M, Aissa AB, Mili A (2012) A cybersecurity model in cloud computing environments. J King Saud Univ Comput Inf Sci

Red E, French D, Jensen G, Walker S, Madsen P (2013) Emerging design methods and tools in collaborative product development. Trans ASME J Comput Inf Sci Eng 13(3):031001-031001-14. doi:10.1115/1.4023917

Rosen DW, Schaefer D, Schrage D (2012) GT MENTOR: A high school education program in systems engineering and additive manufacturing. In: Proceedings of the 23rd Annual International Solid Freeform Fabrication Symposium—An Additive Manufacturing Symposium (SFF 2012). Austin, Texas, 6–8 Aug 2012

Schaefer D, Thames JL, Wellman RD, Wu D, Yim S, Rosen D (2012) Distributed collaborative design and manufacture in the cloud—motivation, infrastructure, and education. In: American Society for Engineering Education Annual Conference, Paper #AC2012-3017, San Antonio, USA

Vaquero LM, Merino LR, Caceres J, Lindner M (2009) A break in the clouds: towards a cloud definition. ACM SIGCOMM Comput Commun Rev (archive) 39(1):50–55

Wu D, Thames JL, Rosen DW, Schaefer D (2013a) Enhancing the product realization process with cloud-based design and manufacturing systems. Trans ASME J Comput Inf Sci Eng 13(4):041004-041004-14. doi:10.1115/1.4025257

Wu D, Greer MJ, Rosen DW, Schaefer D (2013b) Cloud manufacturing: strategic vision and state-of-the-art. J Manufac Syst. http://dx.doi.org/10.1016/j.jmsy.2013.04.008

Wu D, Schaefer, Rosen DW (2013c) Cloud-based design and manufacturing systems: a social network analysis. In: International Conference on Engineering Design (ICED13), Seoul, Korea

Wu D, Thames JL, Rosen DW, Schaefer D (2012) Towards a cloud-based design and manufacturing paradigm: looking backward, looking forward. In: Proceedings of the ASME 2012 International Design Engineering Technical Conference & Computers and Information in Engineering Conference (IDETC/CIE12), Paper Number: DETC2012-70780, Chicago, USA

Wu D, Rosen DW, Schaefer D (2014a) Modeling and analyzing the material flow of crowdsourcing processes in cloud-based manufacturing systems using stochastic petri nets. In: Proceedings of the ASME 2014 International Manufacturing Science and Engineering Conference (MSEC14), Paper Number: MSEC2014-3907, Detroit, Michigan

Wu D, Rosen DW, Wang L, Schaefer D (2014b) Cloud-based manufacturing: old wine in new bottles? In: Proceedings of the 47th CIRP Conference on Manufacturing Systems, Windsor, Canada

Multi-User Computer-Aided Design and Engineering Software Applications

Edward Red, David French, Ammon Hepworth, Greg Jensen and Brett Stone

Abstract This chapter will introduce multi-user computer-aided engineering applications as a new paradigm for product development, considering past collaborative research and the emerging wave of cloud-based social and gaming tools. In a historical context, computer-aided design and engineering models have become much more complex since their inception in the middle of the twentieth century. However, the way design teams approach these models has, at least in one sense, not changed much; a given model can still only be accessed by one user at a time, despite the fact that the entire design team needs to evolve the model. Single user applications have become a productivity bottleneck and do not provide interfaces or architectures for simultaneous editing of models by a collaborative team. Single user applications convert any hope for process concurrency into a serial sequence of design activities. When the single user designer experiences difficulties, the process halts until the designer can reach out to other experts to resolve the problems, which usually requires some form of external collaboration. Unfortunately, single user applications are deficient when it comes to complex and globalized product development. The chapter herein will consider how multi-user architectures will change the single user paradigm from serial to simultaneously collaborative, promote new on-demand access methods like cloud serving, and bring long hoped for efficiencies to product development. We will investigate three research areas of

E. Red (✉) · D. French · A. Hepworth · G. Jensen · B. Stone
Brigham Young University, Provo, UT, USA
e-mail: ered@byu.edu

D. French
e-mail: davidfrench11@gmail.com

A. Hepworth
e-mail: ammon.hepworth@gmail.com

G. Jensen
e-mail: cjensen@byu.edu

B. Stone
e-mail: brettstone87@gmail.com

D. Schaefer (ed.), *Cloud-Based Design and Manufacturing (CBDM)*,
DOI: 10.1007/978-3-319-07398-9_2,

importance to this emerging paradigm: (1) multi-user CAx architectures, including cloud serving; (2) multi-user CAx requirements; and (3) multi-user CAx standards. Of these three, architectures are most investigated, with numerous proof-of-concept prototypes, while requirements and standards, the least investigated, partially explain the reason for non-adoption and non-commercialization of this powerful new paradigm.

Keywords Computer-aided design · Multi-user · Cloud serving · Organizational management · Collaborative design · Concurrent engineering

1 Introduction

Decades ago, engineers and other technical personnel gathered around large drafting tables to review and blend their design contributions, and consider technical requirements and pressing schedules, Fig. 1. Collaboration was natural and transparent in this hands-on environment, at least until the ensuing digital age of desktop workstations and computer-aided applications (CAx) encouraged new methods of decomposing product design among technical individuals, where one person is assigned a part model design. Today, file-based control systems like product lifecycle management (PLM) track and maintain model file version changes, using secured check-in and check-out procedures.

Collaboration has become increasingly difficult as multiple designers are unable to simultaneously enter an editing session, although they can screen share the model. As products become increasingly complex, profit incentives have decomposed the various system (airplane, ship, tank, etc.) components among globally distributed suppliers. These suppliers often use different vendor supplied CAx applications to produce their contracted component, resulting in model variations and inconsistent file formats, referred to as noninteroperable data. Distribution and integration of these models among the supplier chain usually results in many model file conversions because of installed CAx heterogeneity. Because of cascaded file conversion errors, suppliers may find it necessary to replicate the model in their native CAx application. The annual cost associated with interoperability is on the order of hundreds of millions of dollars (USD) (Brunnermeier et al. 1999).

Computer-aided engineering applications like CAD/CAE/CAM (CAx) have continued to grow in complexity and capability, but have remained single user while the concurrent demands of building a new class of technically evolved products like modern transportation systems have escalated. Concurrency is further complicated by business practices where large product companies now depend on distributed global supplier chains to build their system components. Each supplier's CAx tools may vary and secondary CAx model file translations introduce errors and extend the development cycle (Contero et al. 2002).

Fig. 1 Collaboration in 1979

Fig. 2 Gaming session

Client-server gaming and related cloud serving architectures have demonstrated that teams can closely and simultaneously collaborate in complex, dynamically changing landscapes, using distributed cloud servers that manage changes to model data (Kim 2002; Fig. 2).

While cloud-based applications centralize data and make it accessible over great distances, even across continents, CAx applications isolate and protect data for a single user. A single designer creates and details a component model using modern computer-aided design (CAD) applications. A single analyst applies computer-aided engineering (CAE) tools to determine whether a component can withstand the structural loading, or elevated temperatures, or whether the component is aerodynamically stable. Finally, single users apply computer-aided manufacturing (CAM) and computer-aided process planning (CAPP) to determine and program the manufacturing processes and machines to make the component.

Surveys (Red et al. 2013a, b) have shown that technical personnel engage in some form of collaboration at least 50 % of each day (Fig. 3), either in formal or ad hoc meetings or by applying a number of social media tools like conference

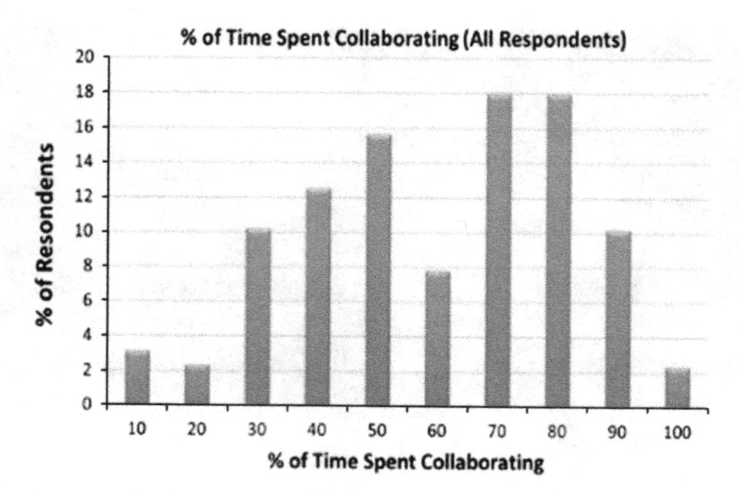

Fig. 3 Daily % collaboration

Fig. 4 Conference call design meeting

calls (Fig. 4), email, texting, instant messaging, and mobile phones. These social tools necessarily compensate for the collaborative deficiencies in modern CAx applications.

Multi-user CAx applications provide real-time collaboration and thus enable Cloud-Based Design and Manufacturing (CBDM), although expanding the concepts described by Wu et al. Wu defines CBDM as follows (Wu 2012):

> Cloud-Based Design and Manufacturing refers to a product realization model that enables collective open innovation and rapid product development with minimum costs through a social networking and negotiation platform between service providers and consumers. It is a type of parallel and distributed system consisting of a collection of inter-connected physical and virtualized service pools of design and manufacturing resources (e.g., parts, assemblies, CAD/CAM tools) as well as intelligent search capabilities for design and manufacturing solutions.

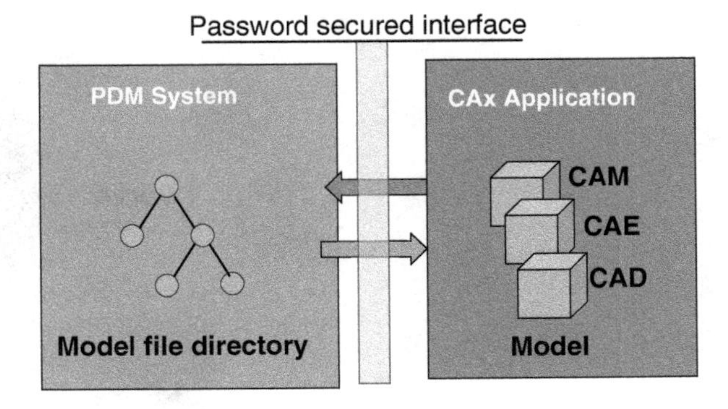

Fig. 5 PDM to CAx file exchange

Although Wu's pooling of CBDM resources refers primarily to cloud computing, we generalize CBDM to include human resources and data consistency resources, in addition to computer hardware and software. Multi-user CAx directly supports this generalized vision by distributing and parallelizing human resources to increase design process efficiency. We suggest that the CDBM vision will not be realized until the product model data becomes interoperable (CAX agnostic), regardless of consumer resource application of choice. Otherwise, the model's data format and integrity will be compromised by conformal translations to each user's application environment.

To understand the power of multi-user CAx applications, this chapter will consider three research areas: (1) multi-user CAx architectures; (2) multi-user CAx requirements; and (3) multi-user CAx standards. The last two have not been well researched, but will be important to consider nevertheless.

2 Multi-user CAx Architectures

We will consider three architectural areas: (1) collaborative networks; (2) collaborative interfaces; and (3) model sharing and conflict resolution.

2.1 Collaborative Networks

Today, network concerns in single user CAx relate to how fast a design model file can be exchanged between a local workstation and PDM server. The PDM server secures and manages the model file revisions, and other product-related data, Fig. 5. Speed is less of a problem when the network is behind company firewalls where high speed LAN's can distribute complex model files in less than an hour. But when companies engage in around the clock, 24 h global design, FTP file delivery may take hours to transfer encrypted models of large size (100's of MB) over the Internet.

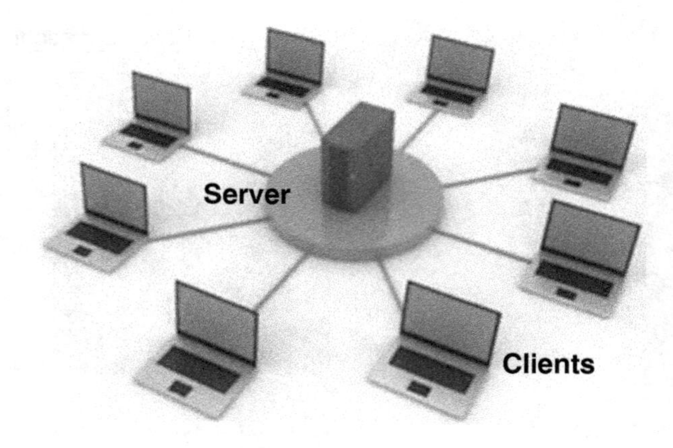

Fig. 6 Client-server architecture

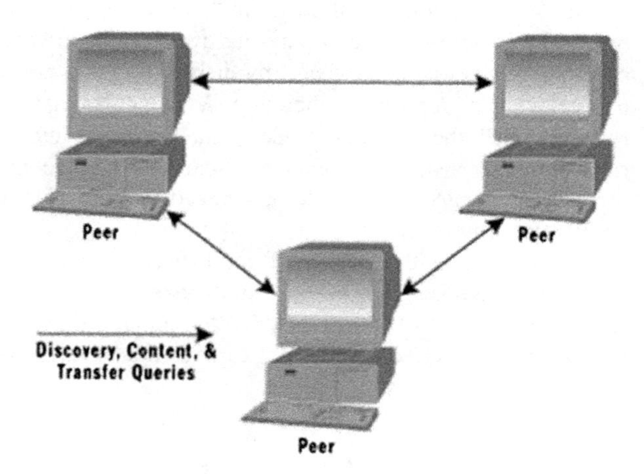

Fig. 7 Peer-to-peer architecture

Contrast single user file exchange with a multi-user network where several (or many) users, referred to as clients, simultaneously edit the model. Rather than files being passed over the network, model changes are passed between clients as small data packets, easily accommodated by even the slowest networks. With this perspective we note that two multi-user architectures dominate the collaborative networks: (1) client-server (CS, Fig. 6) and peer-to-peer (P2P, Fig. 7).

2.1.1 Client-Server

Most collaborative multi-user prototypes use client-server networks. The client-server variations configure the client workstations as thick clients or thin clients. Li's et al. review paper notes client-server dominance, lack of data security, and the difficulty of real-time interaction over networks (Li et al. 2005). Li distinguishes thick clients having a full model from thin clients where a partial model is mainly used for visualization.

Thick clients will run the CAx application on the each client's workstation and integrate software plug-ins to detect and transmit client model edits to the server where they are then reflected to other collaborating clients. These methods are referred to as "transparent adaptation" by Sun et al. (2006, only applied to MS Word and PPT) and Zheng et al. (2009, Co-AutoCAD) because the CAx application source is not modified.

BYU's client-server multi-user prototypes—Several multi-user CAx prototypes have been developed at Brigham Young University's NSF Center for e-Design site. These prototypes were built on native application API's (using C++, C#, Visual Basic) as software plugins to verify that mainstream CAx applications could function in multi-user mode, and to discover architectural limitations that might impede multi-user functionality. These prototypes allow for modeling in both single and multi-user modes. In multi-user mode, users visualize contributions from collaborators in real-time. This allows users to simultaneously contribute to a model and respond in real-time to input from others. Table 1 presents their functionality and limitations, while Figs. 6 and 7 show their architectures, although CS was most dominant. The CUBIT CAE prototype has access to the source code for CUBIT Core and thus does not entirely depend on API's for the implementation. The limitations of Table 1 will be considered later in Sect. 3 on *Multi-User CAx Requirements*.

We have implemented the Fig. 8 architecture on a laboratory LAN, on a college cloud server, and across an Internet WAN, all with similar effectiveness because small data packets easily transmit across networks. When using a cloud server, we applied Hewlett-Packard's Remote Graphics Software (RGS) to effectively turn the local workstation into a terminal screen, with the CAx application instances and multi-user application server running on remote server blades. We will discuss the security implications of these architectures later.

The architecture in Fig. 9 differs somewhat in implementation because we had access to the CUBIT Core source code. This allowed us to thread client interactions with the server using Windows Named Pipes (NP) (Microsoft Developer Network 2012) for inter-process communication (IPC) and TCP/IP sockets for network communications. Two networking clients (External and Internal Clients) reside on each local computer, with the Internal Client embedded into the source code.

The External Client (EC) organizes the different type of messages through a serialization process. The multi-user actions require that client ID's be appended to any GUI instruction as a message structure. Examples of message structures established in the EC are command message, master trigger, and database reset. Commands messages are generated from the CUBIT GUI, Fig. 9b, and delivered to the CUBIT core to process (e.g., "create sphere" or "mesh volume 1").

Master trigger messages are messages initiated by the user to resynchronize their model with the server database for unusual situations such as network down, or when clients engage the design process asynchronously and thus join the work environment at different times. Message structures help the server distinguish between different CUBIT Connect operations so it can respond accordingly.

Table 1 BYU's v-CAx site prototypes

Vendor	Type	Name	Function	Limitations
Siemens NX	CAD	NX Connect	CAD	Single user architecture (kernels, single-threads, GUI's, API's configured for a single user) Limited access to event callbacks Single API thread Single user GUI Memory handles to entity data Few social media tools Some API functionality errors Single user undo/redo Models stored as files
Dassault Systèmes CATIA	CAD	CATIA Connect	CAD	Same as NX except CATIA API runs on separate thread
AutoDesk Inventor	CAD	Inventor Connect	CAD	Same as NX
Sandia CUBIT	CAE	CUBIT Connect	CAE structural analysis	Same as NX, except greater access to event handler Multi-user synchronicity limited by meshing times Meshing time user variability P2P model updating latencies, thus move to client-server version Entity ID's computer fixed

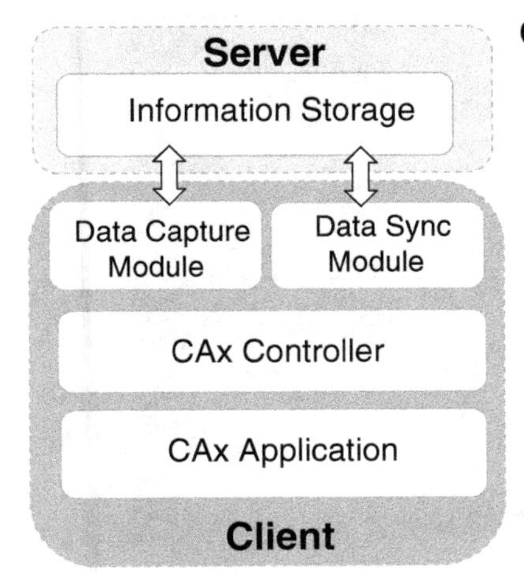

Client-server module functions:

- **Information Storage Module** – uses a relational database for data storage and a hierarchical structure to sync the part features and data changes.
- **Data Capture Module** - monitors the CAx session for changes to the part file and then passes the change information to other users through the server.
- **Data Sync Module** - monitors the information storage module for changes uploaded by another user, using these changes to alert the CAx Controller.
- **CAx Controller** - converts all model edit information into primitive values for database storage, translating the primitive data and parameters back into the API constructs required on each user's computer.

Fig. 8 Client-server architecture for BYU's CAx (CAD) multi-user prototypes

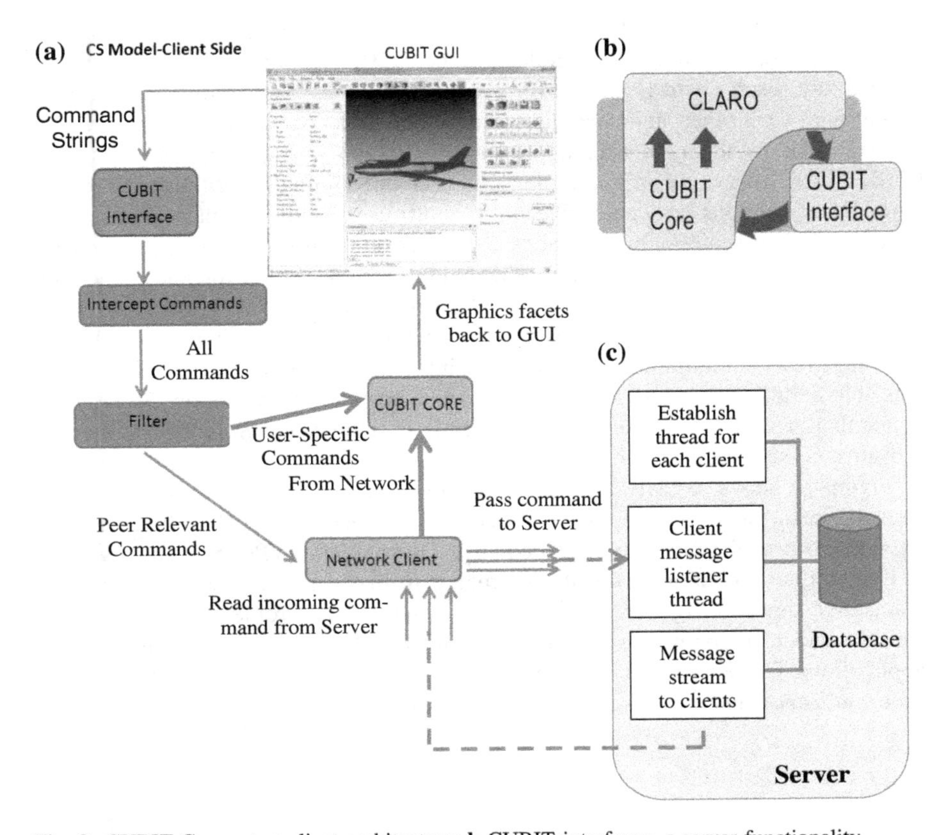

Fig. 9 CUBIT Connect: **a** client architecture; **b** CUBIT interfaces; **c** server functionality

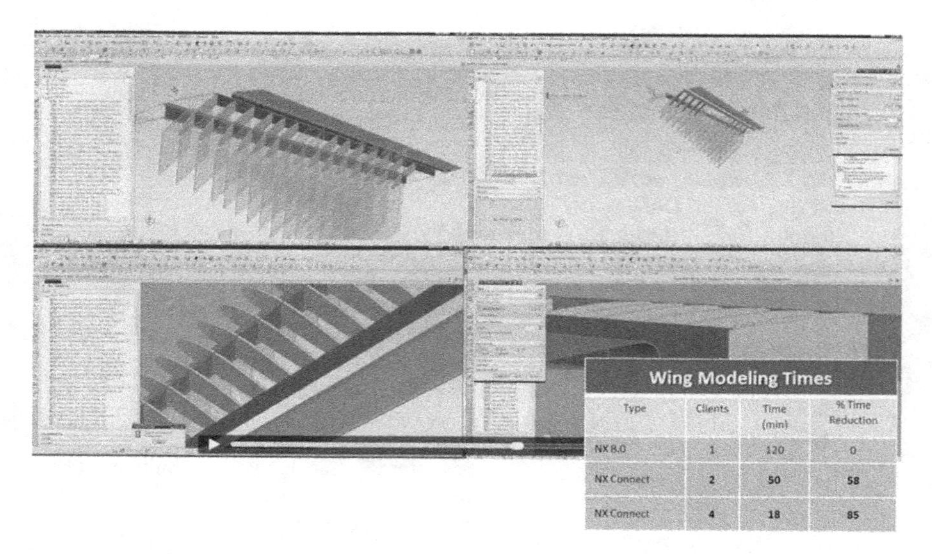

Fig. 10 NX Connect with 4 clients developing a wing structure simultaneously

It should be mentioned that meshing algorithms, unlike CAD algorithms, can take minutes or even hours to complete. Since CUBIT Connect uses a thick client architecture, each client has to perform those same mesh operations to stay consistent with each other, and users must wait while those operations are performed. CUBIT, like most single user applications, runs on a single thread; thus, the CUBIT GUI freezes while a complex and time-consuming algorithm runs in the core. These actions are untenable in a multi-user environment and required changes to our server/client methods to permit multi-users to control when to update from clients, along with passing CPU meshing times for each meshing command among the clients for informed decision making.

In contrast, Fig. 10 shows an instance in a NX Connect modeling session where four multi-users simultaneously design a wing structure. Because the team leader was able to decompose tasks based on expertise, actual time reduction (85 %) was better than T/N (80 %), where T is the single user time and N is the number of multi-users; see the inset table.

Figure 11 shows a CUBIT Connect design session where three users simultaneously mesh three components of a race car developed in collaboration with 26 universities over 4 years as part of the PACE Global Vehicle Project. The question of how to filter and/or stack update commands from other multi-user clients is still being investigated. Algorithmic delays, not a substantial problem in CAD applications, and a minor problem in CAM applications, can be significant in CAE applications. Current research is considering operational vectors that are tagged to each multi-user and that can be applied at optimal times.

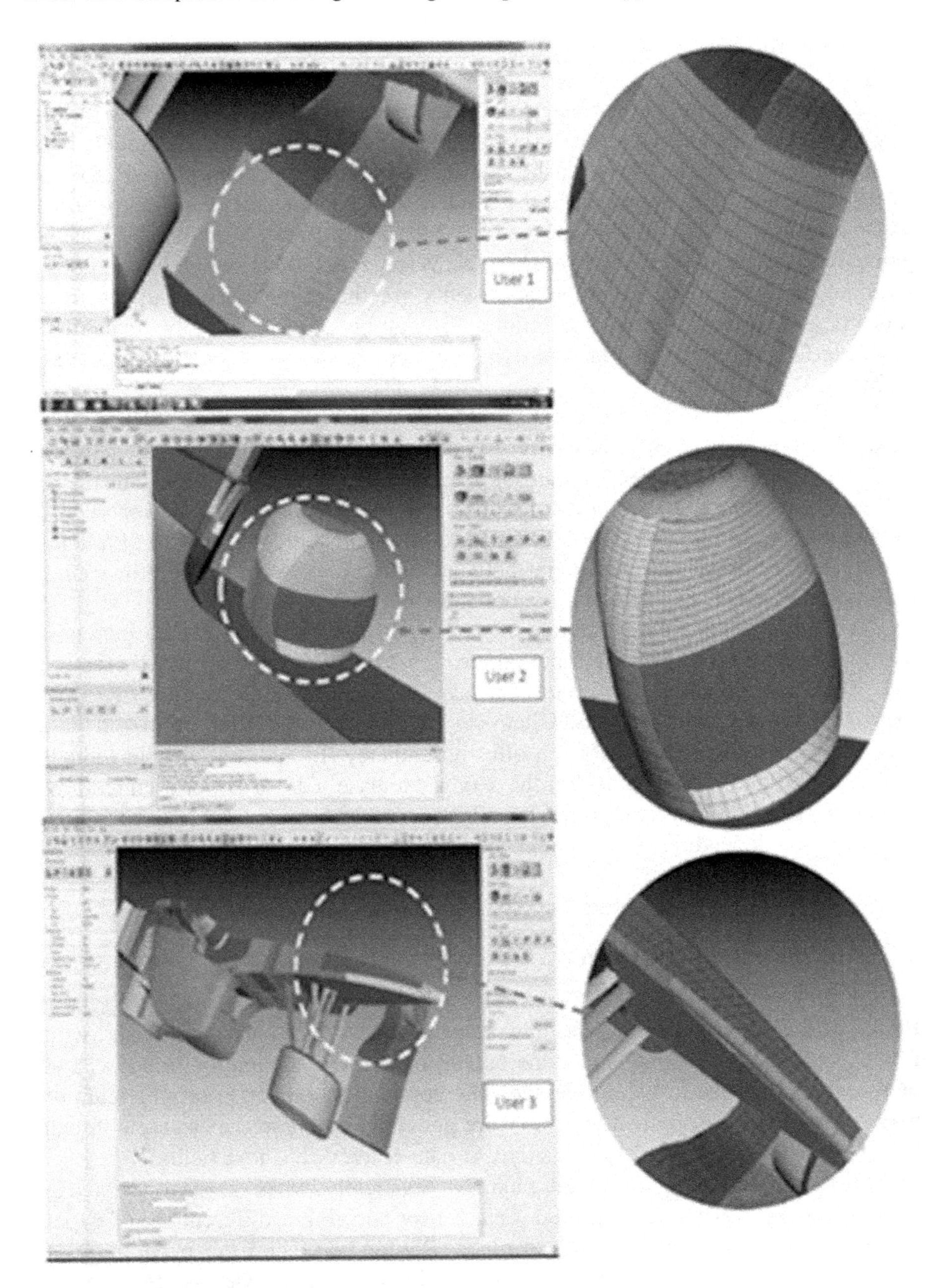

Fig. 11 CUBIT Connect session with 3 clients simultaneously meshing a race car

2.1.2 Peer-to-peer

An alternative architecture connects clients directly as peers, with each workstation replicating the software used to manage the change control for the model (Jing et al. 2009; Fan et al. 2008). In the gaming community, peer-to-peer (P2P) has many adherents, partially because the model data (game space) can accept some weak inconsistencies (Bharambe et al. 2006, http://www.cs.cmu.edu/~ashu/gamearch.html; GauthierDickey et al. 2004; Wang et al. 2004; Douglas et al. 2005; Endo et al. 2007). The common arguments are that client-server suffers from robustness and single point of failure (Bharambe et al.), lack of scalability (GauthierDickey et al.), and lack of dynamic reconfigurability (Ramakrishna et al. 2006).

P2P architectures also have weaknesses: (1) data inconsistency; (2) variable network latencies; (3) weak security; (4) additional client software; (5) network congestion; and (6) requires Internet service provider (ISP) cooperation. Thus, P2P solutions are often cast as hybrid solutions using traffic density allocated servers to maintain the gaming data. The solutions are often clustered as hybrid P2P/client-server architectures, with additional network activity software to monitor traffic, and measure client responsiveness among the players.

CAx applications cannot tolerate any ambiguity in the form of data inconsistency. Any entity data error will propagate disastrously through a model's feature tree. Data centralization and protection has long been a stable feature of PDM/PLM file management systems. Client-server architectures offer improved data security while mirrored servers can offer redundancy. In addition, design models can be stored and protected within the server database. Client-server architectures can be deployed behind company firewalls as a server-centered LAN or as a local cloud server, depending on the client to server network configuration.

2.2 Collaborative Interfaces

Real world CAx applications and their respective interfaces are single user, with few exceptions like GoogleDocs: one cursor, one active application window, and one mouse or mouse-like input device. The application expects a serial stream of user actions, and the application GUI's are generally configured as a single thread to react to mouse (or touch) events and to data entered into text fields.

The collaborative environment changes drastically when several users can simultaneously edit a common model, each user carrying a different set of experiences, capabilities, and cultural attitudes. Researchers echo the need for user awareness. Liu et al. (2008) propose that intelligent agent software be deployed at each client workstation to support awareness interactions among multi-users, such as capturing and transmitting audio and video streams for each user. Another agent might capture design changes, and update these changes among users, according to user privileges.

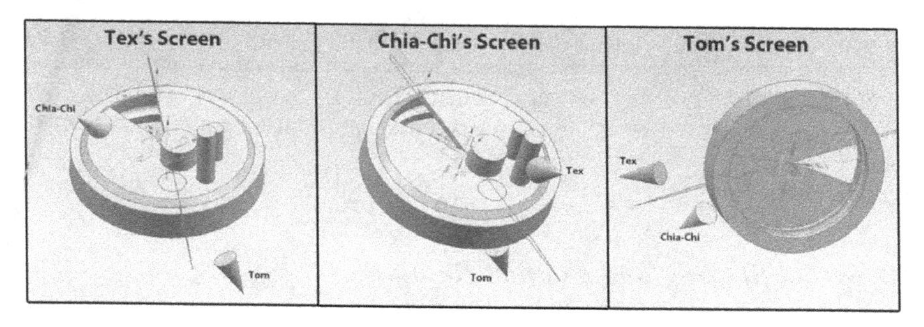

Fig. 12 Three client perspective visualization demonstration

Table 2 Flexible context filters

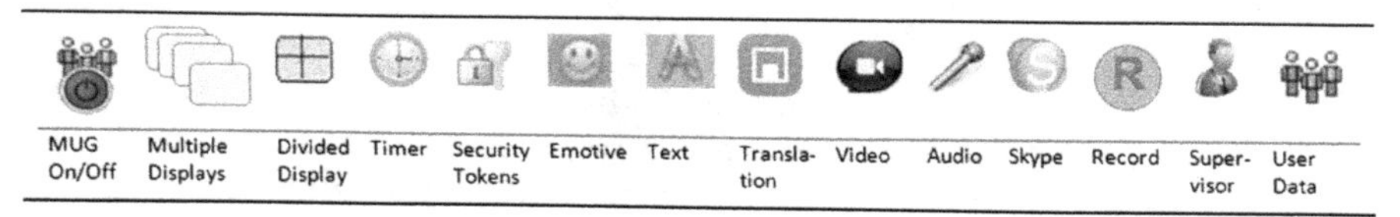

MUG On/Off	Multiple Displays	Divided Display	Timer	Security Tokens	Emotive	Text	Transla-tion	Video	Audio	Skype	Record	Super-visor	User Data

The BYU NXConnect prototype allows users to share their view perspective with other users who are in the same model (Nysetvold and Teng 2013). This improves the quality of their collaboration while keeping bandwidth usage low compared to existing screen share technologies. Since each NXConnect client has an up-to-date model, the only additional data that needs to be sent to facilitate view sharing is the view translation, rotation, and scale information, which can be described with as few as twelve double-precision floating point numbers. In addition, not only can users choose to follow another user's viewing perspective, they can also see cones that represent the view perspectives of other users (Fig. 12).

This capability enhances each user's awareness of the viewing perspective of other users within the same model. Knowledge of what another user is looking at may even allow a user to infer what the other user is currently working on.

Xu (2011) proposes a flexible context interface between users working on the same part at distributed locations, and defines the context as options in a multi-user GUI (MUG), Table 2. MUG uses agent software filters to render the user prototype outside of NX and NX Connect. The interface can then be used with other engineering applications where several users wish to collaborate. NX MUG, as configured for Siemens NX, enables users to view a collaborating user's workspace, send/receive messages between multi-users, and is capable of translating text interactions into different languages. Although Xu did not implement all the filters shown in Table 2, multiple icon selection sequences could be used to vary context preferences for each user in a collaborative session.

Other researchers have experimented with multiple cursors or touch control, but default to time slicing a single cursor event (see http://sourceforge.net/projects/multicursor-wm/#; http://lifehacker.com/5080196/teamplayer-enables-multiple-input-device). Focusing on the importance of GUI collaboration, Baomin Xu et al. (2009) proposed sending Windows GUI events to multi-users rather than model changes made through the CAx API, but this seems difficult in the face of so many GUI versions. Although GUI's change constantly, they are built on a fairly stable API library.

2.3 Model Sharing and Conflict Resolution

Having multiple users concurrently contribute to the same model introduces data consistency challenges that are not present in the single user scenario. Syntactic and semantic conflicts can occur, and the order in which features are created can vary between clients. Approaches for preventing data inconsistency include, spatial decomposition (Red et al. 2010, 2011, 2012, 2013b), collaborative constraints (Panchal et al. 2007; Lai 2009; Chen et al. 2004; Marshall et al. 2011; Ram et al. 1997), locking and blocking of model features (Bu et al. 2006; Moncur et al. 2012; Hepworth et al. 2013a), user negotiated feature access control (Zheng et al. 2009), and role-based and lean model editing access (Cera et al. 2003; Wang et al. 2006).

Figure 13 shows how simple planar constraints can be used to confine multi-users to a certain space, so that mouse events are only effective for the user's assigned space (Marshall 2011). Once fully implemented, these methods will work for even the most complex models. Objects that cross boundaries will require negotiation among adjoining users for editing rights. Extending these concepts to other constraint surfaces simply require that the entity boundaries be compared against the equations in Table 3, or for other possible constraint surface types.

Marshall (2011) implements the constraint equations by a selection filtering tool. The selection filtering portion of the implementation is integrated within the CAD system (mouse cursor combined with feature selection ray cast normal to the viewing window) and has a single dialog window that allows for selection among different multi-users. Depending on the user, a selection filter is applied to all possible selections based on four constraint planes. This early prototype allows for selection of edges and faces, which make up a model P where any spatial point $\mathbf{p}$ is described by coordinates x, y, and z.

$$\mathbf{p} \in P \tag{1}$$

Let X, Y, and Z represent the x, y, and z ranges of points $\mathbf{p}$ in P such that

$$x \in X;\ y \in Y;\ z \in Z \tag{2}$$

For the model of Fig. 10, the constraints in (3)–(6) have been implemented using inch units. ACCEPT P means a feature on the model is selectable by the multi-user. A selectable feature can be edited by the multi-user.

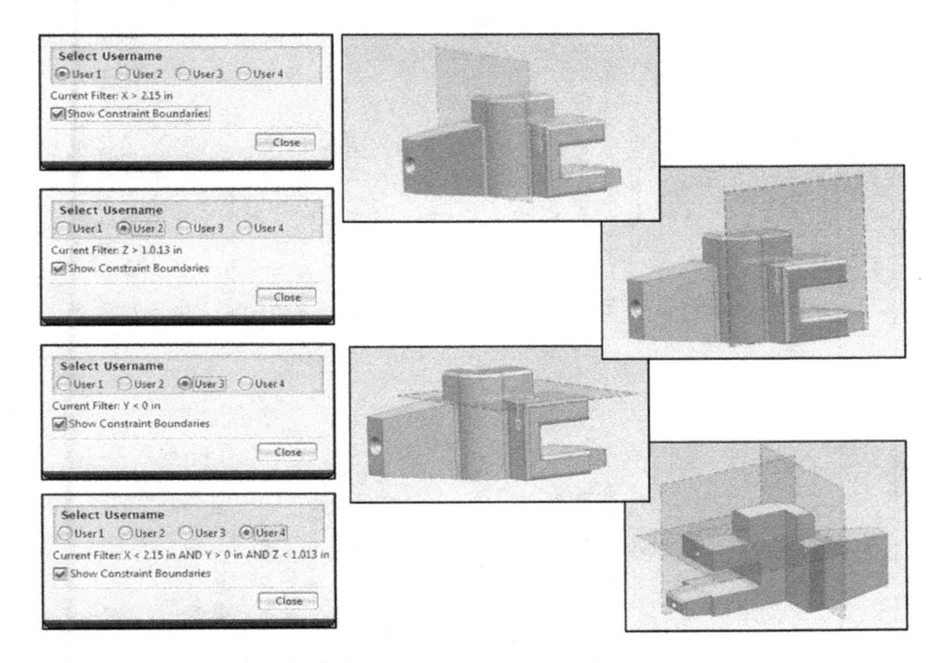

Fig. 13 Geometric constraint limiting of user feature selection

User 1. only select edges and faces for which

$$\text{if any } x \in X > 2.15, \text{ ACCEPT } P \tag{3}$$

User 2. only select edges and faces for which

$$\text{if any } z \in Z > 1.013, \text{ ACCEPT } P \tag{4}$$

User 3. can select edges and faces for which

$$\text{if any } y \in Y < 0, \text{ ACCEPT } P \tag{5}$$

User 4. can select edges and faces for which

$$\text{if any } x \in X < 2.15 \text{ and } y \in Y > 0 \text{ and } z \in Z < 1.013, \text{ ACCEPT } P \tag{6}$$

Normally, a feature would highlight as the mouse hovers over it to show what would be selected if the user were to click the mouse. However, if a feature is not selectable, the feature will not highlight when the mouse hovers over it. The constraint boundaries can be toggled on/off for visibility and are colored differently for each user.

Table 3 Constraint boundaries for spatial decomposition

Constraint	Graphical	Constraint Surface	Constraint forms (IE = Inequality; EQ = Equality)
Plane (unbounded)		$\mathbf{n}^T\mathbf{p} = d$ $\mathbf{n}$ = outward plane normal $\mathbf{p}$ = point in plane d = plane distance	$\mathbf{u}$ = user selected point $\mathbf{n}^T\mathbf{u} < d$ (inward, IE) $\mathbf{n}^T\mathbf{u} \leq d$ (inward, EQ) $\mathbf{n}^T\mathbf{u} > d$ (outward, IE) $\mathbf{n}^T\mathbf{u} \geq d$ (outward, EQ)
Cylinder (unbounded)		$(\mathbf{p} - \mathbf{v})^T\mathbf{e} = r$ $\mathbf{n}$ = cyl axis unit vector $\mathbf{e}$ = unit vector normal to cyl axis unit directed at $\mathbf{p}$ $\mathbf{v}$ = point on cyl axis $\mathbf{p}$ = point on cyl surface r = cyl radius	$\mathbf{u}$ = user selected point $(\mathbf{u}-\mathbf{v})^T\mathbf{e} < r$ (inward, IE) $(\mathbf{u}-\mathbf{v})^T\mathbf{e} \leq r$ (inward, EQ) $(\mathbf{u}-\mathbf{v})^T\mathbf{e} > r$ (outward, IE) $(\mathbf{u}-\mathbf{v})^T\mathbf{e} \geq r$ (outward, EQ)
Conical Frustum (bounded, reduces to cone if $r_2 = 0$)		$r_c = (\mathbf{p}-\mathbf{v})^T\mathbf{e}$ $(r_2 \leq r_c \leq r_1)$ $\mathbf{n}$ = cone axis unit vector $\mathbf{e}$ = unit vector normal to cyl axis directed at $\mathbf{p}$ $\mathbf{v}$ = point on cone axis at base where $r_c = r_1$ $\mathbf{p}$ = point on conical surface r_c = cone radius at $\mathbf{p}$ h = frustum length	$\mathbf{u}$ = user selected point *Step 1:* $d = (\mathbf{u}\text{-}\mathbf{v})^T\mathbf{n}$ *Step 2: Inward* IE: if $(0 < d < h)$ and $r_c = r_1 + \mathrm{d}(r_2 - r_1)/\mathrm{h}$ $r = \lvert\mathbf{u} - \mathbf{v} - \mathrm{d}\mathbf{n}\rvert < r_c$ EQ: if $(0 \leq d \leq h)$ and $r_c = r_1 + \mathrm{d}(r_2 - r_1)/\mathrm{h}$ $r = \lvert\mathbf{u} - \mathbf{v} - \mathrm{d}\mathbf{n}\rvert \leq r_c$ *Step 2: Outward* IE: if $(d < 0)$ or $(d > h)$ or $r = \lvert\mathbf{u} - \mathbf{v} - \mathrm{d}\mathbf{n}\rvert > r_c$ given $r_c = r_1 + d(r_2 - r_1)/h$ EQ: if $(d \leq 0)$ or $(d \geq h)$ or $r = \lvert\mathbf{u} - \mathbf{v} - \mathrm{d}\mathbf{n}\rvert \geq r_c$ given $r_c = r_1 + d(r_2 - r_1)/h$

Fig. 14 Front frame easily decomposed

Spatial decomposition, a divide and conquer approach, will be useful for complex and large models, or for models that have obvious shape independencies, such as the jet engine front frame in Fig. 14. Three design regions (inner case, outer case, radial vanes) can be decomposed by cylindrical and conical constraint surfaces. But decomposition may not always involve just looking for different model shapes, since users may be assigned based on model feature expertise as discussed by Moncur (2013).

2.3.1 Abstract Decomposition

Figure 15 shows an abstract representation of a complex design space divided into multi-user regions that we will refer to as design regions, d_i ($i = 1, n$), and where design space model D is the sum of these regions. Region d_i is the set of features, attributes, operations, and/or geometry associated with the design region. If only spatial decomposition is used, then D represents the volume which bounds the model geometry. If only feature decomposition is used, D represents all features that can be edited in the CAx application. Thus, the design space can be represented by $D = \Sigma\, d_i$, assuming that the decomposition is full, i.e., the set of design regions span the entire design space. Because of regional dependencies, the design space representation can be more complex as will be shown later.

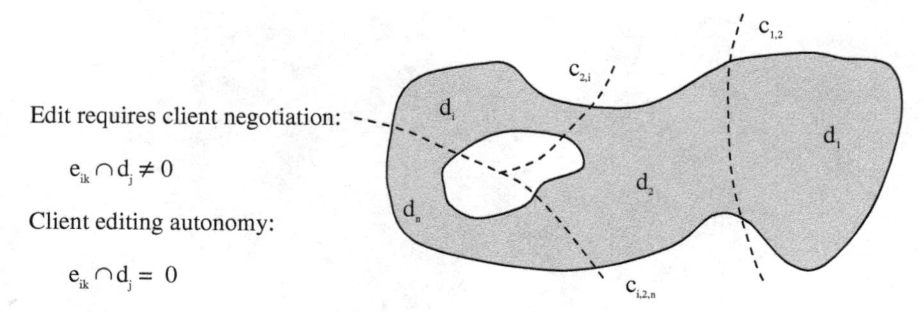

Fig. 15 Spatial decomposition

The set of elements k that define region i is represented as $e_{i,k}$. In multi-user *autonomous* decomposition, it is desired that all k elements reside within region i as depicted by $e_{i,k} \cap d_j = 0$ for all $j \neq i$. A simpler algorithm compares $e_{i,k}$ against all constraints $c_{i,j,k}$ bounding region i. If $e_{i,k}$ satisfies all constraints, then the client can edit the element without constraint. If $e_{i,k}$ does not satisfy all constraints, then element $e_{i,k}$ is locked and multi-users assigned to the intersected regions must be notified, where negotiation allows the now unlocked element to be edited.

The boundaries between two design regions are represented by user constraints $c_{j,k}$, where j, k refers to the constraint between multi-user regions j and k and where $c_{j,k} = c_{k,j}$. User constraints can be represented as geometric equations or feature sets, depending on the design space. Some design regions are defined by only one constraint equation, such as d_1 (constrained only by $c_{1,2}$) and d_n (constrained by $c_{\mathrm{i},2,n}$), while others may require comparison against several constraints, such as d_2 (constrained by $c_{1,2}$, $c_{2,\mathrm{i}}$, $c_{1,2,\mathrm{n}}$) and d_i (constrained by $c_{2,i}$, $c_{1,2,n}$).

As a simple example consider the problem of developing a manufacturing process plan to machine a part. The part model could be comprised of several feature sets that encompass operations required to manufacture a part. The feature set could be used to decompose the process plan into regions represented by operations like these: (1) roughing tool paths; (2) semi-roughing tool paths; (3) finishing tool paths for surface features; (4) pocketing and slotting features; (5) drilling, tapping, and threading features; (6) profiling features; (7) fixturing hardware and setup; and 8) tooling. Some of these operations are reasonably independent of other operations (e.g., pocketing as compared to drilling/tapping/threading) and could be used to decompose the process planning among several multi-users for simultaneous process planning. When a design region is totally independent of other regions, it will not need a constraint relationship to constrain the assigned user actions and thus $c_{i,..} = 0$. Independent regions, or mildly independent regions, are the best candidates for multi-user decomposition. Regions that are strongly dependent, i.e., connected by dependent features, will require cooperation and intense interaction between the multi-users to simultaneously edit the design space.

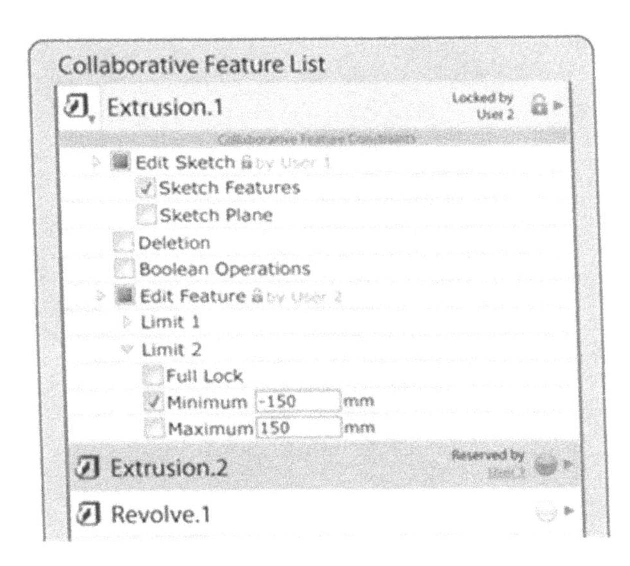

Fig. 16 Collaborative feature list

Some constraints may be common to more than one multi-user region, e.g., $c_{\mathrm{i},2,n}$ as shown in Fig. 15. For example, a multi-user assigned to region i will be constrained by those relationships that contain i in the constraint subscripts: $c_{2,i}$ and $c_{\mathrm{i},2,n}$. Thus, user design region i is defined by the associated feature set associated and constrained by those functions $c_{i,j,}$ with included i subscript.

Since modern CAx applications use feature representations, Moncur displays a representation of a part's feature list, Fig. 16 (Moncur et al. 2013). A user can change a 3-state toggle to reserve, lock, or release the selected feature. When changed, the new feature state is propagated to the other multi-users. When a feature is reserved, it means that the user who reserved the feature is able to edit or delete the feature. A *reservation* can be made by any user at any time, even if the feature is already reserved by another user (although this does not hold true if the feature is locked by another user). Because a reservation can only be held by a single user at a time in a collaborative design session, it prevents two users from editing the same feature at the same time. A reservation could be thought of as a per-feature edit token, or a "soft" lock for edit permissions for a given feature. This allows for open collaboration between users and allows them to work on a model while preventing update and delete conflicts.

A color scheme could be applied to the rendered features inside the CAx application to help other users see which features are available for editing, Fig. 17. Lock symbol decals could also be applied to features to convey editing state to other users.

Features can also be locked or reserved in groups. Thus, an entire branch of features can be locked by locking the parent-feature in the tree. As new features are added to a model, they can inherit any lock/reservation properties from their parent feature. This provides a way to effectively partition a CAx model by feature rather than shape.

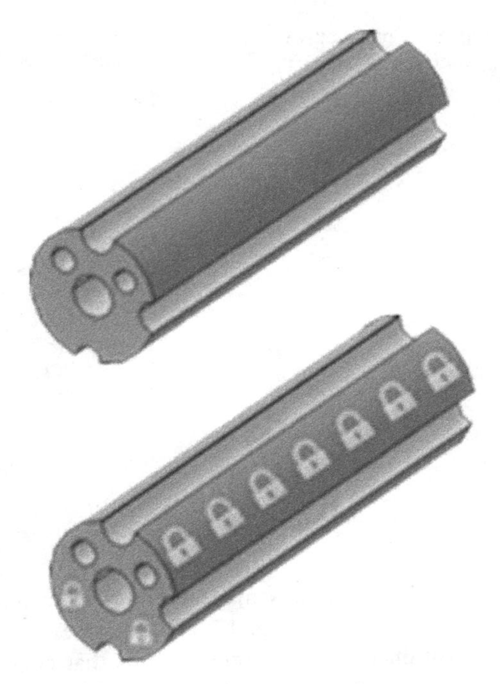

Fig. 17 Reserving/locking using colors/decals

Hepworth et al. (2013b (Automated…)) extend this approach by automatically reserving features during a feature edit operation. This prevents multiple users from simultaneously making conflicting edits to the same feature (feature-self conflicts). Feature-self conflict prevention is expressed mathematically by (7)–(10), given that F is the set of all features in part P, and in time interval Δt, and where f is a feature in part P, Δt. U is the set of all users in part P, Δt, and u is a user in part P, Δt. A part contains:

$$f \in F(P, \Delta t) \tag{7}$$

$$u \in U(P, \Delta t) \tag{8}$$

To prevent feature-self-conflicts for a given edit operation in part P, there must exist a unique set E, Δt, which contains only one feature f and one user u where

$$E(P, \Delta t) = \{f, u\} \tag{9}$$

interval

$$t_{\text{beginEdit}} \leq t \leq t_{\text{endEdit}} \tag{10}$$

forced by allowing only a single instance of set E to exist in part P, Δt, the time interval from the beginning to the end of the edit operation.

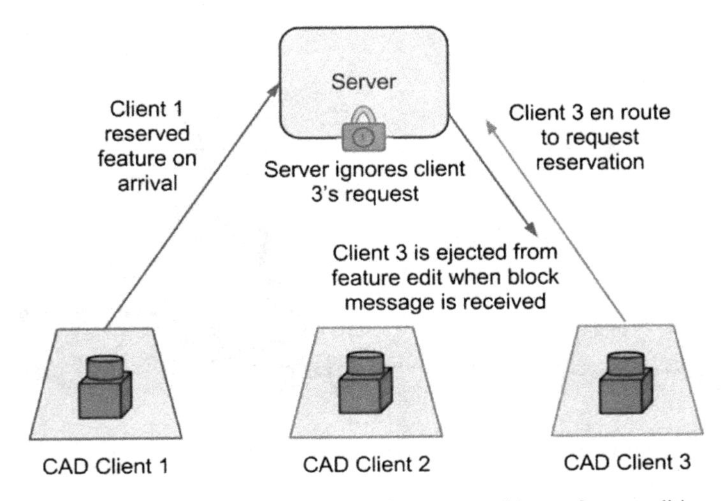

Fig. 18 Server reservation method prevents simultaneous, multi-user feature editing

Otherwise, feature-self conflicts occur if users on separate clients simultaneously edit the same feature.

Hepworth implements a method to prevent feature-self-conflicts by reserving a feature on the server on a first come first serve basis. When a user begins an edit operation, a message is automatically sent to the server requesting reservation of that feature. If the feature is not already reserved, the user is allowed to continue editing the feature. However, if the feature is already reserved, any other client request will be denied. When the user completes the edit operation, another message is automatically sent to the server removing the reservation. Figure 18 is a diagram depicting Client 1 reserving a feature while Client 3 makes a request and is ignored.

In addition to the reservation on the server, a client blocking mechanism is implemented to apprise users that a feature is reserved and prevent users from editing reserved features. When a feature is reserved by a user, all other users receive a message, blocking the feature. A block forcibly prevents users from editing a feature and changes the color of the feature to a predefined blocking color. The color helps to communicate the reservation and also show where users are currently working in the part. Figure 19 shows an example a one user editing a feature while a second user has the feature blocked.

Users may also request that reservations be removed on blocked features; see Fig. 19. This prevents users from having features blocked for an unnecessarily long period of time. Reservation request removal is done by sending a message to the reservation owner requesting reservation removal. The user has the opportunity to accept or reject the request (Fig. 20). The user has a limited time to respond before the request is automatically granted to the requester. This allows users to remove reservations on blocked features even if the owing user is unavailable to respond to the message (Hepworth et al. 2013b (Automated…)).

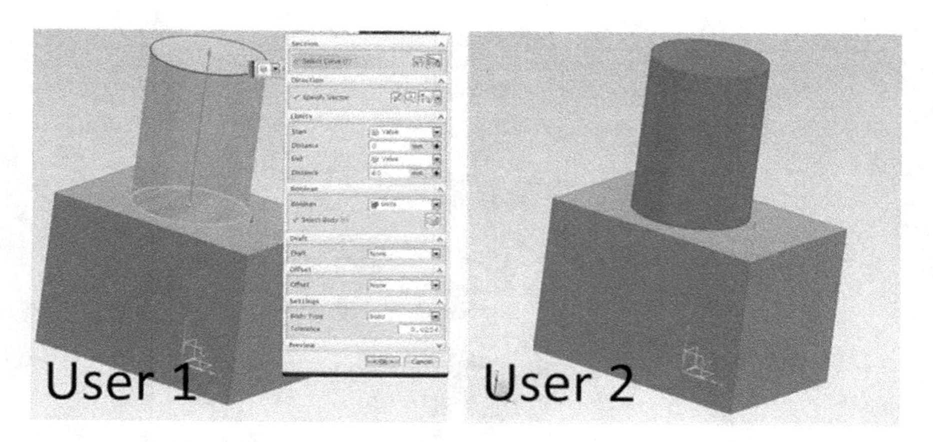

Fig. 19 User 1 edits an extrusion and it is blocked on user 2 (shown in *red*)

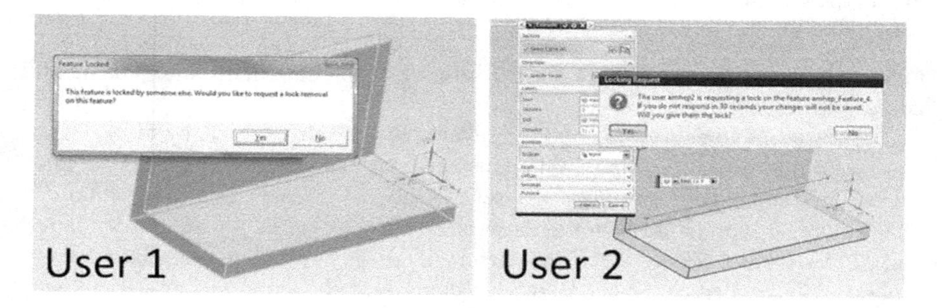

Fig. 20 User 1 requests reservation removal and user 2 receives the request

2.3.2 Undo/Redo in Multi-user CAD

Modern CAx applications allow the user to undo/redo their last actions (often implemented as CTRL + Z = undo; CTRL + Y = redo). But in a multi-user session, undo actions cascade through the feature tree which may have been constructed by several users through dependent operations. Li et al. (2008) note that "a feature of a product model might depend on other features and modifying an early feature may cause later features to become invalid." Gao et al. (2009) consider multi-user undo intent and use local history buffers to record the changes made by local clients. Because any user in a multi-user session can apply a feature undo function, dependencies can cascade and cause chaotic collaboration.

We have successfully applied an undo command method to our NX Connect prototype. The undo method we implemented only allows users to undo their own recent commands—not the commands of other users. Each model-changing command a user performs creates an undo action construct that contains both the data required to undo the command and the data required to redo the command.

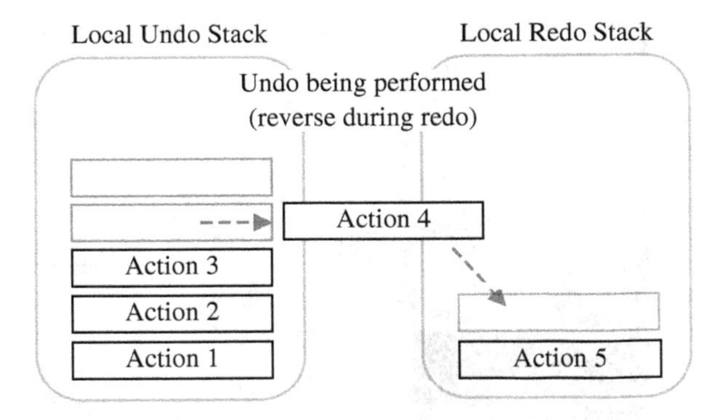

Fig. 21 Local Undo/Redo stacks

A timestamp records the time at which the command is performed. The action constructs are then placed on top of the local user's undo stack (see Fig. 21). None of the local user's action constructs or undo/redo stacks are sent to the server or to other clients. Each user has their own undo and redo stacks which only store that user's actions.

When a user performs an undo command, the action construct on the top of their undo stack is used to perform a local undo. This action construct contains the data to perform the undo operation. After undo is completed locally, the action construct is transferred from the user's undo stack to their redo stack as shown in Fig. 21. After the undo succeeds on the local client, a message is sent to the server, which forwards it to the other clients so they can perform the same modeling operation.

Figure 22a, b, and c show a multi-user engine block design session with three users. User 1 performs a Boolean operation to create the piston chambers, see upper left of Fig. 22a. Figure 22b shows the Boolean operations reflected to users 2 and 3. Figure 22c shows an undo operation performed on the piston Boolean in the local history tree. The undo operations have not yet been sent to the other two users, but this happens in a succeeding step.

3 Multi-user CAx Requirements

This section considers three main areas: (1) multi-user CAx functionality requirements, considering the deficiencies currently existing in single user CAx applications; (2) administration and management of product requirements in multi-user collaborative environments, and (3) human elements in multi-user team forming. It is important to note that the multi-user functionality already demonstrated in early prototypes like NX Connect and CATIA Connect exceeds single user functionality when it comes to reducing design times and sharing design rationale. Current CAx API's can be utilized to make CAx applications behave in a multi-user mode, but with extensive programming and API workarounds.

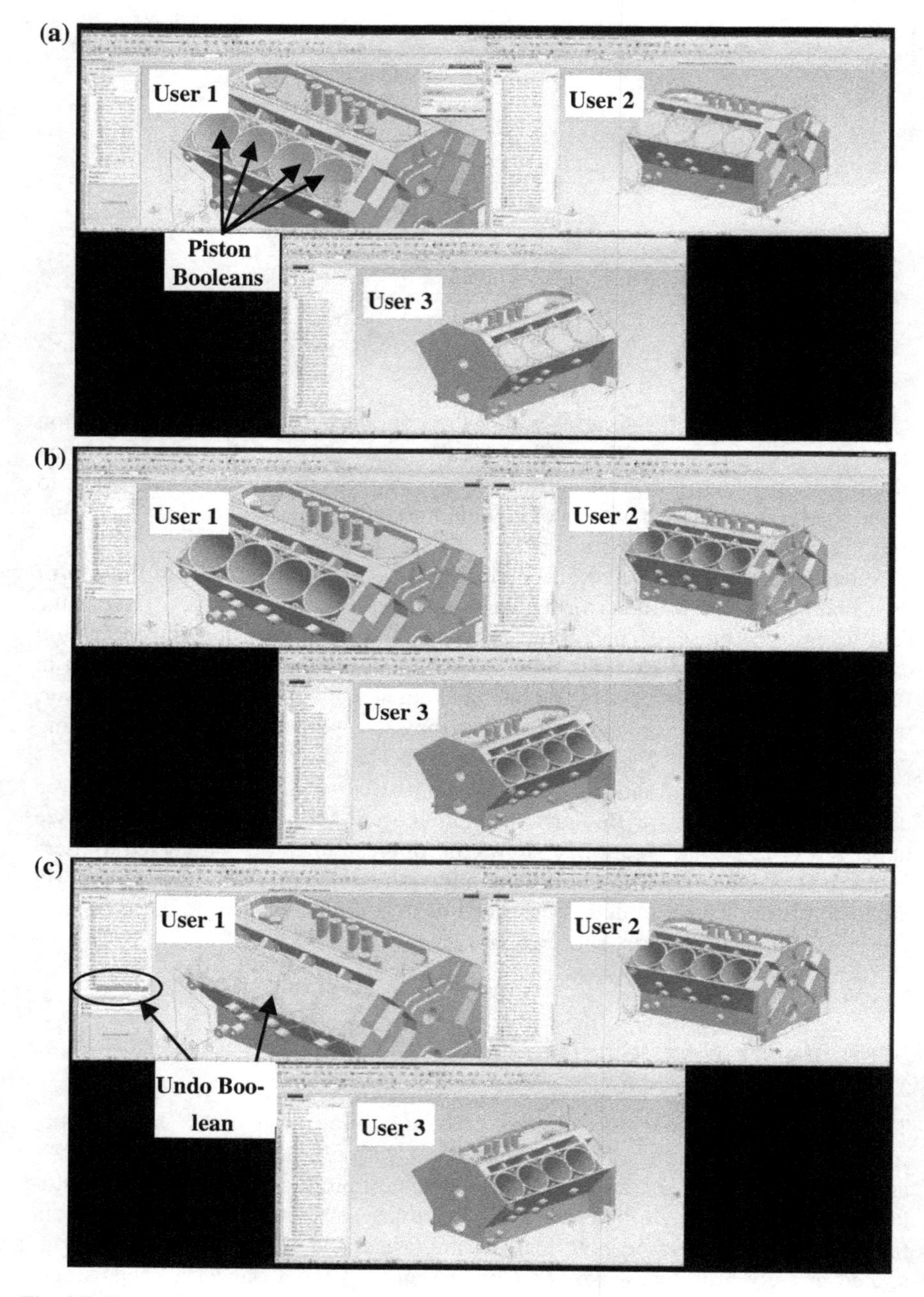

Fig. 22 Create pistons undo example. **a** Create pistons. **b** Reflect edit. **c** Undo pistons

Table 4 Desired multi-user features

Single user limitations	Multi-user implications	Importance (1 most)
Single threaded kernels*, GUI's, API's (*Parasolid, ACIS, etc.)	The advantage of threading is that multi-users can share a kernel, GUI, or API, based on computational and network resources. Threading would reduce the number of site licenses required for multi-user sessions. Presently, each client must have a separate CAx license.	2
API entity memory "handles"	Handles, i.e., memory addresses, are allocated differently in each computer. Current multi-user plug-ins require extensive programming to discover the data structure format for each entity and related operation that must be passed between clients. API's should provide object/data structure copies for data exchanged among clients in multi-user sessions.	1
API functional errors	Many of the less used API's have functional errors, and should be fixed.	2
Single user undo/redo	Single user undo/redo operations do not work in multi-user CAx, but we have demonstrated that simple methods such as tracking of local user actions can provide acceptable undo/redo functionality. Single user CAx applications will have to be converted to client-server architectures because a database must track operations performed by each user in the session.	1
Models file-based	Client-server architectures with server databases provide more accessibility to all product related data, and permit multiple users/clients to access and edit models simultaneously.	1
No access to event handler	Without access to a CAx event handler, prototypes must track temporary session files associated with undo markers, or track Windows GUI events. A simple event handler API to trigger alerts as users edit model data would replace more complex plug-in programming and replace polling timers.	1

(continued)

Table 4 (continued)

Single user limitations	Multi-user implications	Importance (1 most)
CAx algorithmic time variability	Algorithms with significant latency can hinder the multi-user process. Possible work-arounds or solutions include providing time-to-completion estimate for each operation, requesting that software providers improve algorithm speeds (e.g. by parallel processing), and multi-threading the application core/GUI/API such that the local user could continue to work while another user's command is being applied to the local user's model. Multi-user CAx applications faced with operational latencies will also require asynchronous functionality so that users can enter and leave an active session as desired. The users must also be able to control when they wish other client operations to update their model. For CAx application with severe algorithmic latencies, autonomous decomposition approaches will be most effective.	3
Entity numbering schemes	Entity numbering schemes are not uniform across CAx applications, with entity numbering often order dependent. In such cases the server must maintain each user's modifications as ordered and then reflect them to the clients in the same sequence. More uniform and predictable numbering algorithms conforming to multi-user model architectures would simplify server implementations.	2
Limited social tools	Social tools for collaborative awareness are critical to effective multi-user teaming and should be integrated into CAx applications. Surveys have revealed that product development personnel are using many such tools to collaborate because these tools are not prevalent in CAx applications.	1

3.1 Multi-user CAx Functionality Requirements

Table 4 lists several important single user CAx architectural limitations, some of which are inherent to a particular CAx application, like CAE/FEA preprocessing of meshes. Table 4 recommends some changes to improve multi-user viability. Many of these limitations are not difficult fixes and would reduce multi-user server complexity, such as entity copies rather than memory address handles, integrated social tools, fixing API errors, predictable entity numbering schemes, and exposing an API to the primary event handler. Multiple threading and algorithmic latencies will prove more challenging.

3.2 Multi-user Administrations of Product Requirements

Liu et al. (2008) recognize the reluctance to use collaborative CAD, which he says can be related to network latencies, security, and cultural issues, and because software companies have not adopted collaborative advances. We suggest that nonadoption is more closely related to potential disruption of normal business processes; stable companies are very reluctant to experiment with their internal processes.

Regardless of potential, industry will not adopt new processes unless the administrative issues are understood and easily implementable. Refer to Hannan and Freeman's (1984) popular paper to review entrenched principles of industrial inertia. We suggest that any company which has applied virtual teaming approaches can be comfortable with multi-user CAx processes. And multi-user tools simply provide an opportunity when appropriate, and do not need to be applied. Structurally, they require reliable networks, task assignments, schedules, administration, and team leadership, not really different from localized process teams. With predetermined workspaces and boundaries, collaboration decreases the product development time in proportion to the number of multi-users (Jensen 2012).

A less trumpeted advantage of multi-user CAx is the knowledge transfer and training that occurs among cross-functional virtual teams. Presently, experts train novice engineers in a time-consuming one-on-one process. Since multi-users, even technical personnel from different disciplines, can view and edit each other's work on their screen, potential problems are more easily identified early on and fixed.

Red et al. (2012) state that:

> The advent of multi-user computer-aided applications (CAx)……will change personnel/organizational assignment processes in product development…..collaborating personnel/organizations will enter design sessions and simultaneously edit/review design spaces. This paradigm shift will require new methods to be developed that decompose development tasks over personnel/organizations at both local and global locations. Experiential data will not be restricted to suppliers, organizations, or sites, or other grouping types, but reflect a different granularity where a particular group of individuals from a variety of organizations might be collected into design teams for optimal collaboration.

Currently, there are no process-specific or model-specific multi-user decomposition principles generally accepted or in standard form, although a few have been suggested in this chapter, such as decomposing a model based on geometric or feature independencies. But spatial or feature independence recognition is of no help if you lack multi-user expertise for the observable independencies.

Consider the following steps employed in current product development practices:

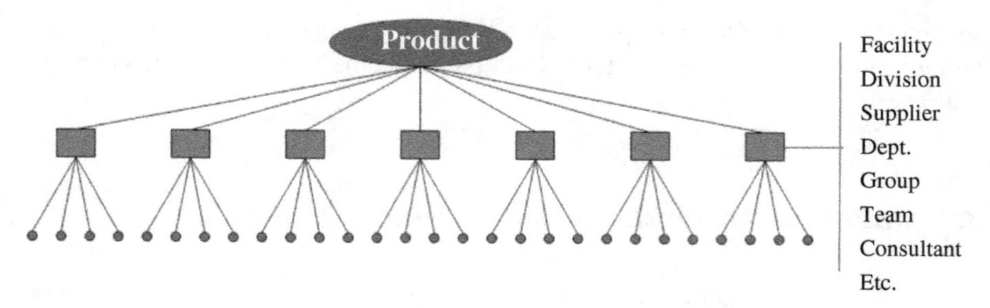

Fig. 23 Product decomposition

Step 1 *Product idea*. A product idea is conceived; requirements are specified, along with related specifications

Step 2 *System decomposition*. Product development is decomposed into elemental forms, like systems, subsystems, components, parts, etc., with more detailed requirements and specifications.

Step 3 *Organizational decomposition*. These elements are then assigned and developed according to a set of decomposed tasks (and related schedules), which recognize organizational capabilities (company, division, contractor, supplier, group, team, etc.), resource availability (raw materials, OTS/off-the-shelf components, tools, workstations, facility, personnel, etc.), schedule, logistics (moving things around), and sometime local politics (unions, laws, etc.) which may affect all the previous. Much of the decision making involves experienced personnel, similar or competitive products, and past development histories.

Step 4 *Resource decomposition*. Assignment of tasks is usually based on experiential information and resource availability, and subject to other constraints, such as government regulations, or competitive response from competing companies.

Step 5 *Completion*. Decomposed product and related development assignments are then completed according to assignments, resources, and schedule. Because current CAx applications are single user, the process is serial with corrections made as development errors are discovered in the stage gate process, which requires time-consuming iterations to the CAx models.

Figure 23 shows the typical decomposition hierarchy used in product development industries. Different organizational principles may be applied to manage the product team, such as matrix, pyramid, group, or even virtual teaming where team members are not necessarily co-located.

Now, consider the emerging paradigm of multi-user CAx. From the viewpoint of multi-user decomposition the same five steps are amended to consider possible multi-user administrative principles in Table 5. An associated asterisk (*) denotes that the tool/principle needed is poorly developed or not formalized.

Table 5 Postulated multi-user administrative principles

Step	Name	Multi-user implementation
1	Product idea	Collaborative principles* are used to review and tag the requirements and specifications for potential multi-user teaming opportunities, according to personnel and supplier availability and capability.
2	System decomposition	Conventional practice decomposes a product based on performance histories of organizations making similar products, by group or supplier capabilities, or by location, logistics, government regulations, etc. Using personnel and supplier multi-user experiential data*, and multi-user possibilities from Step 1, decomposed elements can be directed towards multiple organizational entities, for potential virtual teaming.
3	Organizational decomposition	Multi-user decomposition requires comparison of available user backgrounds against model specifications and features to determine whether personnel are available and suitable for multi-user mode. It would seem that new databases* would maintain relevant experiential data about personnel based on their educational background, technical and model space experiences, and expertise both in technical matters and in human-interaction and management. It is likely that department or group level resources, i.e., experiential databases*, could be searched to establish expertise and experience, and existing schedules can also be used to determine personnel availability.
4	Resource decomposition	Multi-user decomposition tools* would examine the model design space to determine model complexity and whether there several regions that are independent or mildly independent, based on proposed model features, or similar products previously developed. If personnel resources are available, and model is sufficiently complex, multi-user decomposition of the design model among several users may be desirable. In the case where multi-user training* of novice users by more experienced personnel is desired, model complexity may be relaxed.
5	Completion	Multi-user sessions are scheduled and managed by a team leader*, using integrated and networked multi-user CAx GUI's* and client-server architectures to store user actions and design rationale histories in a central database*.

Figure 24 shows how multi-user virtual teams can be organized from a community of personnel, suppliers, or other organizational entities, given access to experiential histories and capabilities. Also note that one individual could be assigned to sit on several multi-user teams, as could personnel from different functional areas, like designers, analysts, testers, marketers, production personnel, etc.

Not all decomposition tasks are compatible with multi-user development. Tang et al. (2000) notes three types of relationships between design tasks: uncoupled, coupled, and decoupled. Uncoupled tasks have low interdependency and can be performed in parallel. These qualify for multi-user collaboration. Coupled tasks are highly dependent and utilize iterative cycles to solve conflicts.

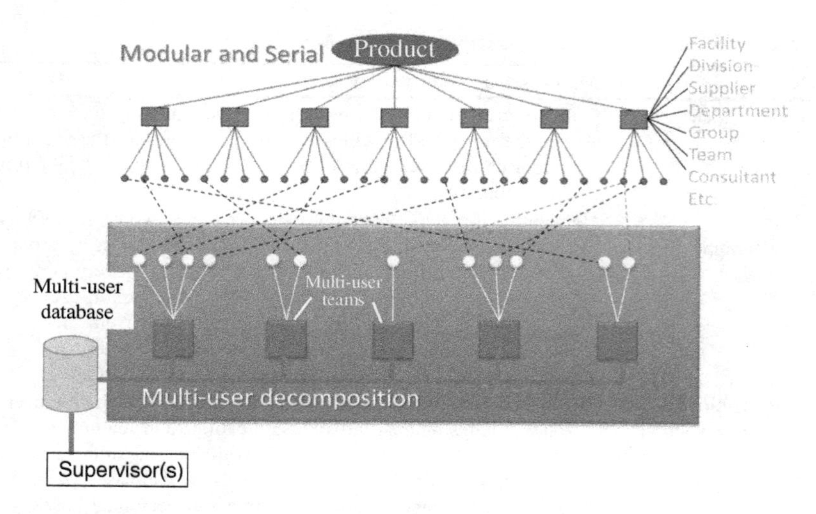

Fig. 24 Product decomposition: *top* is conventional; *bottom* shows organized multi-user teams (*red squares*)

Tasks that are decoupled can be performed sequentially and contain only one way dependencies. Holyoak (2012) extended the methods of Tang to apply a Design Structure Matrix (DSM) to correlate design tasks specifications and determine task dependencies. From these multi-user assignments can be appropriately made.

3.3 Human Elements in Multi-user Team Forming

Step 3 of Table 5 presents an opportunity to more optimally allocate human resources within product development communities. By adapting forming techniques based on personnel experiences and capabilities, product organizations can organize employees into multi-user teams that take full advantage of their experiential backgrounds.

"Customer Relationship Management" and "Data Mining" (Rygielski et al. 2002) are techniques that identify and extract information about individual customers. Using this data, an organization can apply algorithms to predict future behavior and improve customer relationships. Customer data is parsed by automated algorithms to predict customer behavior, needs, and desires. Vendors like Amazon.com participate in a Customer Relationship Management interaction.

When an Amazon webpage suggests another purchase, it does so automatically based on data it has gathered about that specific customer's preferences using "collaborative filtering" algorithms and Search-Based Methods. Additional information such as customer age, gender, geographic location, and more (gathered through surveys, coupon offers, etc.) allow the vendor to compare an individual

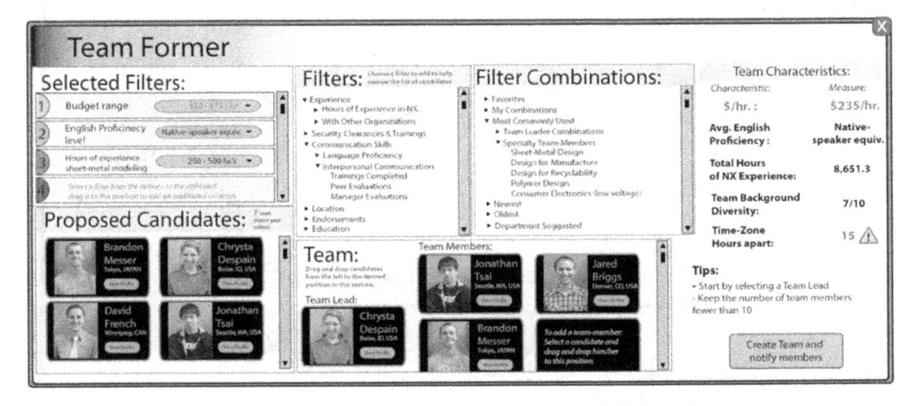

Fig. 25 Artistic rendition of a multi-user experiential search interface

customer to other similar customers. Cluster algorithms (Linden et al. 2003) can then be used to display advertisements and product suggestions that will attract a specific customer.

Now consider the gathering of user experiential data for multi-user CAx environments. Multi-user architectures require methods to organize personnel into local (or co-located) or virtual project teams. To do this optimally, project leaders, or supervisors, need dynamic data representing an individual's capability and experience in applying CAx applications, as appropriately mapped to project requirements and specifications. We suggest that both a user's CAx experiences and their social/cultural markers could be mapped into experiential databases; Fig. 25 is an interface rendition. This data could then be searched by applying search algorithms and automated suggestions similar to the marketing algorithms discussed earlier.

It would not be necessary for a manager to personally know or have interacted with an employee to accurately predict personnel skills, or motivation to work on a given project. Managers can quickly filter employees from multiple organizations for a project group, allowing organizations to more fully utilize the abilities and talents of their employees. Managers will access a much larger pool of candidates. Employees will work on projects for which they are most qualified. Other potential benefits include a decrease in need to relocate employees or travel long distances to work as part of a project team.

4 Multi-user CAx Standards

Multi-user architectures, by providing uniform methods for interacting with distributed suppliers (or even within company divisions using different CAx vendor applications), will spawn standards. These standards will seek to make CAx data interoperable, regardless of vendor application.

Brunnermeier and Martin (1999) note interoperability deficiencies for the automotive industry in the executive summary of a 1999 report prepared for National Institute of Standards and Technology (NIST): "This study estimates that imperfect interoperability imposes at least $1 billion per year on the members of the U.S. automotive supply chain. By far, the greatest costs relate to the resources devoted to repairing or reentering data files that are not usable for downstream applications. This estimate is conservative because we could not quantify all sources of interoperability costs."

Another interesting statement can be found in the Committee (2008) summary of the report *Pre-Milestone A and Early-Phase Systems Engineering: A Retrospective Review and Benefits for Future Air Force Acquisition*: "There has been about a *threefold increase in delivery time* for most major systems...puzzling given enormous productivity advantages conferred by...Internet and e-mail...advances in knowledge-management and collaboration tools such as...(CAD)...(CAM)..."

It is not difficult to see that "imperfect interoperability" is extending schedules and costs for complex products. Simply consider the increasing supplier chains for large product companies. For example, the Boeing 787 Dreamliner schedule slipped several years, attributed 67 % to outsourcing to suppliers. Boeing uses approximately 6,450 suppliers; Goodrich (now UTAS) and PPG Aerospace are two Boeing Tier 1 suppliers each with supply chains listed in the 1,000's. Ultimately the *leaves* of the Boeing supply chain touch small manufacturing firms with minimal CAx tools. Model file conversions in the millions and schedule delays are required to get suppliers access to Boeing (or UTAS or PPG) data. File-based *uniformity* translations like IGES, STEP, JT, etc., are architecturally not the solution, because vendor model representations and their data files are not standardized (Choi et al. 2002; Li et al. 2012).

Researchers have targeted solving this problem as early as the mid-1990s where Hoffman et al. (1993) proposed Erep, an editable representation which serves as a global schema for data exchange. Anderson (1998) proposed the Enabling Next GENeration (ENGEN) project to develop a data model to exchange design intent information. He later proposed Solid Model Construction History to exchange history-based information (Anderson 2002). Several papers have discussed the macro-parametrics approach in which a macro records the modeling commands of the parametric feature history. Data is exchanged through a standard set of commands used as a neutral format, which acts as an intermediary between the various commercial CAD macro files (Choi 2002; Mun 2003; Yang 2004; Kim 2007; Han 2010).

Rappoport (2003) recognized the incompatibilities between CAD systems so he proposed Universal Product Representation (UPR) architecture as a potential solution. This is a union of all common data and operation types between CAD systems. Data types not compatible are rewritten as another data item. In this manner, compatible data is transferred between CAD systems parametrically. Li et al. (2009) use a two stage process to recover the entire modeling process of a user and use this to simulate a real human in the target system. Although the

Fig. 26 Entity record

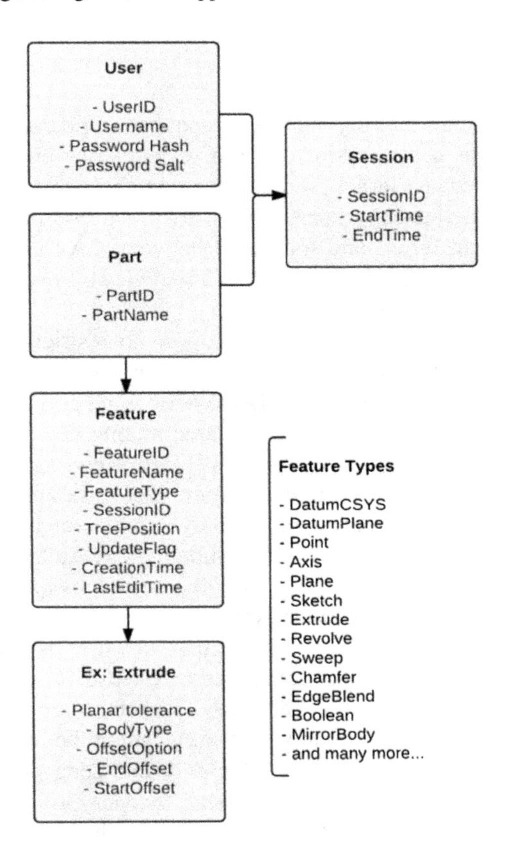

parametric data exchange problem has yet to be completely solved, much progress has been made in this area.

Translation aside, multi-user architectures (e.g., client-server) naturally provide for standardization, because model data can be stored in a neutral, easily accessible format on server databases. Replacing proprietary model files with database representations provides an architectural framework that inherently supports data interoperability. If the database representations can be normalized and standardized, something like that in Fig. 26, then any compatible CAx application can edit the model data regardless of location, assuming that security provisions only allow authorized clients to view and edit the data. Neutral CAx databases will permit organizations up and down the supply chain to collaboratively interact in the same design session to avoid errors caused by passing information "over the wall". In the proposed system data translation is automatic, the supply chain process becomes more collaborative, and data integrity is maintained.

4.1 Data Security

Model sharing security has not been a primary research area (Red et al. 2013a, CIS), but it is a major concern to industries engaged in global product development. Because larger product development industries use many distributed suppliers, their intellectual property (IP) is always at risk, particularly when the model data must be transferred and translated between CAx applications. But even client-server and P2P architectures used in the reviewed research prototypes expose IP to multiple collaborators.

Wang et al. (2006) propose to restrict model sharing by providing selective information based on collaborator need-to-know. Encryption can be used to secure the information among the networked collaborators. Role-based viewing methods, where data is partially shared among designers, can deter reverse engineering.

Mensah and Teng (2013) performed an initial investigation into security principles and solutions for multi-user CAx applications. They proposed solutions for solving three of the primary security challenges, namely (1) authentication, (2) authorization, and (3) confidentiality. Authentication involves validating the client computer's identity with the server using the Transport Layer Security (TLS) protocol, and then validating the user's login credentials. An authorization mechanism then ensures that the user is only permitted to see or modify the appropriate CAx data based on the user's security clearance and other permissions. Confidentiality is necessary to ensure that messages being passed between network nodes (e.g., clients, servers) cannot be intercepted/deciphered or modified by unauthorized entities. Mensah and Teng propose that the TLS security protocol would provide confidentiality by encrypting messages that can only be deciphered by clients who have successfully authenticated. They asserted that further research is required to address the security challenges of nonrepudiation, auditing, and message integrity checks.

An authorization mechanism has been implemented into the NX Connect prototype. It consists of a dialog that allows users to manage permissions for the active part, with client-side logic that enforces those permissions, and additional columns in the database for storing each part's permission settings. Upon creating a part, the user who created it is defaulted to be the part manager. In addition to manager level permission, there are also contributor, viewer, and no-access permission levels. Registered users can be assigned to any of these levels for a given part. Users with no access to a part cannot even see that the part exists. Users with viewing access can only view the part; they cannot contribute to, modify, or delete the part. Users with contributor-level access can modify and contribute to the part, but only managers (who have full access) can delete the part, modify the part's permission settings, or perform other part-management types of operations. View-only restrictions are enforced by rejecting all part modification commands from the restricted user via the NX API.

As new multi-user collaborative methods evolve, security architectures will also evolve (Kushwaha and Roy 2010). Cloud serving architectures offer an alternative to placing CAx applications and model data on local workstations. It seems inevitable that CAx applications will be deployed on company secured cloud servers, and thus offer some security advantages, if client access can be managed and intrusions blocked. When models or editing deltas must be moved between mirrored servers to promote global teaming efficiencies, security becomes more challenging (Reddy and Reddy 2011).

5 Multi-user CAx Cloud Service

By implementing a cloud-based CAx agnostic database, CAx cloud serving becomes perfectly compatible with the multi-user CAx architectures presented in this chapter. Cloud computing is growing in popularity for a number of reasons: (1) less need for local deployment of expensive computing technologies; (2) organizational IT infrastructures can be reduced; (3) smaller businesses (e.g., suppliers) can access sophisticated CAx applications that were previously too expensive to maintain; and 4) cloud servers can be scaled according to user demand (Lynn 2012).

Of course, there are hurdles to overcome, some unique to a product development environment where IP protection is critical and where downtime is not tolerated. One of the advantages of widespread and local deployment of CAx applications is redundancy. Depending on the PDM implementation and permission system, design models may be accessible locally even if PDM systems or networks go down, so that work is not stopped. If that is not the case, a cloud server or network outage could shut down an entire development division or suppliers dependent on a particular set of CAx applications. Another solution to this is to replicate (mirror) cloud servers so that if one server goes down, clients can simply access the desired data from another server.

TeamPlatform and AutoDesk 360 are two commercial CAx applications that have placed CAx on the cloud (Dobrzynski 2013). Both applications offer the collaborative access and on-line advantages expected for cloud serving, but remain file-based solutions. Thus, these early applications seem to be more compatible with simpler models, and thus smaller files. In the near future, cloud serving of multi-user CAx applications will provide much greater efficiency and reduce product development costs.

Acknowledgments The National Science Foundation, the Center for e-Design, and BYU's industrial members and research students are acknowledged for their funding and conducting of this research and Center Site. Most importantly, our member companies, by providing access to facilities and personnel, have helped us assess their current collaborative environments and the potential for multi-user CAx improvements.

References

Anderson B (1998) ENGEN data model: a neutral model to capture design intent, PROLAMAT98

Anderson B (2002) Implementor's guide—solid model construction history

Bharambe A, Pang J, Seshan S (2006) Colyseus: a distributed architecture for online multiplayer games. NDSI '06: 3rd Symposium on Networked Systems Design & Implementation 155–168

Brunnermeier SB, Martin SA (1999) Interoperability cost analysis of the U.S. automotive supply chain final report, Research triangle Institute Center for Economics Research, RTI project number 7007–03

Bu J, Jiang B, Chen C (2006) Maintaining semantic consistency in real-time collaborative graphics editing systems. Int J Comput Sci Netw Secur 6(4):57–61

Cera CD, Braude I, Comer I, Kim T, Han JH, William CR (2003) Hierarchical role-based viewing for secure collaborative CAD. In: Proceedings of DETC'03 2003 ASME design engineering technical conferences

Chen L, Zhijie SZ, Feng L (2004) Internet-enabled real-time collaborative assembly modeling via an e-Assembly system: status and promise. Comput Aided Des 36:835–847

Choi G, Mun D, Han S (2002) Exchange of CAD part models based on the macro parametric approach. Int J CAD/CAM 2:13–21

Committee (2008) Pre-milestone A and early-phase systems engineering: a retrospective review and benefits for future Air Force acquisition, National research council of the National Academy of Science

Contero M, Company P, Vila C, Aleixos N (2002) Product data quality and collaborative engineering. IEEE Comput Graph Appl 22:32–42

Dobrzynski D (2013) Comparison of cloud-based CAD collaboration services: TeamPlatform versus Autodesk 360, CADdigest. http://www.caddigest.com/exclusive/MCAD/teamplatform/050213_teamplatform_vs_autodesk_360.htm

Douglas S, Tanin E, Harwood A, Karunasekera S (2005) Enabling massively multi-player online gaming applications on a P2P architecture. In: Proceedings of the international conference on information and automation, Colombo, Sri Lanka, pp 7–12

Endo K, Kawahara M, Takahashi Y (2007) A distributed architecture for massively multiplayer online services with peer-to-peer support. Int Fed Inf Proces 229:147–158

Fan LQ, Kumar AS A, Jagdish BN, Bok SH (2008) Development of a distributed collaborative design framework within peer-to-peer environment. Comput Aided Des 40:891–904

Gao L, Lu T, Gu N (2009) Supporting semantic maintenance of complex undo operations in replicated Co-AutoCAD environments. In: Proceedings of the 2009 13th international conference on computer supported cooperative work in design, Santiago, Chile

GauthierDickey C, Zappala D, Lo V (2004) A fully distributed architecture for massively multiplayer online games, ACM Netgames workshop draft

Han S (2010) Macro-parametric—an approach for the history-based parametrics. In: Han S (ed) Int J Prod Lifecycle Manage 4(4):321–325

Hannan MT, Freeman J (1984) Structural inertia and organizational change. Am Soc Rev 49:149–164

Hepworth A, Tew K, Nysetvold T, Bennett M, Jensen G (2013a) Automated conflict avoidance in multi-user CAD. Comput-Aided Des Appl. accepted for publication, CAD & A and presentation at CAD14, 25 June 2014, Hong Kong

Hepworth A, Nysetvold T, Bennett J, Phelps G, Jensen G (2013b) Scalable integration of commercial file types in multi-user CAD. Comput-Aided Des Appl. accepted for publication, CAD & A and presentation at CAD14, 25 June 2014, Hong Kong

Hoffmann C, Juan R (1993) Erep, an editable, high-level representation for geometric design and analysis. Geometric and product modeling pp 129–164

Holyoak VL (2012) Effective collaboration through multi-user CAx by implementing new methods of product specification and management, Master's thesis, Brigham Young University, Dec

Jensen G (2012) Collaborative multi-user synchronous and asynchronous modeling, analysis and design, Keynote presentation, Defense manufacturing conference (DMC), 26–29 Nov, Orlando, Florida

Jing S, He F, Han S, Cai X, Liu H (2009) A method for topological entity correspondence in a replicated collaborative CAD system. Comput Ind 60(7):467–475

Kim B, Han S (2007) Integration of history-based parametric translators using the automation APIs'. Int J Prod Lifecycle Manage 2(1):18–29

Kim S, Kuester F, Kim HK (2002) A global timestamp-based scalable framework for multi-player online games. In: Proceedings of the IEEE fourth international symposium multimedia software engineering, pp 2–10

Kushwaha J, Roy BN (2010) Secure image data by double encryption. Int J Comput-Aided Appl 5(10):28–32

Lai YL (2009) A constraint-based system for product design and manufacturing. Robot Comput-Integr Manuf 25(1):246–258

Li WD, Ong SK, Fuh JYH, Wong YS, Lu YQ, Nee AYC (2004) Feature-based design in a distributed and collaborative environment. Comput Aided Des 36:775–797

Li WD, Lu WF, Fuh JYH, Wong YS (2005) Collaborative computer-aided design - research development status. Comput Aided Des 37:931–940

Li M, Gao S, Fuh JYH, Yang YF (2008) Replicated concurrency control for collaborative feature modeling: a fine granular approach. Comput Ind 59:873–881

Li X, He F, Cai X, Chen Y, Liu H (2009), Using procedure recovery approach to exchange feature-based data among heterogeneous CAD systems, CSCW in design

Li J, Kim B, Han S (2012) Parametric exchange of round shapes between a mechanical CAD system and a ship CAD system. Comput Aided Des 44(2):154–161

Linden G, Smith B, York J (2003) Amazon.com recommendations: Item-to-item collaborative filtering. IEEE Computer Society, Jan–Feb, pp 76–80

Liu Q, Cui X, Hu X (2008) An agent-based intelligent CAD platform for collaborative design. ICIC CCIS 15:501–508

Lynn S (2012) 20 Top cloud services for small businesses. PCMAG.com. http://www.pcmag.com/article2/0,2817,2361500,00.asp

Marshall F (2011) Model decomposition and constraints to parametrically partition design space in a collaborative CAx environment, Master's thesis, Department of Mechanical Engineering, Brigham Young University

Mensah F, Teng C (2013) Security mechanisms for multi-user collaborative CAx. In: Proceedings of the 2nd annual conference research information technology (RIIT '13), 59-60

Moncur RA (2012) Data consistency and conflict avoidance in a multi-user CAx environment, Master's thesis, Department of Mechanical Engineering, Brigham Young University

Moncur R, Jensen C, Teng C, Red E (2013) Data Consistency and Conflict Avoidance in a Multi-User CAx Environment. Comput Aided Des Appl 10:1–19

Mun D, Han S, Kim J, Oh Y (2003) A set of standard modeling commands for the history-based parametric approach. Comput Aided Des 35:1171–1179

Nysetvold T, Teng C (2013) Collaboration tools for multi-user CAD. In: IEEE 17th international conference on supported cooperative work in design (CSCWD), 2013, pp 1–5

Panchal J, Fernandez M, Paredis C, Allen J, Mistree F (2007) An interval-based constraint satisfaction (IBCS) method for decentralized, collaborative multifunctional design. Concurrent Eng 15(3):309–323

Ram DJ, Vivekananda N, Rao CS, Mohan NK (1997) Constraint meta-object: a new object model for distributed collaborative designing. IEEE Trans Syst Man Cybern Part A Syst Hum 27(2):208–221

Ramakrishna V, Robinson M, Eustice K, Reiher P (2006) An active self-optimizing multiplayer gaming architecture". Cluster Comput 9(2):201–215

Rappoport A (2003) An architecture for universal CAD data exchange. In: Proceeding of ACM symposium on solid modeling and applications 2003. ACM Press, pp 266–269

Red E, Holyoak V, Jensen G, Marshall F, Ryskamp J, Xu Y (2010) A research agenda for collaborative computer-aided applications. Comput-Aided Des Appl 7(3):387–404

Red E, Jensen G, French F, Weerakoon P (2011) Multi-User architectures for computer-aided engineering collaboration. In: 17th international conference on concurrent enterprising, Aachen, Germany

Red E, Marshall F, Weerakoon P, Jensen G (2012) Considerations for multi-user decomposition of design spaces. CAD12, Niagara Falls, Canada, June (to be published in Journal of Computer-Aided Design and Applications, 10(5))

Red E, French D, Jensen G, Walker S, Madsen P (2013a) Emerging design methods and tools in collaborative product development (to be published in J. Computing and Information Science in Engineering, 13(3))

Red E, Jensen G, Weerakoon P, French D, Benzley S, Merkley K (2013b) Architectural limitations in multi-user computer-aided engineering applications (to be published in the Journal of Computer and Information Science, 6(4))

Reddy VK, Reddy LSS (2011) Security architecture of cloud computing. Int J Eng Sci Technol 3(9):7149–7155

Rygielski C, Wang JC, Yen DC (2002) Data mining techniques for customer relationship management. Technol Soc 24(4):483–502

Sun C, Xia S, Sun D, Chen D, Shen H, Cai W (2006) Transparent adaptation of single-user applications for multi-user real-time collaboration. ACM Trans Comput-Hum Interact 13(4):531–582

Tang D, Zheng L, Li Z, Li D, Zhang S (2000) Re-engineering the design process for concurrent engineering. Comput Ind Eng 38:479–491

Wang Y, Tan E, Li W, Xu Z (2004) An architecture of game grid based on resource router.Grid and cooperative computing. Springer, Berlin 3032

Wang Y, Ajoku PN, Brustoloni JC, Nnaji BO (2006) Intellectual property protection in collaborative design through lean information modeling and sharing. J Comput Inf Sci Eng 6(2):149–159

Wu D, Thames JL, Rosen DW, Schaefer D (2012) Towards a cloud-based design and manufacturing paradigm: looking backward, looking forward. In: Proceedings of the ASME 2012 International design engineering technical conference and computers and information in engineering conference (IDETC/CIE12), paper number: DETC2012-70780, Chicago, US

Xu B, Gao Q, Li C (2009) Reusing single-user applications to create collaborative multi-member applications. Elsevier, Adv Eng are 40:618–622

Xu Y, Red E, Jensen G (2011) A flexible context architecture for a multi-user GUI. Comput-Aided Des Appl 8(4):479–497

Yang J, Han S, Kim B, Cho J, Lee H (2004) An XML-based macro data representation for a parametric CAD model exchange. Comput-Aided Des Appl 1(1):153–162

Zheng Y, Shen H, Sun (2009) Leveraging single-user AutoCAD for collaboration by transparent adaptation. In: 13th international conference on computer supported cooperative work in design, Santiago, Chile, 22–24 April. ISBN:978-1-4244-3534-0

Distributed Resource Environment: A Cloud-Based Design Knowledge Service Paradigm

Zhinan Zhang, Xiang Li, Yonghong Liu and Youbai Xie

Abstract Design can be viewed as a knowledge intensive process, which requires more and more collaboration between design resources within and without an enterprise for product innovation. A company's capacity for product innovation essentially means the ability to discover, use, and manage different kinds of design resources. However, design resources are distributed unevenly within or across organizational boundaries. In order to benefit from the outsourcing of design knowledge within different design resources, with lower costs and within a shorter time, is a great challenge for enterprises. The design knowledge must flow quickly and reliably from when and where it is located to when and where it is needed for design activity. Unfortunately, there are many barriers having a negative influence on quick and reliable knowledge flow between knowledge owners and knowledge demanders. The lack of supporting mechanisms (e.g., a knowledge service platform) between service providers and consumers is one of the barriers. Thus, there is a need to develop mechanisms to overcome the barriers and thereby improve the performance of knowledge flow. This chapter introduces a cloud-based knowledge service environment, i.e., a distributed resource environment, which enables companies to utilize collective open innovation and rapid product development with reduced costs. The definition, function, structure, and characteristics of a distributed resource environment are presented. Then, the concept of a cloud-based

Z. Zhang (✉) · X. Li · Y. Xie
School of Mechanical Engineering, Shanghai Jiao Tong University,
200240 Shanghai, China
e-mail: zhinanz@sjtu.edu.cn

X. Li
e-mail: li.xiang@sjtu.edu.cn

Y. Xie
e-mail: ybxie@sjtu.edu.cn

Y. Liu
SANY Group Co., Ltd, 410100 Changsha, China
e-mail: yhong.liu@gmail.com

D. Schaefer (ed.), *Cloud-Based Design and Manufacturing (CBDM)*,
DOI: 10.1007/978-3-319-07398-9_3,

knowledge service framework is proposed to organize the knowledge sources in a distributed resource environment. The knowledge sources are composed of design entities' knowledge cloud and resource units' knowledge cloud. The former is the knowledge consumer and latter is the knowledge provider in a distributed resource environment. Next, a cloud-based knowledge service framework is presented for the well effective operation of a distributed resource environment. Two agents, i.e., a knowledge service publishing agent (KSPA) and a knowledge service consuming agent (KSCA) are developed to implement the online knowledge service. KSPA can be used by knowledge providers to encapsulate and publish their design knowledge as a service into the service market, whereas KSCA can be used by knowledge consumers to request knowledge service from the service market. Finally, an inner-enterprise distributed resource environment is implemented to verify the proposed knowledge service paradigm.

Keywords Distributed resource environment · Cloud-based design · Knowledge cloud · Knowledge service · Knowledge flow

1 Introduction

The highly competitive nature of business in a global market economy is forcing companies to make continuous efforts to leverage their product innovation in order to gain or maintain their competitive edge. The design and development of products are becoming increasingly complex. In this scenario, manufacturing enterprises have to utilize just-in-time widespread knowledge resources to meet various customer requirements. There are many knowledge acquisition resources (e.g., experienced engineers, software, knowledge base, all kinds of test equipment and instruments), which belong to different organizations or enterprises. Generally, these knowledge acquisition resources are geographically distributed, and usually unknown to the seekers (i.e., enterprises or organizations who have a desire to utilize the outsourcing resources) outside organizational boundaries. This scenario makes the sharing of knowledge acquisition resource across organizational boundaries very complicated.

In recent years, new concepts about the cloud have emerged from computer science, such as cloud computing (Zhang et al. 2010), cloud service, service-oriented architecture (SOA), and grid computing. A constant infiltration of new techniques within a manufacturing industry, the concept of cloud-based design and manufacturing (CBDM) has become more and more popular as an enterprise-level design and manufacturing model (Wu et al. 2013a, b, c). As pointed by Dirk Schaefer in the *call for chapter proposals* of the book "*Cloud-Based Design and Manufacturing (CBDM): A Service-Oriented Product Development Paradigm for the 21st Century,*" cloud-based design and manufacturing (CBDM) is "a new product realization model that supports open innovation through social networking and negotiation platforms between service providers and consumers. It is a type of parallel and distributed system consisting of a collection of inter-connected physical and virtualized service pools of design and manufacturing resources as well as intelligent search capabilities for design and manufacturing solutions." In this context, knowledge service bridges the gap between service providers and consumers, and thus facilitates knowledge flow and sharing among them.

Knowledge service is a knowledge flow intensive process to satisfy the users' requirement of specific knowledge (Li et al. 2009). As product design becomes more and more complicated and increasingly multidisciplinary knowledge dependent, any individual or enterprise will find it difficult to own all kinds of needed knowledge in the design process. Many researchers have done a lot of work in the knowledge service area. For example, in order to provide a semantically interactive environment for users, Li et al. (2009) proposed an approach of service-oriented knowledge modeling on an intelligent topic map. Chen and Chen (2009) proposed a novel product knowledge service model based on the product life cycle and its supply chain. Meng and Xie (2011) presented the concept of an embedded knowledge service to embed the independent technical activities of external resource units as a part of the complete product development process of design entities. Zhuge and Guo (2007) proposed a virtual knowledge service market based on a knowledge grid. Mentzas et al. (2007) introduced the concept of knowledge trading and analyzed an ontology-based approach for trading knowledge services using semantic technologies. In addition, according to Li et al. (2009), Schaefer et al. (2012), Wu et al. (2012, 2013a, b, c), Ding et al. (2012), Tao et al. (2011), and Xu (2012), scholars are working on defining, understanding, and discussing the framework of CBDM. For example, Wu et al. (2012, 2013a, b, c) and Schaefer et al. (2012) introduced a comprehensive definition and concept of CBDM as well as a first step towards understanding the key characteristics and fundamentals of it. As summarized by Wu et al. (2013a, b, c), they have proposed four kinds of services to cloud-based service consumers. The services are: (1) Hardware-as-a-service (HaaS) which delivers hardware sharing services (e.g., milling, lathe machines, CNC machining centers, hard tooling, and manufacturing processes) to cloud consumers, (2) Software-as-a-service (SaaS) which delivers software applications, e.g., CAD/CAM, FEA tools, and Enterprise Resource Planning (ERP) software to cloud consumers, (3) Infrastructure-as-a-service (IaaS) providing consumers and providers with computing resources, e.g., high performance servers and data centers, and (4) Platform-as-a-service (PaaS) offering a social networking media for cloud consumers and providers to communicate and collaborate, providing a new channel for mass collaboration and open innovation. Zhang et al. (2010) presented a survey of cloud computing, which provided a better understanding of the key concepts, architecture, and implementation of cloud computing. Ding et al. (2012) proposed an ontology-based cloud service integration model for the representation of manufacturing resources. A three-layer collaborative manufacturing resource sharing platform is proposed for manufacturing service integration. Tao et al. (2011) proposed a four-stage cloud manufacturing model where manufacturing resources are controlled through the internet via intelligent monitoring systems. Xu (2012) discussed the typical characteristics, architecture, core enabling technologies for cloud manufacturing, and the relationships between cloud manufacturing and cloud computing.

As mentioned above, current knowledge service work focuses on how to supply consumers with organizations' existing knowledge in a plug-in and plug-out pattern. In fact, since design is full of uncertainty, knowledge service providers have to do certain knowledge acquisition activities ahead of providing knowledge to meet the consumers' knowledge needs. Therefore, a design knowledge service should be a knowledge flow intensive iterative process. To the best of our knowledge, a few studies on knowledge service and CBDM focus on how to organize knowledge acquisition resources to form a distributed resource environment (DRE) and represent their knowledge service capability for providing design knowledge services. The demands of a distributed resource environment are more owing to today's competitive nature of the global marketplace and increasing complexity of products. This phenomenon requires that companies design products right-the-first-time and for shorter time-to-market. A distributed resource environment can help companies by reducing the cost of product design and development, accelerating time-to-market for new products, and improving product performance. This is an evolution of the design transition away from traditionally depending on research and design (R&D) in-house towards more open innovation by integrating knowledge from distributed resources. Core competencies, i.e., knowledge integration competency for design entities (i.e., knowledge service consumers) and knowledge acquisition competency for resource units (i.e., knowledge service providers), rather than physical assets, increasingly define competitiveness of companies in the market.

In this study, a DRE aims at realizing a service-based design knowledge flow and paid sharing among enterprises or organizations. A well operated distributed resource environment means that design resources can be integrated and the flow of knowledge is manageable. Manufacturing enterprises can rapidly integrate different kinds of knowledge from professional design resources based on their dynamic and various design requirements. In short, the manufacturing enterprises and individuals enabled by the more efficient cooperative capability of enterprise cloud service systems could obtain diverse design and manufacturing services from the internet and intranet more conveniently. This chapter introduces a cloud-based knowledge service environment, i.e., a distributed resource environment, which enables companies to utilize collective open innovation and rapid product development with reduced costs. The outline of the chapter is as follows. Section 1 presents the background of this study. Section 2 presents the concept of the distributed resource environment. Section 3 proposes a cloud-based knowledge service framework. Section 4 introduces the implementation of the proposed framework via an illustrative case study. Finally, the conclusions are presented in Sect. 5.

2 Distributed Resource Environment

2.1 Concept of Distributed Resource Environment

The aim of a distributed resource environment is to effectively organize all kinds of knowledge service providers (e.g., design entities) and knowledge services consumers (e.g., resource units) through a knowledge service platform. Based on a distributed resource environment, any design entity or resource units could easily obtain the knowledge to solve their problem by knowledge services. Its concept is introduced as follows.

The Oxford English Dictionary http://oxforddictionaries.com/page/oxford englishdictionary defines the noun *environment* as

> The surroundings or conditions in which a person, animal, or plant lives or operates.

The essentials of this definition are *surroundings* or *conditions*, and *object* (referring to person, animal, or other things). Thus, the concept of *distributed resource environment* also includes elements as surroundings and object. Therefore, the key factor to define the concept of a *distributed resource environment* is to clarify the surroundings and the object.

In order to solve the problem, a brief review of the development of this concept should be discussed first. The concept of a distributed resource environment is presented in the background that design plays a critical role for improving the competitive power of Chinese manufacturing enterprises, and that most of Chinese manufacturing enterprises lack knowledge and knowledge acquisition resources to help design products with shorter time-to-market. The concept of a distributed resource environment was first presented by Xie (1998), which aimed at organizing distributed resources in China through service-based knowledge flow for the purpose of improving their competitive power http://www.chinamoderndesign.com .

As mentioned above, the *surroundings* are distributed knowledge service providers and consumers, i.e., design entities and resource units, respectively. The *object* is an intelligent, sustainable internet-based platform (e.g., a kind of knowledge) that enables design entities (fulfilling the role of knowledge service or knowledge acquisition service consumers) and resource units (fulfilling the role of knowledge service or knowledge acquisition service providers) to effectively capture, publish, share, and manage knowledge services, or knowledge acquisition service resources. Therefore, the distributed resource environment is the *Internet of Knowledge*.

Thus, a DRE can be defined as an ecosystem of design entities and resource units, supporting networks and infrastructure that is designed to be: competitive, satisfy customer needs, and create value for both design entities and resource units. This ecosystem refers to companies in the system where orderly competition instead of the current disorderly competition is exhibited.

For resource units (*knowledge service providers*), DRE means a shift from producing and offering products to providing services and system solutions that will intensively meet customers' needs. With traditional R&D cooperation in China, R&D means one organization that provides services for one company for a certain period, and the fruit will be owned by the company or shared by both. However, DRE emphasizes on open innovation that has the potential for one resource units servicing many design unities and many resource units to service one design unit. This requires resource units to continuously improve their knowledge service and knowledge acquisition (abbreviated as knowledge service later) service level to maintain their competitive edge.

For design entities (*knowledge service consumer*), DRE means a shift from long time-based R&D cooperation to buying just-in-time services and system solutions that have a potential to minimize the risk for short-to-market product development due to failed or delayed R&D projects.

2.2 Function of a Distributed Resource Environment

For both resource units and design entities (*provider and consumer*), a distributed resource environment is a knowledge flow facilitator designed to support cross-organizational intense flow of knowledge. Both resource units and design entities will benefit from knowledge flow. On the one hand, design entities can fulfill their design-led innovation by integrating knowledge from outside knowledge resources. On the other hand, resource units can provide knowledge services to different design entities to enlarge their profit, and this will in turn encourage permanent investment in maintaining the knowledge acquisition edge.

A paramount purpose of the DRE should be to minimize the impact of lack of knowledge acquisition resources or insufficiently utilizing existing knowledge in improving the design-led competitiveness of manufacturing enterprises, especially Chinese manufacturing enterprises.

2.3 Structure of a Distributed Resource Environment

As is shown in Fig. 1, a distributed resource environment is the *internet of knowledge*, which includes three main constitutional elements, i.e., design entities, resource units, and an internet-based knowledge service platform. Figure 1 may provide a cognitive understanding of the distributed resource environment. It is the platform that bridges design entities and resource units and fulfills the function as knowledge flow facilitator.

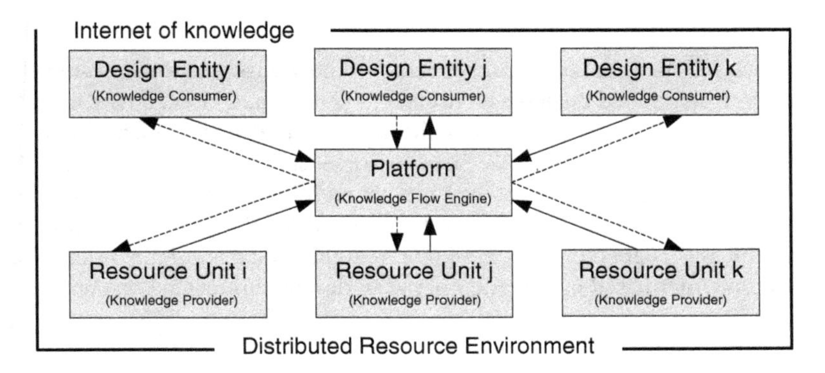

Fig. 1 Structure of distributed resource environment

- *Design entity: knowledge service consumer*

Design entities refer to organizations, which have the capacity to integrate domain knowledge to perform the innovative product design and development process, and which know what knowledge is needed to accomplish their tasks. Within DRE, many design entities focus on the integration of knowledge, and in their product design process, they will consume knowledge services and knowledge acquisition services to overcoming the barriers of lack of time and money spent on new knowledge acquisition. For example, internal combustion engines OEMs are design entities. In their engine design and development process, design entities will consume knowledge services from outside organizations, for example, they may ask universities to do fundamental research on a frontier field, or they may ask component suppliers (e.g., piston OEM, bearing OEM, etc.) to develop engine elements to meet their needs.

- *Resource unit: knowledge service provider*

Resource units are organizations or individuals, which have a competitive advantage in certain domains and which may offer knowledge services and have the ability to acquire new knowledge to meet the need of knowledge service requirements from design entities. Many resource units (having the capacity of acquiring new knowledge in given regions) will provide knowledge services and new knowledge acquisition services for design entities in order to obtain their rewards. As for the engine design case, if an Engine OEM is a design entity, then suppliers, such as a piston OEM and piston ring OEM are resource units, which will provide service to the engine OEM. On the other hand, if the piston OEM asks for knowledge services (e.g., evaluate piston skirt tribological performance) provided by others, then the role of the piston OEM is as a design entity. Therefore, there are no clear boundaries between design entity and resource unit. To name an organization as a design entity or a resource unit depends on the functional role of the organization.

- *Knowledge service platform*

As shown in Fig. 1, the platform is a hub for knowledge service consumers and providers in a distributed resource environment (Zhuge and Guo 2007; Choy et al. 2007). Design entities and resource units in design related domains find their needed information and knowledge here to solve their problem. They therefore help both design entities and resource units through knowledge services to obtain design-led competitiveness, which is at the heart of the platform. A platform with many instruments is good enough to support a highly efficient, highly reliable, and very cheap internet-based cooperation between design entities and resource units.

2.4 Characteristics of a Distributed Resource Environment

- *Internet of knowledge*

The internet of knowledge refers to design entities and resource units that have the potential in knowledge services, and both design entities and resource units are taken as knowledge flow nodes being connected by the internet within a distributed resource environment. The concept of internet of knowledge means that the internet bridges the time and space gaps between design entities and resource units, and meets the need for the just-in-time supply of knowledge and information. The Internet of Knowledge can possibly enable manufacturing enterprises to compete in the knowledge economy, especially Chinese manufacturing enterprises. DRE can help manufacturing enterprises to strategically manage its limited knowledge resources and focus on core capabilities for sustainable development. On the other hand, manufacturing enterprises may depend on knowledge services to gain faster insights into new business opportunities, and compete on knowledge to drive productivity and innovation.

- *Partial self-organization of design entities and resource units*

Distributed resource environments emphasize an open innovation instead of a closed innovation through knowledge service. A key element of DRE is a well operated platform. As mentioned above, the platform is a hub for integrating information about distributed design resources. Hence, the platform fulfills the role of organizer and manager for organizing and managing distributed resources. However, both design entities and resource units have the right to determine whether joint DRE or not.

- *Power of the many*

One resource unit serves for one design entity on a traditionally project-based R&D pattern, and all the knowledge acquired should be transferred to the project founder. A resource unit is forbidden to use the knowledge acquired in one project to serve another consumer. Therefore, this pattern will not allow improvements in design-led Chinese manufacturing enterprises' competitive power due to limited knowledge flow. On the other hand, using a one-to-one service pattern, design entities may lose the chance to ask more powerful resource units to provide better knowledge service or knowledge acquisition services.

Within a distributed resource environment, any design entity or resource unit has the chances to ask many to provide or consume knowledge services or knowledge acquisition services. The one which can provide better service will have more chances to serve the many and thus to make more profits.

3 A Cloud-Based Knowledge Service Framework

The cloud-based knowledge service framework consists of knowledge service, working platform, and resources. The key technology of this framework mainly includes the service management technology, service providers and consumers management technology, safety and reliable security technology, and cloud technology. This study focused on the development of knowledge service platform and related key technology, e.g., development of tools to help knowledge providers encapsulate and publish their design knowledge as a service (i.e., SaaS). The platform consists of a collection of knowledge service pools, i.e., simulation related knowledge service, test-related knowledge service, consulting-related knowledge service, and component of system design related knowledge service.

3.1 Architecture of Knowledge Service Platform

According to the definition and structure of the distributed resource environment, the architecture of a knowledge service platform is shown in Fig. 2. The platform is also a type of knowledge service provided by a certain resource unit. The platform has three key components. The universal discovery, description, and integration of knowledge (i.e., KUDDI) component deals with the publishing and searching of knowledge service. All the publishing services and requirements will be included in the service and requirements list, respectively. Using this component, both knowledge service providers and consumers will obtain business opportunity. The negotiation component helps both the knowledge service provider and demander to negotiate and produce contracts to protect the intellectual property and patent rights within the knowledge flow. It should be noted that an inner-enterprise distributed resource environment is able to avoid the intellectual property rights disputes. The trade management component deals with the management of complaints, payment, and trust for services and offers a system to evaluate the knowledge service provider. The functions of the management component can be refined with reference to *www.taobao.com*.

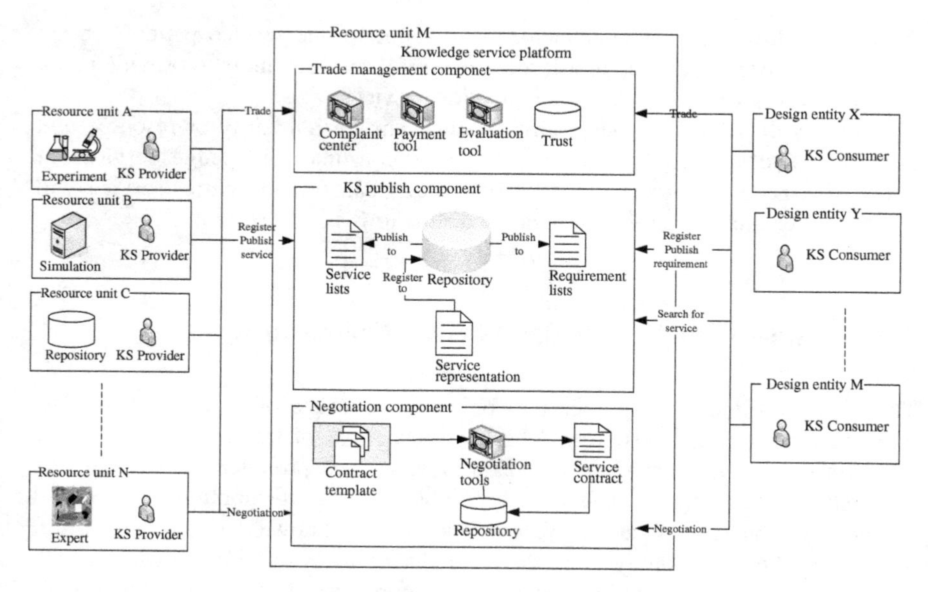

Fig. 2 Architecture for a knowledge service platform in DRE

3.2 Represent the Knowledge Service Capacity of a Resource Unit

In the architecture of a knowledge service platform, one of the key issues is the representation of the knowledge service capacity of a resource unit and the service requirements of a design entity. In this subsection, the capacity model of a resource unit and an ontology based modeling approach was proposed to represent the knowledge service providers' knowledge service capacity and the consumers' knowledge service requirement.

- *Capacity model of resource unit*

The capacity model of a resource unit consists of four main aspects: (1) People with knowledge and experience that are able to acquire extensive knowledge in a certain domain. (2) The knowledge base that stores the knowledge and experience of a resource unit. (3) Instruments including software and hardware, which supports knowledge acquisition. (4) Sufficient investment of time and money to improve and maintain a resource unit's innovation and competitiveness. With the support of its knowledge acquisition resource, a resource unit has to develop its own broad knowledge acquisition methodology and be able to develop technology-based knowledge services and/or test-based knowledge services. Figure 3 illustrates a schematic diagram of the capacity model of a resource unit.

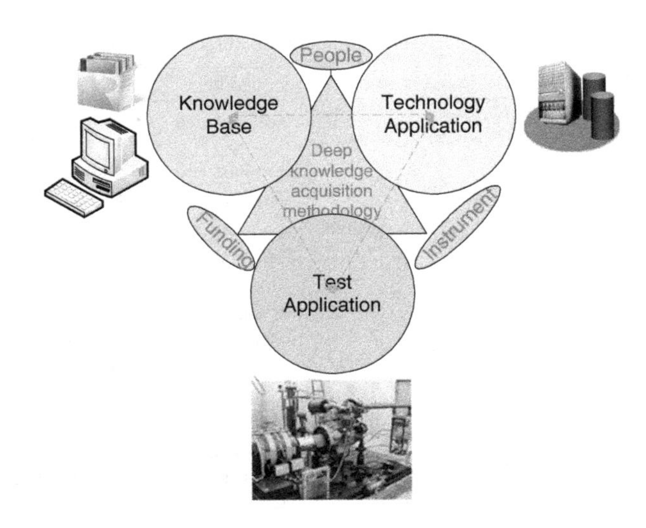

Fig. 3 Capacity model of a resource unit

- *Modeling of service providers' capacity*

As shown in Table 1, the provider, product, discipline, tools, and provision compose the ontologies of resource units, and each ontology consist of one or more attributes. For instance, the provider ontology has the attributes of contact information, finance information and human information, and all the attributes have determined descriptions. Using the provider ontology, the basic information of a knowledge service provider will be represented and ready for being discovered in the KUDDI component. Suppose each resource unit provides two kinds of knowledge services, i.e., product-oriented services and discipline-oriented services. For the product ontology, all or some of the six attributes have to be quantitative or qualitative descriptions. For the discipline ontology, the provider needs qualitative description application domains and their capacity of new knowledge acquisition and knowledge services, such as tribology design. Our approach also provides ontology for the description of tools that might be used to acquire new knowledge. The provision ontology includes the trust attribute, where intellectual property and the quality of service are defined; the price ontology, which specifies price and payment issues; and the delivery ontology, where transfer issues are represented.

- *Capacity model of resource unit*

Table 2 provides the design entities with four kinds of ontologies, i.e., the demander ontology, the product ontology, the discipline ontology and the provision ontology, to represent their knowledge requirements. As the attributes and descriptions of those ontologies are similar with those presented in Table 1, those will not be detailed here.

Negotiation among providers and demanders will be assisted by a conference system or other negotiation tools, such as Skype, Microsoft MSN, Tencent QQ et al.

Table 1 Ontology based modeling of resource units' knowledge service capacity

Vendor	Ontology	Attributes	Description
Resource units (Service providers)	Provider	Contact	Name, address, tel, fax, e-mail, reputation, and so forth
		Finance	Financial information
		Human	Human resource
	Product	Function	What is for?
		Structure	For physical products
		Constraint	Must or Must not
		Context	In which context the product can be used
		Process	Describes how to use the service object
	Discipline		Describes the application domain that can offer knowledge service, i.e., tribology design
	Tools	Test instrument	Name, vender, specialty, purpose
		Software	Name, vender, specialty, purpose
		Hardware	Name, vender, specialty, purpose
	Provision	Trust	How to protect the rights of both parties
		Price	Price for different kinds of service
		Delivery	How to transfer the knowledge object
		Time	Time to customer

Table 2 Ontology based modeling of design entities' knowledge requirements

Vendor	Ontology	Attributes	Description
Design entities (Service consumers)	Demander	Contact	Name, address, tel, fax, e-mail, reputation, and so forth
		Finance	Financial information
	Product	Function	Requirement for function
		Behavior	Requirement for behavior
		Structure	Requirement for structure
		Constraint	Must and must not
	Discipline		Describes the requirement of discipline knowledge service
	Provision	Price	Price for the service
		Delivery	How to transfer the knowledge object
		Time	Time to demander

After successful negotiation, a contract will be produced to protect the intellectual property and patent rights within the knowledge flow.

Using knowledge flow, even if a company lacks some domain knowledge, it can also design a new product through integrating domain knowledge. Thus, to facilitate knowledge flow within DRE by the proposed architecture, both service provider and service demander will benefit from the flow of knowledge. In the next section, a brief case is given to show how a DRE facilitates the flow of knowledge and how a manufacturer benefits from KF.

3.3 Modeling Knowledge Service Process

The process of knowledge service is shown in Fig. 4, which includes two kinds of stakeholders (i.e., knowledge service provider and knowledge service consumer), nine kinds of actions and five kinds of states.

Figure 5 shows the actions and the variation of states in a knowledge service process. After the knowledge service provider published his knowledge service, the status of the service becomes *ready*. If knowledge service consumer requests knowledge service from a knowledge provider, the status of service enters *proposed*. The knowledge service provider needs to evaluate its capability to decide whether it can provide knowledge service. If not, the knowledge service provider rejects the request, and gives feedback to the knowledge service consumer; the status of service will be *closed*. If yes, the status of service becomes *active*. The active status indicates the service is beginning. When a knowledge service provider solves the requested problem, it will give the solution to the knowledge service consumer. The status of service becomes *resolved*. The knowledge service consumer evaluates the result to decide whether the service satisfied his requirements. If not, it will ask the knowledge service provider to rework; the status of service becomes *active* again. If yes, it indicates that the solution provided by the knowledge service provider meets the requirement of knowledge demander, the service is *closed*, and then a total knowledge service process is done.

3.4 Classification of Knowledge Services

In a distributed resource environment, there are four major kinds of knowledge services. The detailed information about each type of knowledge service is as follows.

- *Simulation-related knowledge service* Knowledge service providers have simulation tools, methods, and engineering applications. Based on these resources, they can provide simulation-related knowledge services to the service consumers.
- *Test-related knowledge service* Knowledge service providers have instruments, equipment, and experiment devices, and the knowledge of how to use these resources to acquire new knowledge.
- *Consulting-related knowledge service* Knowledge service providers have domain experts and fruitful knowledge and experience. Providers can use these resources to solve consumers' problems.
- *Component of system design related knowledge service* Knowledge service providers are often component suppliers with the capabilities to offer the service of design and development of new component to consumers.

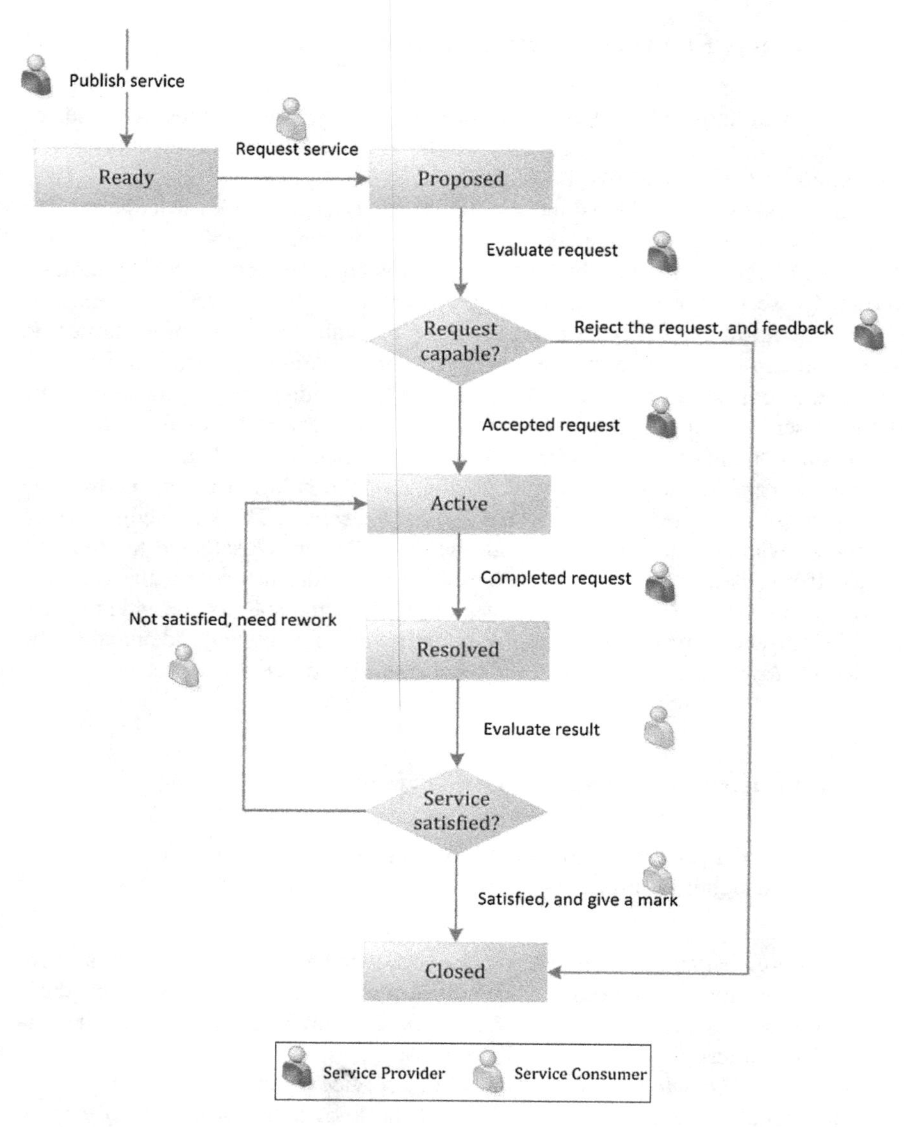

Fig. 4 The planning for knowledge service process

3.5 The Framework of Knowledge Service Agents

Two knowledge service agents are designed for supporting the cloud-based knowledge service environment, i.e., knowledge service publishing agent (KSPA) and knowledge service consuming agent (KSCA). The function of KSPA is to help knowledge providers encapsulate and publish their design knowledge as a service.

Action	State	Role	Info.
Publish service	Ready		SP is ready for providing service to customers.
Request service	Proposed		SC wanna request a relay test service.
Evaluate request			SP will evaluate your request.
- Accepted	Active		Okay, SP decides to accept SC's request.
- Reject	Closed		Sorry, SP decides to reject SC's request.
Completed service	Resolved		SP told SC that he has completed the request, and waiting for SC's evaluation.
Evaluate result			SC evaluates the result of service.
- Satisfied, and give a mark	Closed		Very good, SC satisfies the result of service, and give a five-star mark to SP's service.
- Not Satisfied, need rework	Active		Not very good, SC does not satisfied with the result of service, and tell SP to rework.

Service Provider (SP) Service Consumer (SC)

Fig. 5 The state variation during knowledge service process

KSCA will help the knowledge consumers use the knowledge service. An overall framework of the two knowledge service agents is shown in Fig. 6. The framework has five layers, i.e., knowledge encapsulation layer, knowledge representation layer, knowledge service layer, knowledge contract layer, and knowledge application layer.

In the knowledge encapsulation layer, design knowledge is encapsulated as the knowledge application part with its input and output files. The knowledge application part includes design knowledge, such as finite element analytical files, or procedures written in any programming language, or which kinds of know-how knowledge to apply and use to obtain the design result. At the same time, knowledge providers can always extract necessary knowledge elements from the input and output files regardless of the format of them. For example, a CAE analysis process generally includes input and output files, where the input files include the three-dimensional model, and the output files include analytical result.

In the knowledge contract layer of knowledge providers, KSPA provides a file parsing function to extract input and output knowledge into input and output frame sets. The frame set is a parameter set, which is based on the frame-based knowledge representation. Each parameter in the input and output frame set has its name, data type, assigned value, and location where it is in the knowledge input and output file. The function of the knowledge representation layer, so the service consumers can give the values for their structured questions according to the input frame set, then KSPA would rewrite these values into the knowledge input file. The knowledge application part is conducted by knowledge providers based on the input knowledge that comes from the knowledge consumers. KSPA would extract the result knowledge from the output files into an output frame sent back to the

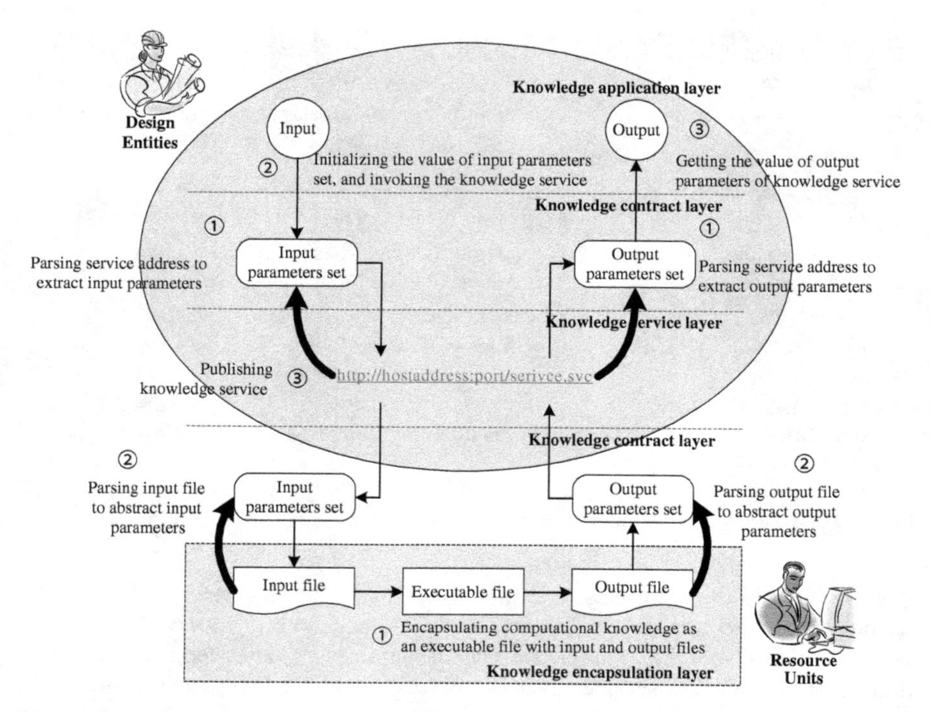

Fig. 6 Overview framework of knowledge service agent (from Li et al. 2012)

service consumers. The goal of this layer is to transform the irregular knowledge from the input and output files to the regular input and output frame set.

In the knowledge service layer, the knowledge providers can publish the knowledge encapsulated in the bottom layer as a service. Each knowledge service has its service address, which comprises a host address, port number, and service name. The service address is a bridge for connecting knowledge suppliers and knowledge demanders.

Knowledge suppliers are responsible for receiving the input knowledge from knowledge demanders, then performing knowledge applications, and giving the output knowledge back to the knowledge demanders.

In the knowledge contract layer of knowledge consumers, KSCA helps knowledge consumers parse the input and output frame set from the service address that published by knowledge suppliers. The frame set is a knowledge interface transferring from knowledge providers to knowledge consumers, which include the slot name, data type, value, and data structure, etc. KSCA will help knowledge consumers understand what knowledge he must give to the knowledge providers, and what knowledge it will get back from the knowledge providers.

In the knowledge application layer, knowledge consumers will set input knowledge according to the input frame set for requesting service from the knowledge suppliers. KSCA will obtain the output knowledge what he wanted through the service address of the knowledge providers. The knowledge service has been completed through KSCA and KSPA.

4 Implementation and Illustrative Case

The implementation of an inner-enterprise distributed resource environment is employed to illustrate how the proposed approach can be applied to real industrial practices. For the sake of commercial secrets, the name of the enterprise (abbreviated as AA later) is omitted here. Since AA is very large and its subcompanies are geographically distributed all over the world, AA is selected as a focal scenario for developing a cloud-based knowledge service paradigm to support knowledge service driven open innovation. The content of the construction of a distributed resource environment includes two main parts, i.e., the construction of a distributed resource environment and the development of a knowledge service platform.

4.1 Background

AA is a global company in the construction machinery industry with a vast product range of concrete machinery, excavators, hoisting machinery, pile driving machinery, road construction machinery, port machinery, and etc. The company is comprised of nine product divisions, one central R&D institute and 31 product R&D institutes. The divisions and R&D institutes are geographically distributed throughout China and other countries. The knowledge and knowledge acquisition resources are distributed unevenly within or across organizational boundaries. Generally, these resources are unknown to the seekers (i.e., enterprises or organizations who have a desire to utilize the resources) outside organizational boundaries. This scenario makes the sharing of knowledge and knowledge acquisition resources across organizational boundaries very difficult and complicated.

However, the highly competitive nature of business in a global market economy is forcing AA to make continuous efforts to leverage their product innovation in order to gain a competitive edge. In order to benefit from the latest fruit of outsourcing design knowledge in different design resources at lower cost and shorter time is a big challenge for enterprises. The design knowledge must flow quickly and reliably from when and where it is located to when and where it is needed for design activity. Unfortunately, there are many barriers having a negative influence on quick and reliable knowledge flow between knowledge providers and knowledge consumers. The lack of supporting mechanisms (e.g., a knowledge service platform) between service providers and consumers is such a barrier. Thus, there is a need to develop mechanisms to overcome the barriers, thereby improving the performance of knowledge flow.

4.2 Construction of Resource Units

In AA group, we have constructed six different domain resource units. Another five resource units are in the construction. A four-step model is used to frame the construction progress of resource units.

Step 1 *Functional planning of a resource unit* First, long-term goals and short-term goals are primarily determined, and then according to the goals, the investment of time and money, knowledge acquisition capabilities, and knowledge service items are planned.

Step 2 *Construction of a resource unit* Based on the planning schedule, resource units are constructed.

Step 3 *Providing knowledge service to consumers* In this stage, there are two kinds of knowledge service patterns. One is the offline service pattern, in which service providers and service consumers often utilize face-to-face communication and will be based on a project contract to realize a successful knowledge service. The other is the online service pattern. This is a real cloud-based pattern. Providers and consumers will use a knowledge service platform to cooperate on a knowledge service.

Step 4 *Managing the acquired new knowledge* Any resource units have to construct their own knowledge base and establish knowledge management mechanisms.

Based on the above four steps, we have planned 12 resource units, and currently six of them are providing knowledge services now. An overview description of the six resource units is shown in Table 1. A case of the Noise, Vibration, and Harshness (NVH) unit is briefly given for the demonstration of resource units (Table 3).

- *Description of the NVH resource unit*

Planning the long-term goal is built a 20-person professional service group to provide NVH service to the whole enterprise, and the short-term goal is to build an 8–10 people professional service group to provide knowledge service to several divisions.

Construction Currently, two senior engineers with Ph.D. degree and eight engineers with Master's degrees are taking charge of the total knowledge service to the whole group of AA. The investment of money, the NVH stimulation, and testing facility is more than 3.7 million Chinese Yuan (CNY). The knowledge service capacities include CAE software applications, the fault identification of vibration and noise; the design and analysis noise and vibration reduced facility.

Service Advanced Simulation of the cab; Vibration and noise control; NVH control of concrete machinery, excavator, hoisting machinery, pile driving machinery, road construction machinery, port machinery, etc.

Management of knowledge Build database; develop structured documents; organize knowledge community.

Table 3 Overview on the construction of professional resource units

Unit name	Knowledge acquisition resource			Service type
	People	Software	Hardware	
Unit of NVH	2 Ph.D. 8 MS	EDM Sysnoise VA One	e.g. Ncode/NI/LMS data collection system; Sensors	Simulation test
Unit of material R&D	5 MS	ANSYS	e.g. MTS fatigue test machine; Metallographic microscope	Test consulting
Unit of environment simulation	5 MS	None	e.g. Temperature test chamber; Salt spray test chamber	Test consulting
Unit of electronic and electrical	5 MS	EDA PCB	e.g. CAN bus test system; Switch performance test system;	Test consulting
Unit of Industry design	5 MS	CATIA PRO/E	Design review facilitate	Design consulting
Unit of product development	7 MS	PRO/E ANSYS	e.g. Bearing test system; product performance test system	Design consulting

4.3 Development of a Knowledge Service Platform

Based on the architecture of a knowledge service platform, we have developed a knowledge service platform to meet the needs of the inner-enterprise knowledge service in AA Group. Figure 7 shows the login page. Users can login or register on this page.

The platform has three key components, i.e., a service market management component, a resource unit management component, and a business management component. Figure 8 exhibits the elements of all the published services in this platform. The services are broken down into four types, i.e., test service, simulation service, design service, and consulting service. Knowledge service consumers can browse and obtain summary or detailed information about a service.

Figure 9 presents the summary information about the seven registered resource units. These units provide four kinds of knowledge services to the whole enterprise.

Figures 10 and 11 show the key functions of the knowledge service management component. As shown on the left side of Fig. 10, users can create a new resource unit and manage their knowledge service information. Using the *My Service* function module, resource units can review and manage their service profile. The right side of Fig. 10 shows the abstract information of a test service. Detailed information can be found after clicking on the *Detail button*.

Figure 11 shows the detailed information of a service interaction process. The rationale to support decision making in the service process can be recorded, and can be traced in later business.

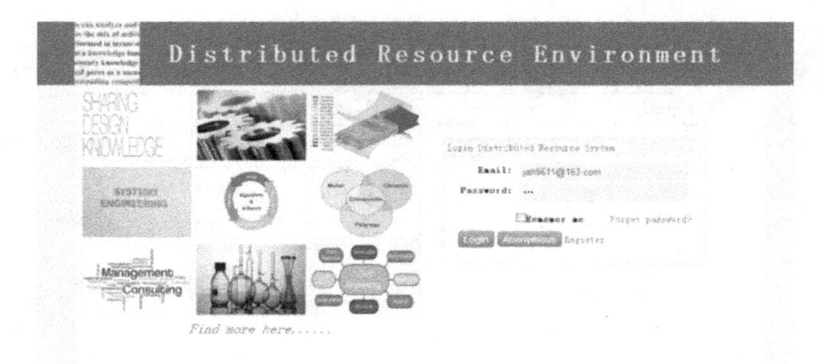

Fig. 7 Distributed resource environment

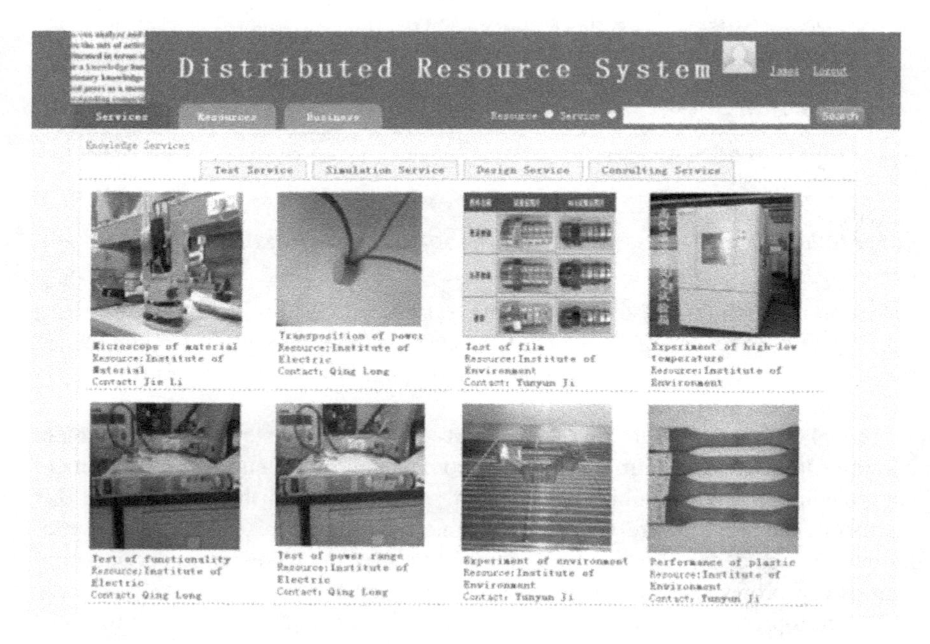

Fig. 8 Knowledge service market

4.4 Results and Discussions

The concept of the distributed resource environment is applied to real industry practice. A one-and-a-half year industry practice has constructed a small scale distributed resource environment in AA group. The environment includes six professional resource units (i.e., knowledge service providers), a knowledge service platform, and numerous knowledge service consumers. The six resource units

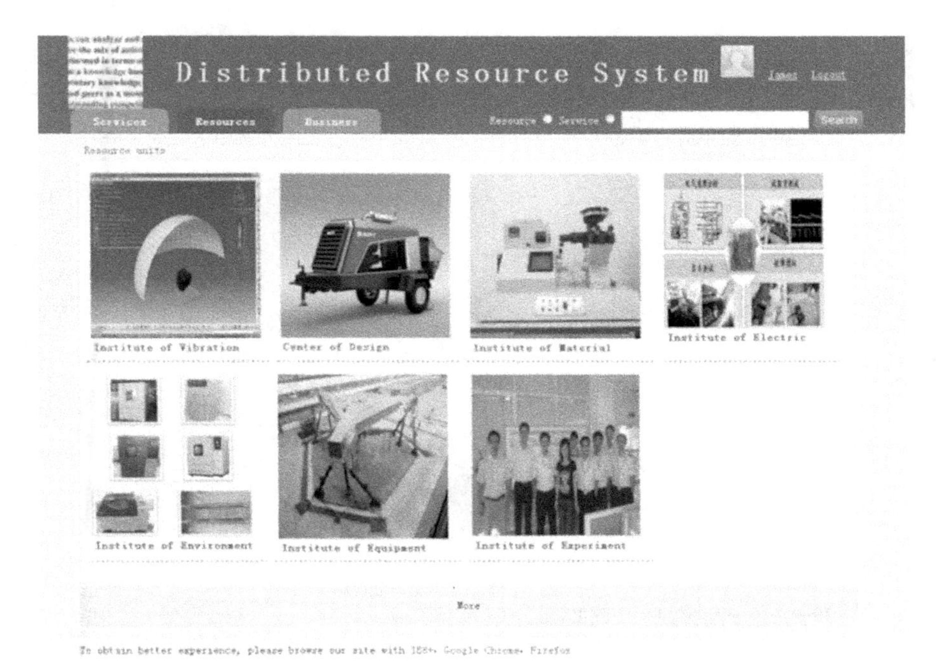

Fig. 9 Resource units management

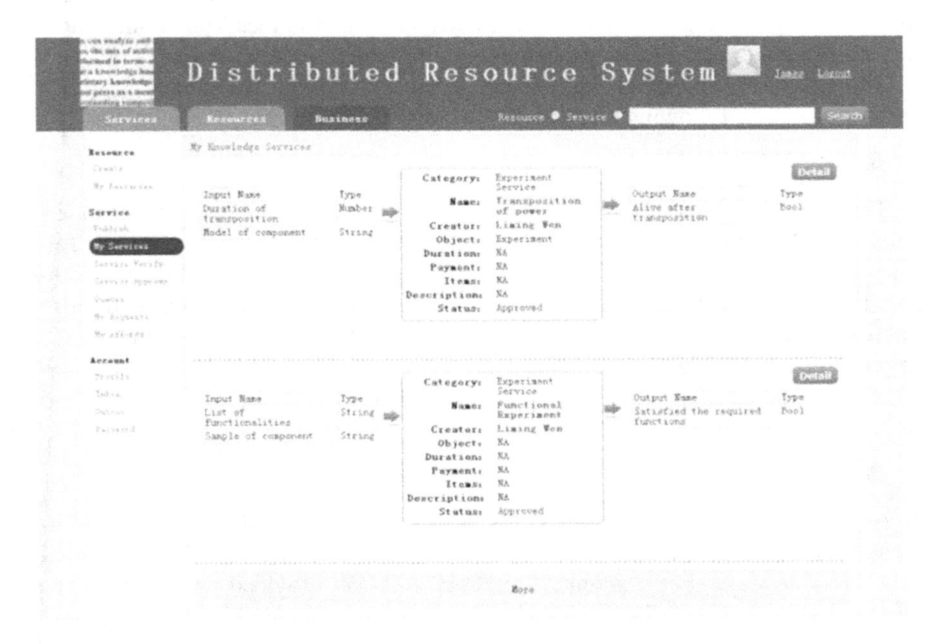

Fig. 10 Knowledge service management

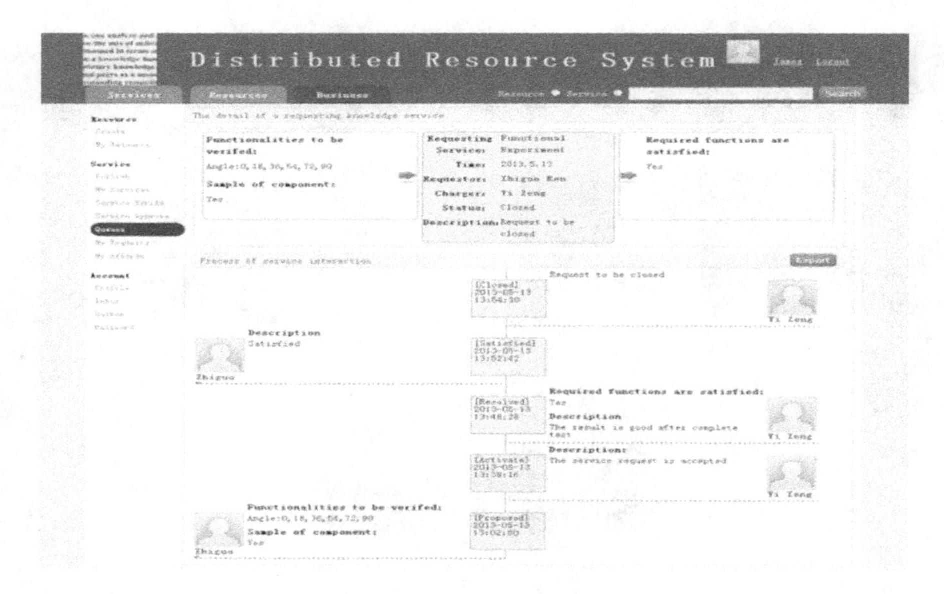

Fig. 11 Process of service interaction

can provide simulation, testing, consulting, and product design services to the whole AA group. The knowledge service platform is a knowledge service market that has bridged the gap of information asymmetry and reduced the communication cost between knowledge consumers and providers in AA group.

Through inner-enterprise knowledge services within a distributed resource environment, the following specific effects were achieved.

- *Better profit*

The outcome of knowledge services shows that the construction of an inner-enterprise distributed resource environment in AA group has generated the following merits: (1) The cost-saving effect is obvious. As shown in Fig. 12, compared with outsourcing knowledge services, a half-year inner-enterprise knowledge service has saved about ¥6.461 million in R&D investment. (2) Improved resource unit deep domain knowledge acquisition capability. (3) Speed enterprise product innovation. (4) Well-managed knowledge for better sharing and reuse.

- *Better social influence*

As shown in Fig. 13, a half-year operation of a distributed resource environment has provided more than 3,100 test services to more than 31 R&D institutes all over the enterprise. For example, the NVH unit has provided 44 noise and/or vibration tests services; the environment simulation unit has provided 1,051 test services. Since July 2012, the industry design unit has provided 23 design services, the other five units have provided more than 7,600 test services and 270 consulting

Unit	People	Income (Completed Services)	Income (Ongoing services)	Spending (Outside service)
NVH	10	617K	360K	1900K
Material R&D	5	282K	170K	900K
Product development	7	630K	50K	1400K
Environment simulation	5	775K	140K	3000K
Electronic and electrical	6	417K	200K	1500K
Industry design	5	428.5K	1695K	2000K
Total	38	3149.5K	1089.5K	10700K

Note: the unit of the currency is CNY, inner-enterprise service will save ¥6461K

Fig. 12 The benefit of a half year inner-enterprise knowledge service

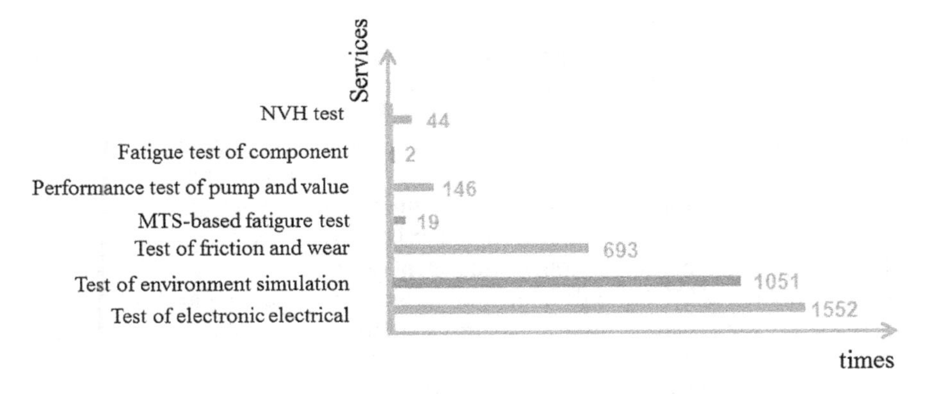

Fig. 13 Overview of half a year knowledge service

services to the whole group. The small scale distributed resource environment has solved a lot of problems within several months. The effect of knowledge service to AA group has been widely acknowledged by knowledgeable service consumers. For example, consumers published thank-you letters through the enterprise's office automation (OA) system. It can be concluded that in the foreseeable future, the distributed resource environment will be a strong driver of product innovation in AA group.

Although we have applied the concept of a distributed resource environment and this small scale distributed resource environment has achieved good results in AA group, there is still a need for further improvement. For example, the concept of a distributed resource environment will be constantly evolving. The development of a theoretical framework for effective design knowledge service will become increasingly important as moving to the next generation of design innovation.

In this case study, there remain several questions still to be answered. For example, (1) How to improve the affordance of the platform? The platform should provide a step-by-step tool for service providers to easily encapsulate their design computing service. The platform should provide an automated mapping mechanism between knowledge service providers and consumers. (2) How to construct more resource units? There are only six design knowledge service providers in this small scale distributed resource environment, therefore how to design organizational mechanisms, which can support the development of more resources units and facilitate knowledge service and knowledge management in AA group would be an interesting direction for future work.

5 Conclusion

In this chapter, the concept of a distributed resource environment (DRE) was presented. The function and structure of a distributed resource environment was proposed and the characteristics of a distributed resource environment were also identified. We proposed a cloud-based knowledge service framework. The architecture of a knowledge service platform was proposed to integrate the knowledge service providers and consumers. A capacity model of a resource unit and the knowledge service process model were presented to support the construction of knowledge service providers and planning their knowledge service processes. Two agents, i.e., a knowledge service publishing agent (KSPA) and a knowledge service consuming agent (KSCA) are developed to implement the online knowledge service. KSPA can be used by knowledge providers to encapsulate and publish their design knowledge as a service into the service market, whereas KSCA can be used by knowledge consumers to request knowledge service from the service market. The concept of a distributed resource environment was applied to a real industry practice. The efficiency and effectiveness of a distributed resource environment as a knowledge flow and innovation facilitator was validated by the industry practice. The results of this study provide a background for setting future research directions on developing a distributed resource environment. However, there are some important questions unanswered and also some disadvantages that need to be overcome. For example, why should resource units provide their knowledge to the platform, when providing their know-how to only one customer is more profitable? Why resource units should provide their knowledge to customers via the platform, even if there is a strong protection of intellectual property? There is also a need to further explore issues like how to reduce the resistance to develop a distributed resource environment from the view of design and how to develop such an environment as a social, technical, and cognitive integrated process.

Acknowledgments This research is partially supported by the National Natural Science Foundation of China (Grant No. 51205247, 50935004). The Research Project of State Key Laboratory of Mechanical System and Vibration (Grant No.MSVZD201401). The authors are grateful to Mr. Qing Long and Li Xiao for their continuous support throughout our case study. Special thanks to Mr. Reynolds and Dr. Bo Xing for proofreading and improving the quality of this paper. The authors also gratefully acknowledge the helpful comments and suggestions of the reviewers, which have improved the presentation.

References

Chen YJ, Chen YM (2009) Development of a distributed product knowledge service system. In: International conference on complex, intelligent and software intensive systems (CISIS2009)

Choy KL, Tan KH, Chan FTS (2007) Design of an intelligent supplier knowledge management system—an integrative approach. Proc Inst Mech Eng Part B: J Eng Manuf 221(B2):195–211

Ding B, Tu XY, Sun LJ (2012) A cloud-based collaborative manufacturing resource sharing services. Inf Technol J 11(9):1258–1264

Li G, Lu H, Ren W (2009) Service-oriented knowledge modeling on intelligent topic map, In: 1st international conference on information science and engineering (ICISE2009), 2009

Li X, Zhang Z, Xie Y (2012) The frame-based Knowledge Service Agent. In: Special Workshop: Knowledge as a Service (CSSS 2012)

Meng X, Xie Y (2011) Embedded knowledge service in mechanical product development. Int J Adv Manuf Technol 53:669–679

Mentzas G, Kafentzis K, Georgolios P (2007) Knowledge services on the semantic web. Commun ACM 50: 53–58

Schaefer D, Thames JL, Wellman RD, Wu D, Yim S, Rosen D (2012) Distributed collaborative design and manufacture in the cloud-motivation, infrastructure, and education, In: American society for engineering education annual conference, 2012

Tao F, Zhang L, Venkatesh VC, Luo Y, Cheng Y (2011) Cloud manufacturing: a computing and service-oriented manufacturing model. Proc Inst Mech Eng, Part B: J Eng Manuf 225(10):1969–1976

Wu D, Thames JL, Rosen DW, Schaefer D (2012) Towards a cloud-based design and manufacturing paradigm: looking backward, looking forward. In: Proceedings of the ASME 2012 international design engineering technical conference and computers and information in engineering conference (IDETC/CIE12)

Wu D, Greer MJ, Rosen DW, Schaefer D (2013) Cloud manufacturing: strategic vision and state-of-the-art. J Manuf Syst 32: 564–579 http://dx.doi.org/10.1016/j.jmsy.2013.04.008

Wu D, Greer MJ, Rosen DW, Schaefer D (2013) Cloud manufacturing: drivers, current status, and future trends. In: Proceedings of the ASME 2013 international manufacturing science and engineering conference (MSEC13), 2013

Wu D, Schaefer D, Rosen DW (2013) Cloud-based design and manufacturing systems: a social network analysis. In: International conference on engineering design (ICED13), 2013

Xie YB (1998) Development and application of distributed design resource. Chin Mech Eng 9(2):16–18 (In Chinese)

Xu X (2012) From cloud computing to cloud manufacturing. Robot Comput Integr Manuf 28(1):75–86

Zhang Q, Cheng L, Boutaba R (2010) Cloud computing: state-of-the-art and research challenges. J Internet Serv Appl 1:7–18

Zhuge H, Guo WY (2007) Virtual knowledge service market for effective knowledge flow within knowledge grid. J Syst Softw 80(11):1833–1842

Research and Applications of Cloud Manufacturing in China

Bo Hu Li, Lin Zhang, Xudong Chai, Fei Tao, Lei Ren, Yongzhi Wang, Chao Yin, Pei Huang, Xinpei Zhao, Zude Zhou, Baocun Hou, Tingyu Lin, Tan Li, Chen Yang, Anrui Hu, Jingeng Mai and Longfei Zhou

Abstract This chapter leads to the achievements of research and applications in Cloud Manufacturing (CMfg) carried out by the authors' team. First of all, the meaning of Big Manufacturing is given, and the challenges and strategies for manufacturing industries as well as the content and development of manufacturing informatization are analyzed. Then the definition, concept model, system architecture, technologies, typical technical characteristics, and service object and content of CMfg are put forward. Moreover, discussions are shown to prove that CMfg

---

*Notation: *Manufacturing* appearing in this chapter refers to *Big Manufacturing* that covers all the activities and processes in the product lifecycle, from the market demand demonstration, the design and simulation, the production and fabrication, the test and experiment, the operating maintenance to the final scraping process.

B. H. Li (✉) · L. Zhang · F. Tao · L. Ren · A. Hu · J. Mai · L. Zhou
School of Automation Science and Electrical Engineering, Beihang University, Beijing 100191, China
e-mail: bohuli@buaa.edu.cn

L. Zhang
e-mail: zhanglin@buaa.edu.cn

B. H. Li · X. Chai · B. Hou · T. Lin · T. Li · C. Yang
Beijing Simulation Center, 52, Yongding Rd., Beijing 100191, China

Y. Wang
China CNR Corporation Limited, 15 Area One, Fangchengyuan Rd., Beijing 100078, China

C. Yin
Institute of Manufacture Engineering, Chongqing University, Chongqing 400044, China

P. Huang
DG-HUST Manufacturing Engineering Institute, Beijing, China

X. Zhao
Beijing ND Tech Corporation Limited, Beijing 100020, China

Z. Zhou
Wuhan University of Technology, 122 Luoshi Road, Wuhan 430070, China

D. Schaefer (ed.), *Cloud-Based Design and Manufacturing (CBDM)*,
DOI: 10.1007/978-3-319-07398-9_4,

is a new paradigm and approach to realize manufacturing informatization, which materializes and extends Cloud Computing in the manufacturing domain. The current status of the technologies, applications, and industries for CMfg in China are presented. Concerns are mostly focused on the 12 key technologies for the CMfg System (also called Manufacturing Cloud), including system overall technology, sensing technology for manufacturing resource and capability, virtualization and service technology for manufacturing resource and capability, construction and management technology for virtual manufacturing environment, operation technology for virtual manufacture environment, evaluation technology for virtual manufacturing environment, trusted service technology for virtual manufacturing, management of knowledge, model and data, pervasive human–computer interaction technology, application technology of service platform, informatized manufacturing technology system, and product service technology. The academic research results of the authors' team are presented. Furthermore, four typical CMfg cases which have been successfully implemented in group enterprises and mid-small enterprise clusters are described as well as cases introducing smart manufacturing into smart city. Finally, future works and development prospect for CMfg are raised.

Keywords Cloud manufacturing · Manufacturing informatization · Cloud computing · Smart manufacturing

1 Introduction

1.1 Big Manufacturing

Traditional manufacturing drives at the production activities and processes from raw material to product. It is shown in Fig. 1.

In this chapter, the meaning of manufacturing is "Big Manufacturing." It has three aspects of meaning: (1) big coverage of manufacturing process,in which demand, design, production, experiment, running, maintenance to the retirement are all involved as shown in Fig. 2; (2) big space coverage of the manufacturing activities, which is from inside enterprises to among enterprises, even to global space; and (3) big coverage of manufacturing pattern, which relates to a wider variety of manufacturing such as discrete manufacturing, process manufacturing and hybrid manufacturing, etc.

1.2 Challenges and Countermeasures of Manufacturing Industry

During the past two decades, many advanced manufacturing *paradigms and approaches* have been proposed in order to realize the aim of TQCSEK (i.e., faster

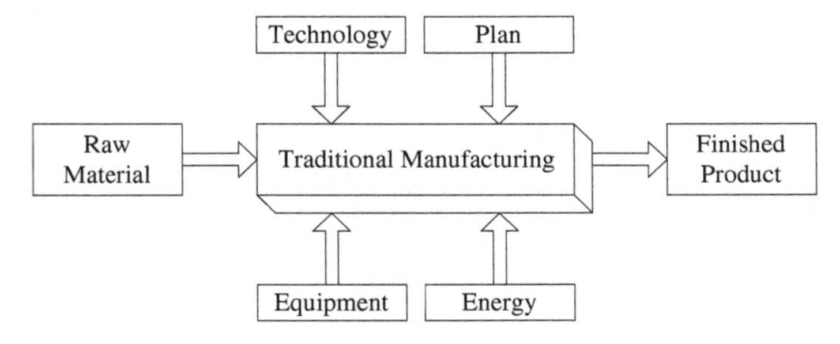

Fig. 1 Traditional manufacturing

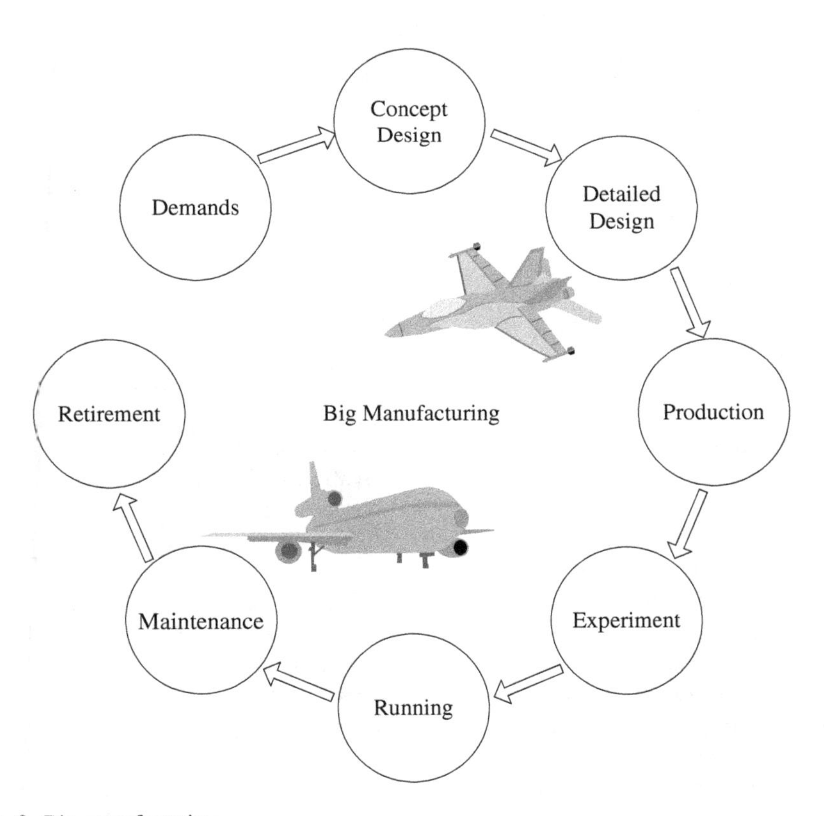

Fig. 2 Big manufacturing

time-to-market, higher quality, lower cost, better service, clean environment, and high knowledge) for manufacturing enterprise as shown in Fig. 3.

"Smile Curve" is a famous curve in the manufacturing industry. It shows the relationship between business process and added value. Currently, many manufacturing enterprises in China locate in the bottom of "Smile Curve" (low profit margin). The problems are low competitive power in product R&D and Service

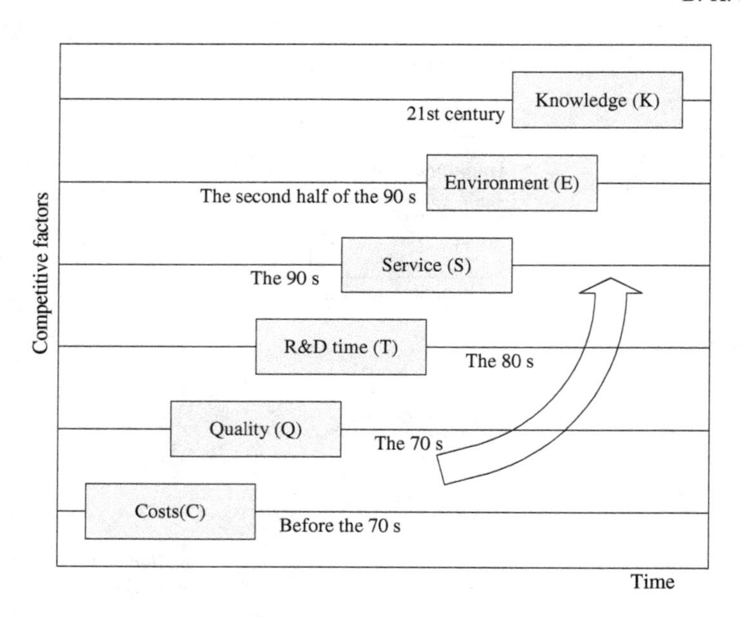

Fig. 3 Challenges of manufacturing

and high energy-consuming in production and high environment pollution. It is shown in Fig. 4.

The manufacturing industry in China is now transforming from production-oriented type to production-plus-service-oriented type, from the lower end of the value chain to the upper end, from major producer to major manufacturing innovator, from "Made-in-China" to "Innovation-in-China." To realize new manufacturing paradigms and approaches, to accelerate the transforming of the economy progressing and to fulfill the TQCSEK will be critical issues for manufacturing in China in next 5–10 years.

The countermeasures promote the overall progress of the manufacturing industry in China under the guidance of "Driven-by-Innovation, Focused-on-Quality, Green-Development, Structure-Optimization" (Zhou 2013). It focused on improving the competitiveness of the enterprises. Among these countermeasures, "manufacturing informatization" is an important measures, and it has been implemented in China for many years.

The manufacturing informatization is a complicated strategic system engineering and important part of the new industrial China. The guidance idea of the manufacturing informatization in China keeps up with the times. It is from "informatization promotes the industrialization" to the "fusion of informatization and industrialization" and then to the "in-depth fusion of informatization and industrialization". The roadmap fuses the information technology, modeling and simulation technology, modern manufacturing technology, system engineering technology, and product-related technology into the life cycle of the production. It sets up a new paradigm and approach. The contents of manufacturing

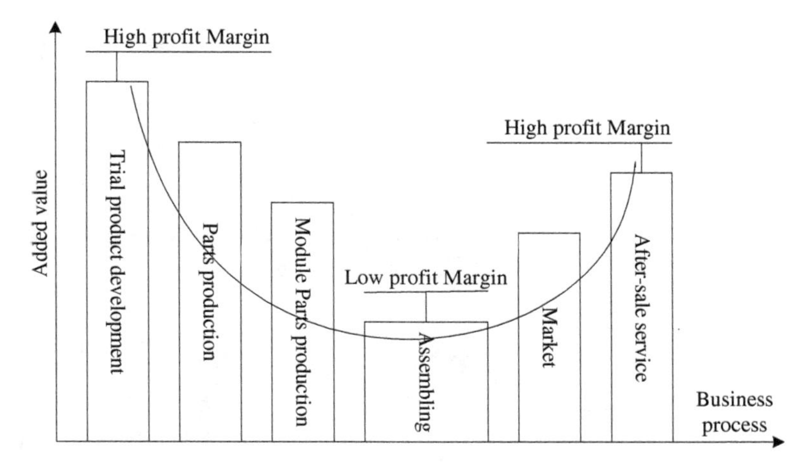

Fig. 4 Smile curve (http://job.workercn.cn/c/2010/07/27/154049780763471.html)

informatization integrates and optimizes the human or organization, management and technology (three elements) together with the information flow, logistics, financial flow, knowledge flow and service flow (five flows) to realize the digital, integrative, collaborative, networked, agile, service-oriented, green, and intelligent manufacturing to improve the agility, flexibility, and robustness of the enterprises, and boost up the competitive ability of the enterprises.

1.3 Development of Manufacturing Informatization

The trends of manufacturing informatization are digital, integrative, collaborative, networked, agile, service-oriented, green, and intelligent. The current focus is "agile, service-oriented, green, and intelligent." It is now becoming key factors of current manufacturing. The main developments are agile manufacturing, service-oriented manufacturing, green manufacturing, and intelligent manufacturing.

Our team proposed the concept of Cloud Manufacturing (CMfg) in 2009, and achieved important breakthrough in the simulation and design in CMfg (Li et al. 2010). Currently, our team cooperates with 28 excellent teams and starts the technical research and application for both group enterprises and minor enterprises, which greatly improves the innovation abilities of those manufacturing enterprises.

Our research is divided into six parts. The first is the common key technology research of group enterprise CMfg service platform. Beihang University is responsible for this matter with other five units. The second is the cloud services platform development, system construction and applications in complex aerospace product manufacturing group enterprise. Second Research Institute of China Aerospace Science and Industry Group is responsible for this matter with other

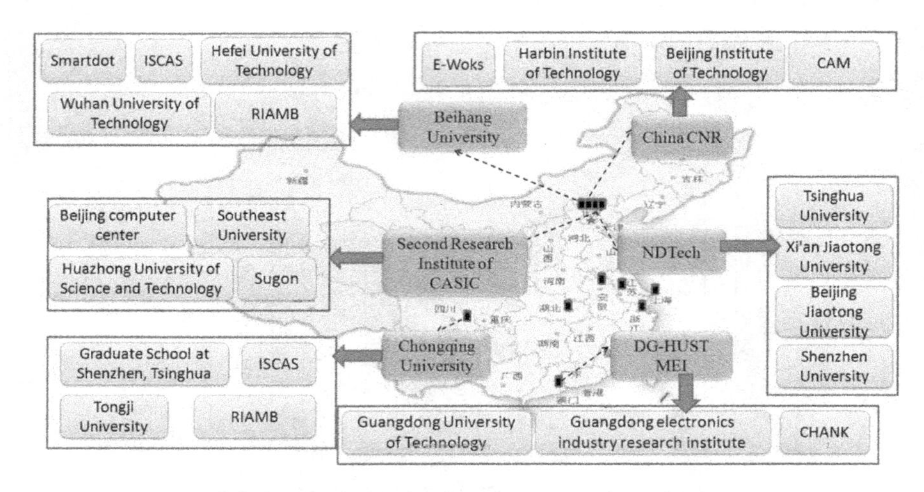

Fig. 5 Research team distribution diagram

four units. The third is CMfg service platform development, system construction and applications in rail transit equipment group enterprise. China CNR is responsible for this matter with other 4 units. The fourth is common key technology research of small and medium-sized enterprise CMfg service platform. Chongqing University is responsible for this matter with other 4 units. The fifth is CMfg service platform development, system construction, and applications in small and medium-sized enterprise supporting inter-enterprise business collaboration. DN tech is responsible for this matter with other four units. The sixth is CMfg service platform development, system construction, and applications in small and medium-sized enterprise supporting industrial cluster collaboration. DG-HUST Manufacturing Engineering Institute is responsible for this matter with other 4 units. The whole project is taken charge by Beihang University. It is shown in Fig. 5.

2 Connotation of CMfg

2.1 Definition of CMfg

CMfg is a new service-oriented smart manufacturing paradigm and approach based on networks (such as Internet, IoT, communication network, broadcasting network, and mobile network). It fuses the current informatized manufacturing technology and new information technology (such as cloud computing, Internet of things, service-oriented computing, modeling and simulation, intelligent science, high performance computing, big data, and e-commerce, etc.) to transform

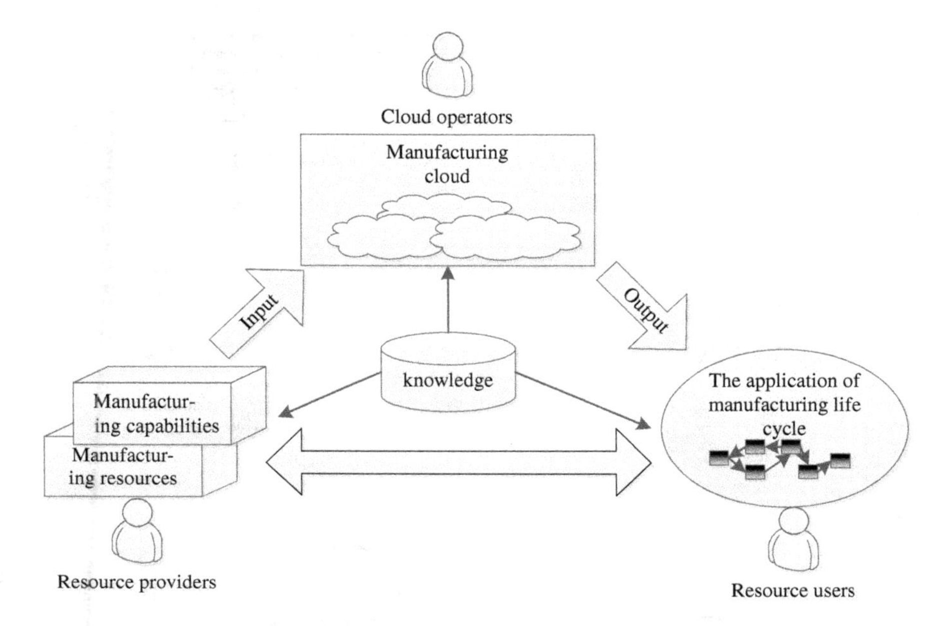

Fig. 6 The concept model of CMfg

manufacturing resources and manufacturing capabilities into manufacturing services and to build a manufacturing service pool, which can be managed and operated in an intelligent and unified way to enable users acquiring safe and reliable, high quality, cheap, and manufacturing services anytime and anywhere on demand for the whole life cycle activities of manufacturing (Li et al. 2010, 2011a, 2012a, b; Zhang et al. 2011, 2012a, 2013; Tao et al. 2011a).

2.2 Concept Model of CMfg

Based on the above definition, the concept model of CMfg is shown in Fig. 6. In this model, there are three parts (manufacturing resources and manufacturing capabilities, manufacturing cloud, or service pool and applications of manufacturing life cycle) corresponding to the three types of users (providers, operators, and demanders) respectively. Therefore, it contains two processes among the three types of users, i.e., input: providers publish the services of manufacturing resources and capabilities to the platform of manufacturing cloud, output: the services in manufacturing cloud are applied into the whole manufacturing lifecycle for demanders. Meanwhile, a core support (knowledge or wisdom) exists covering the whole operation of CMfg.

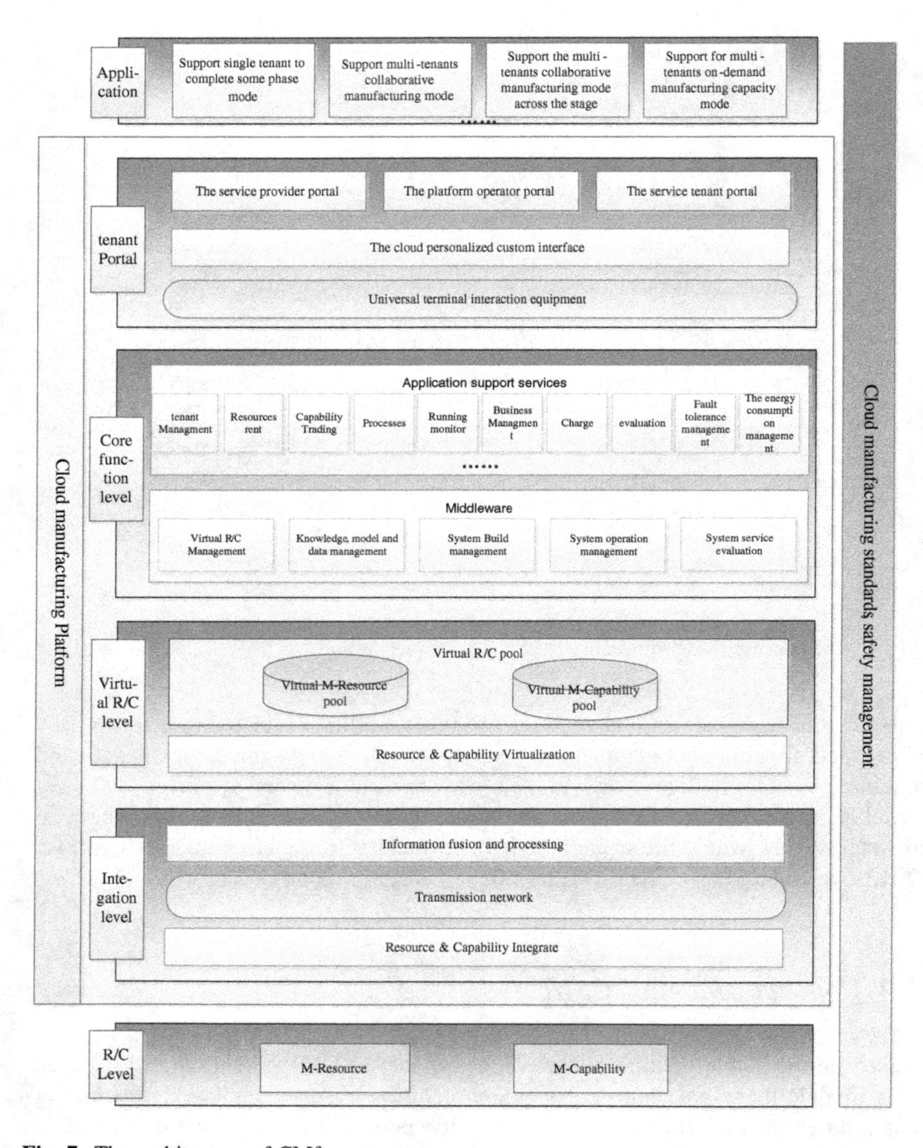

Fig. 7 The architecture of CMfg system

2.3 Architecture of CMfg System

The architecture of CMfg system is shown as Fig. 7, and the key layers involved in this CMfg architecture are as follows:

(1) Resource and capability layer: It is the layer of physical manufacturing resources and capabilities.

(2) Integration layer: It includes resource and capability integration, transmission network and information fusion and processing. It can sense the physical manufacturing resources and capabilities, and realize the overall connection of various resources and capabilities.
(3) Virtual resource and capability layer: It includes two virtual resource and capability pools, namely, Virtual M-Resource Pool and Virtual M-Capability Pool.
(4) Core function layer: It provides core functions and service support for the operation and management of manufacturing cloud services.
(5) User portal: It consists of universal terminal interaction equipment, the cloud personalized customer interface and some portals. These portals include the service provider portal, the platform operator portal, and the service user portal.
(6) Application layer: It provides different specific application interfaces and related end-interaction equipment which consists of 4 kinds of manufacturing modes.

2.4 Technology System of CMfg

The technology system of CMfg can be classified into the following 12 key categories as shown in Fig. 8:

(1) System overall technology,
(2) Sensing technology for manufacturing resource and capability,
(3) Virtualization and service technology for manufacturing resource and capability,
(4) Construction and management technology for virtual manufacturing environment,
(5) Operation technology for virtual manufacturing environment,
(6) Evaluation technology for virtual manufacturing environment,
(7) Trusted service technology for virtual manufacturing,
(8) Management of knowledge, model and data,
(9) Pervasive human–computer interaction technology,
(10) Application technology of service platform,
(11) Informatized manufacturing technical system as shown in Fig. 9,
(12) Product service technology.

2.5 Key Features of CMfg

2.5.1 Technical Features of CMfg

Based on the above technology system of CMfg, the typical technical features of CMfg are as follows:

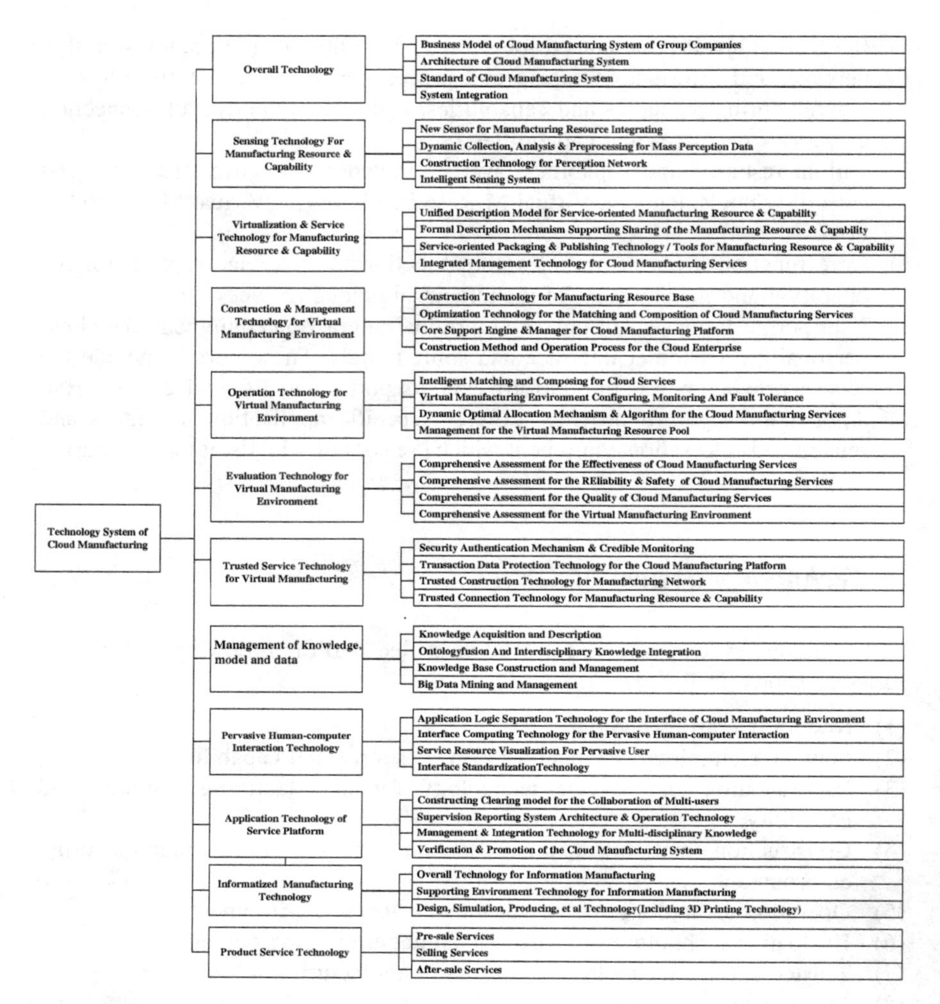

Fig. 8 The technology system of CMfg

(1) Digitalization of manufacturing resource and capability
It refers to convert the attributes, both the static and dynamic behavior of the manufacturing resources and capabilities into digit, data, models, in order to carry out a unified analysis, planning, and restructuring process. The integration of digitalization technology and manufacturing resource and capability can be used to control, monitor, and manage the manufacturing resources systems, such as hardware manufacturing resources (i.e., CNC machine tools, robots, etc.) and soft manufacturing resources (i.e., computer-aided design software, management software, etc.).

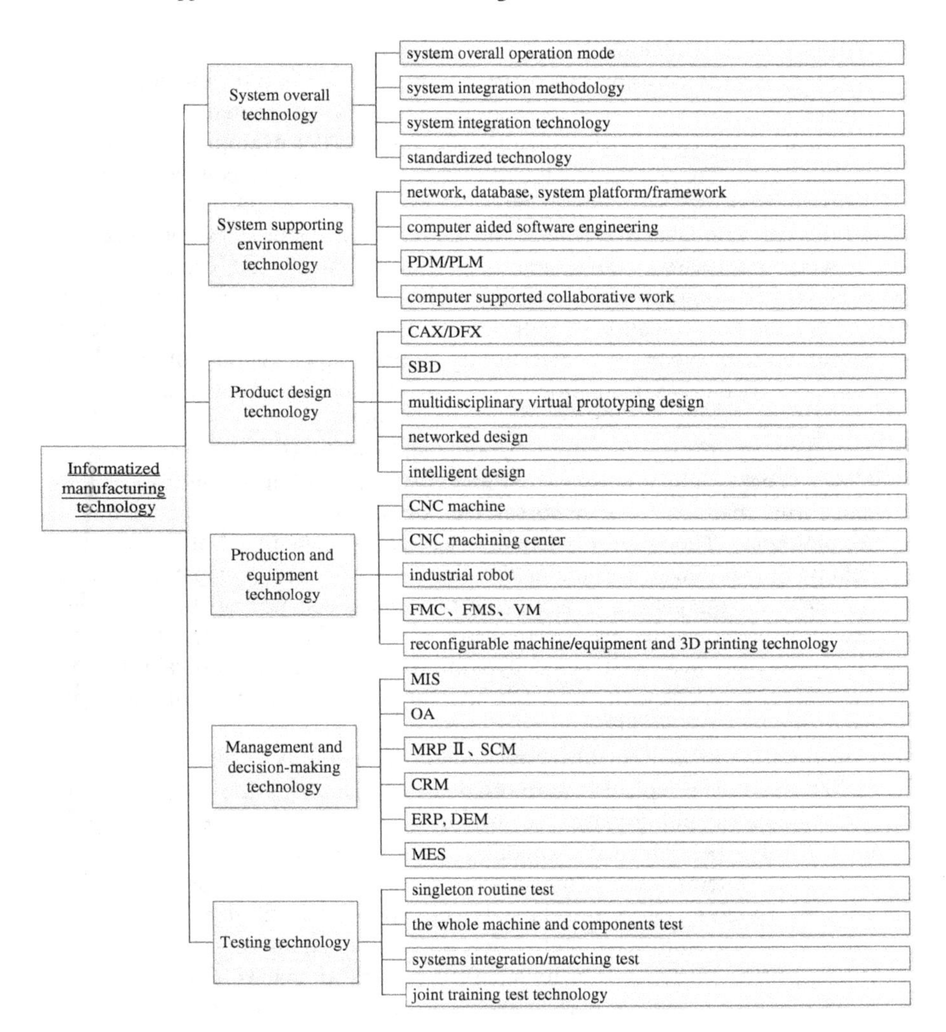

Fig. 9 Informatized manufacturing technology

(2) Connection of things of manufacturing resource and capability

Integrating the newest information technologies such as IoT and cyber physical system (CPS), CMfg achieves the comprehensive and thorough accessing and perception of the system-wide, whole life cycle manufacturing resources and capabilities, especially the hard manufacturing resources such as machine tools, machining centers, simulation equipment, test equipment, logistics of goods and manufacturing capabilities such as human, knowledge, organization, performance, and reputation. To support the integration and optimization of the human, organization, management, and technology in the full life cycle of the manufacturing.

(3) Virtualization of manufacturing resource and capability
It means to provide logic and abstract representation and management of the manufacturing resources and capability, which is not limited to the specific physical instances with virtualization technology. For example, ① a physical manufacturing resource or capability can form multiple isolated packaged "virtual devices," ② multiple physical manufacturing resources and capabilities can aggregate to form a larger-grained "virtual devices" organization, ③ when necessary, virtual manufacturing resources and capabilities can achieve live migration and dynamic scheduling. Virtualization technology enables the simplification of representation and access of the manufacturing resources and capabilities and further optimal management of them uniformly. It is the key technology to realize the service-oriented and collaborative manufacturing resources and capabilities.
(4) Servalization of manufacturing resource and capability
CMfg system gathers mass virtual manufacturing resources and capabilities, and forms them as on-demand services by service package and computing technologies. Then the on-demand service is primarily reflected in: ① achieving centralized service of the distribute resource or capability by the on-demand aggregation, ② achieving distributed services of the centralized resource or capability by the on-demand split. The characteristics of the CMfg services are on demand dynamic architecture, interoperability, collaborative, heterogeneous network integration, fast response, full life cycle smart manufacturing, etc.
(5) Collaboration of manufacturing resource and capability
Collaboration is a typical feature of advanced manufacturing mode. Through information technology such as virtualization, service-oriented and distributed, or high performance computing, and CMfg, forms the flexible, interconnected, interoperable "manufacturing resources and capabilities as a service" module. On the technical level, the cloud service modules can achieves the system-wide, full life cycle, comprehensive interconnection, interoperation, collaboration to satisfy the users' demands. CMfg also provides comprehensive support to the dynamic collaborative management of agile virtual enterprise organizations, to achieve on-demand construction of virtual enterprise organizations by multi-users dynamically, and to achieve seamless integration in the collaboration of virtual enterprise business.
(6) Intellectualization of manufacturing resource and capability
Another typical technique feature of CMfg is to achieve the system-wide, full life cycle, comprehensive intelligence. Knowledge and intelligent science is the core to support the intelligent operation of the CMfg service system. Manufacturing cloud brings together all kinds of knowledge and builds a multidisciplinary knowledge base. Knowledge and intelligent science penetrate all aspects and levels of manufacturing, to support the two dimensions of the life cycle: full life cycle of the manufacturing, and full life cycle of manufacturing resource and capability service.

2.5.2 Relationship Between CMfg and Cloud Computing

As we know, the main service models involved in cloud computing are Infrastructure as a Service (IaaS), Platform as a Service (PaaS), and Software as a Service (SaaS). According to the concept of CMfg, the extension and development of CMfg is based on IaaS, PaaS, and SaaS, as shown in Fig. 10.

In CMfg, the main users of service are manufacturing enterprise user and manufacturing product user, and the contents of service are Argumentation as a Service (AaaS), Design as a Service (DaaS), Fabrication as a Service (FaaS), Experiment as a Service (EaaS), Simulation as a Service (SIMaaS), Management as a Service (MaaS), Operation as a service (OPaaS), Repair as a Service (REaaS), and Integration as a Service (INaaS). And the services in CMfg have the following characteristics:

(1) On-demand dynamic architecture: according to the user needs, provide manufacturing services anywhere, anytime.
(2) Interoperability: support interoperability among manufacturing resources and capabilities.
(3) Collaboration: face to the collaboration of large-scale complex manufacturing tasks and multi-tenants.
(4) Heterogeneous network integration: support the integration of online distributed heterogeneous manufacturing resources and capabilities.
(5) Fast response: compose various services (unlimited) in response to demand quickly and flexibly.
(6) Full life cycle smart manufacturing: serve the manufacturing life cycle and make use of intelligent information technology to achieve the multi-phase smart manufacturing.

In terms of the resources and capabilities sharing, shared resource in cloud computing are mainly computing resources such as storage, CPU, software, and data. However, the shared resources in CMfg contains hard manufacturing resource (i.e., all kind of Models, data, software, knowledge in the process of manufacturing, etc.), soft manufacturing resource (such as machine tools, robotics, machining centers, computing devices, simulation test equipment, etc.), and manufacturing capability resources (such as argumentation, design, simulation, fabrication, experiment, management, operation, repair, integration capabilities in the process of manufacturing).

In terms of service pattern, the service contents of CMfg are in addition to the cloud computing pattern of IaaS, PaaS, and SaaS. It pays more attention and emphasis on the resources and capabilities required in the full life cycle of manufacturing such as AaaS. Regard to service pattern, cloud computing supports submitting job online and interacting with computing resources, furthermore, CMfg can make users submit task online, and use various collaborative services for the whole manufacturing lifecycle.

Finally, in terms of supporting technologies, it includes the following enabling technologies:

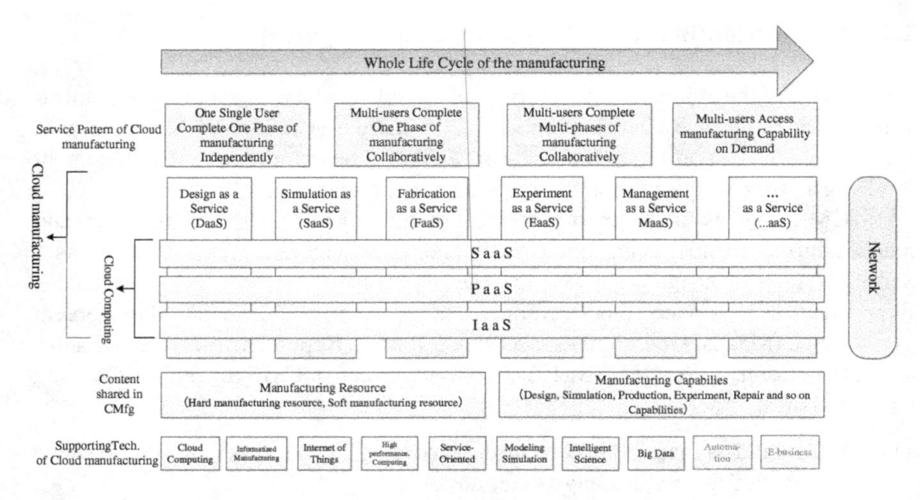

Fig. 10 The relationship between CMfg and cloud computing

(1) Cloud computing: it provides a series of enabling technologies and new manufacturing patterns for manufacturing resources or capabilities in CMfg, as well as the enabling technologies for intelligent information processing and decision in manufacturing activities. It is the core technology to handle all of manufacturing data.
(2) Internet of things: it provides a series of enabling technologies for interconnection between things in manufacturing domain and realization of wisdom manufacturing, that is to say, the enabling technologies for the realization of intelligent perception and effective connection among different things (including men-to-men, machine-to-machine, and men-to-machine). It is for the source and transmission of various manufacturing data.
(3) Service-oriented technology: it provides a series of enabling technologies for intelligently constructing and executing a virtual manufacturing and service environment, such as SOA, web service, semantic web, and so on. It makes the possible from manufacturing resources and capabilities to cloud services.
(4) Intelligent science: it provides the enabling technology for making manufacturing resources or capabilities intelligent, and helps to improve the intelligent operation and using of the CMfg service platform with much more manufacturing related knowledge generation and management.
(5) High performance computing: it provides the enabling technologies for solving large-scale and complex manufacturing problems and carrying out large-scale cooperative manufacturing and work.
(6) Big data: it provides enabling technology for accurate, high-efficiency, intelligence full lifecycle activities of CMfg with the entire realization of manufacturing informatization which brings out massive data.
(7) Modeling and simulation: it provides enabling technology for effective development and running of manufacturing system.

(8) Automation: it provides enabling technology for inspection, control, evaluation of manufacturing system.
(9) E-business: it provides enabling technology for the business of the full lifecycle activities of CMfg and the integration of logistic, information flow and capital flow, and it also makes all production activities and business flows online and much easier for manufacturing enterprises.
(10) Informatized manufacturing technology: it is the basic technology of the CMfg.

2.5.3 CMfg: The Extension of Traditional Manufacturing Informatization Technology

With the evolution from information integration and process integration to enterprise integration in manufacturing, the servitization, creation based on knowledge, the aggregation and collaboration of various manufacturing resources and capabilities, and the friendliness to environment, are more important for enterprise competitiveness and the development of manufacturing informatization. Under the driven of new demands and new technologies, CMfg is proposed based on the advanced characteristics and results of these existing manufacturing modes, as well as digesting such newest developing thoughts in manufacturing industries at home and abroad as IPS2, collaborative networks, crowd sourcing, and so on.

Therefore, CMfg is the extension of traditional informatization manufacturing technology on paradigm and approach:

(1) New manufacturing paradigm:
It is from "production" mode to the "production" plus "service" mode with the feature of "network, collaboration, agility, green, service and intelligence."
(2) New manufacturing approach:
It has the smart technical features of "digitalization", "things networking," "virtualization," "service-oriented," "collaboration" and "intelligence."

2.6 The Comparison Between CMfg Model and Other Manufacturing Model

In order to improve the produce efficiency of enterprises, scholars put forward many manufacturing modes, such as agile manufacturing, gridding manufacturing, and industry 4.0, IPS2, and so on. These manufacturing modes above all make great contributions to manufacturing industry. However, these manufacturing modes have their own limitation in terms of paradigm, approach, and supporting technology, as shown in the following table.

Terms model	Cloud manufacturing	Industry 4.0	Agile manufacturing	ASP/Network manufacturing
Manufacturing paradigm	Smart (digital, integrative, collaborative, networked, agile, service-oriented, green, intelligent) manufacturing in the entire life-cycle, realizing the on-demand individualized, and socialized product-plus-service manufacturing at any time anywhere	Crosswise integration of the full value chain, end-to-end digital integration, networked intelligent manufacturing	Virtual enterprise and collaborative network, establishing the cooperation federation among enterprises to fast response to the market demands	Online rent and usage of business application and management services software as a service
Manufacturing approach	Digital, thing-networked, virtualized, service-oriented, collaborative, intelligent manufacturing systems based on resources and capabilities in pervasive network	Intelligent manufacturing system and intelligent factory based on Internet of things and Internet of services	The multi-enterprise business model integration and work-flow control, the enterprise information integration system(the management of clients, supply chain and enterprise resources)	The network-based application system services, the rent and host of software resources
Supporting technology	Cloud computing, Internet of things, service computing, M&S, automation, intelligent science, HPC, Big data, E-commerce, informatized manufacturing technology	CPS, embedded technology, Internet of things, internet of services, automation, simulation, informatized manufacturing technology	EAI, Supply Chain, ERP, SCM, PDM, Work-flow, PDM	Multi-tenant, web-service, internet, E-commerce

3 Research Works of CMfg in China

Supported by the National High Technology Research and Development Program of China (863 Program), "Key generic techniques of CMfg service platform", our team has studied the CMfg technology and achieved a series of research results, as follows.

3.1 Overall Technology of CMfg

In this technology, the system architecture, technology system, and its technical characteristics were proposed. And four CMfg standards were draft, one of which has been approved and is waiting for release.

To solve complex manufacturing problems and perform larger scale collaborative manufacturing, the impediments to practical application and development of Networked Manufacturing (NM) were analyzed. Technologies, such as cloud computing, cloud security, high performance computing, Internet of things were used to solve the mentioned impediments. The concept of CMfg was defined. And differences among CMfg, application service provider and manufacturing grid were discussed. CMfg architecture was also proposed, key technologies for implementing CMfg were studied (Li et al. 2010).

CMfg was a new service-oriented, high-efficiency and low consumption, knowledge-based and intelligent-networked agile manufacturing model and technology. It enriched and expanded the range of resource sharing and service model in cloud computing, promoted agile, service, green, and intelligent-oriented manufacturing development. Service system and technical architecture of CMfg were constructed by integrating cloud computing, Internet of things, High Performance Calculation (HPC), service computing, Artificial Intelligence (AI), and information-oriented manufacturing (Li et al. 2011a).

3.2 Sensing, Access and Handling for CMfg Resources, and Capabilities

In this technology, the authors have proposed real-time, dynamic, online, distributed new principles, methods, and technologies based on fiber Bragg grating sensor, which has built the manufacturing equipment running state IntelliSense-integrated system based on a line multi-point, dynamic multi-field of FBG sensing. New method using data fusion, resource description, and intelligent retrieval for Intelligent Information Processing has been proposed in CMfg. Physical manufacturing resources Intelligence that oriented sensor network, transmission network and application network have been researched and developed (Zhou et al. 2011; Lou et al. 2012; Xu et al. 2012).

The embedded technology as a technological fundament plays an important role in the intelligent monitoring and diagnosis for modern mechanical equipment (MME). The fiber Bragg grating sensor (FBGS) technology has been rapidly applied to most of the industrial and mechanical engineering fields in recent years. It is valuable to study the new principle and new method of intelligent monitoring and diagnosis system for MME based on the integration of embedded technology and FBGS technology. According to the principle of embedded technology, the embedded network and field bus gateway for the integration of monitoring and diagnosis system are investigated, and the embedded high-speed demodulation is studied. Moreover, the embedded sensing signal processing and data transmission, which can meet the requirements of multi-parameter measurement, synchronous sampling, and long-term intelligent monitoring for MME, are proposed. In addition, by integrating the FBGS technology and embedded technology, the embedded integration system of intelligent monitoring and diagnosis for MME is presented (Zhou et al. 2011).

Multi-agent-based proactive–reactive scheduling for job shops is presented, aiming to hedge against the uncertainties of dynamic manufacturing environments. This scheduling mechanism consists of two stages, including proactive scheduling stage and reactive scheduling stage. In the proactive scheduling stage, the objective is to generate a robust predictive schedule against known uncertainties; in the reactive scheduling stage, the objective is to dynamically rectify the predictive schedule to adapt to unknown uncertainties, viz. the reactive scheduling stage is actually complementary to the proactive scheduling stage. A stochastic model is presented, which concerns uncertain processing times in proactive scheduling stage on the basis of analyzing the deficiencies of a classical scheduling model for a production schedule in practice. For the stochastic scheduling problem, a multi-agent-based architecture is proposed and a distributed scheduling algorithm is used to solve this stochastic problem. Finally, the repair strategies are introduced to maintain the original proactive schedule when unexpected events occur. Case study examples show that this scheduling mechanism generates more robust schedules than the classical scheduling mechanism (Lou et al. 2012).

Due to the continuous growth in the application of networks in manufacturing, quality of service (QoS) has become an important issue. In our research, the concept of QoS for manufacturing networks is discussed. To provide overall performance assurance for manufacturing networks, a service framework integrating the QoS mechanisms of the networked resource service management function and the communication networks is proposed. The novel framework maps an application to resource services and then to communication networks, adopts an intelligent optimization algorithm for QoS management of resource services, and provides QoS schemes for data transfer across communication networks. A prototype implementation has been realized and a set of simulation experiments conducted to evaluate the validity of the framework. The results obtained demonstrate the ability of the framework to satisfy the various performance requirements posed by such applications and provide efficient overall performance assurance for manufacturing networks (Xu et al. 2012).

3.3 Virtualization and Servitization for CMfg Resources and Capabilities

The manufacturing resources and capability must have been preliminarily classified. In this context, the extraction method for connotation and characteristics of manufacturing capability have been given, and formal modeling and representation method of manufacturing capability based on multidimensional information model has been proposed. Besides, processing method for big data of manufacturing services based on Relational Database and Hadoop framework has proposed. And virtual resources optimization scheduling, fault-tolerance and monitoring technologies have been proposed (Cao et al. 2012; Bao et al. 2012; Zhang et al. 2013).

On one hand, performance of data streaming applications is co-determined by both networking and computing resources, and therefore they should be co-scheduled and co-allocated in an integrated and coordinated way. Dynamic control of resource scheduling and allocation is required, because unilateral redundancy in either networking or computing resources may result in the overprovision of it and the other may become a bottleneck. To avoid resource shortage as well as over-provision, a virtualized platform is utilized to implement data streaming and processing. Fuzzy logic controllers are designed to allocate CPU resources. Besides, iterative bandwidth allocation is applied and is processing and storage-aware to guarantee on-demand data provisioning (Cao et al. 2012).

On the other hand, CMfg provides a new concept and model for manufacturing informatization, which has become a hot research topic in modern networked manufacturing area. CMfg systems generate huge amounts of sensor data from distributed manufacturing devices, and how to manage them efficiently becomes one of the main challenges of CMfg. Various software frameworks have been developed in Cloud Computing to handle massive data management, and the Hadoop framework has proved an effective solution. The requirement of massive sensor data management in CMfg and the defects of traditional RDBMS (Relational Database Management System) have been considered, and a framework supporting parallel storage and processing of massive sensor data in CMfg systems based on Hadoop has been presented. This framework can provide a promising solution for massive sensor data management in CMfg (Bao et al. 2012).

In addition, the manufacturing ability of each dimension each layer of qualitative and quantitative property is fully considered and it is based on knowledge of manufacturing ability language so as to make the manufacturing ability of the main body of the service to be stored in CMfg service platform. And dependent service time of each switch is formed producing ability of service network so as to user the intellective searching. A CMfg capability description method supporting use on demand and sharing circulation of manufacturing capability is provided. This method comprises the following steps:

(1) Extracting and classifying fuzzy information and dynamic behavior information in a knowledge-based manufacturing capability description model (namely a multidimension manufacturing capability information model).
(2) Formalizing a fuzzy concept and a fuzzy role relationship according to a fuzzy description logic; describing service state change, and dynamic combination flow according to a dynamic description logic.
(3) Giving an intelligent search and recommendation mechanism of the manufacturing capability based on the description method.

According to the CMfg capability description method, the qualitative and quantitative properties of each dimension and each level of manufacturing capability can be taken into sufficient consideration, and a knowledge-based manufacturing capability language is provided, so that manufacturing capability services are stored in a CMfg service platform in an ontology mode, and a manufacturing capability service network is formed on the basis of various relationships among the services and supports the intelligent search and the use on demand of users (Zhang et al. 2013).

3.4 Construction and Management Technologies for Virtual Manufacturing Services Environment

In this technology, we have given platform architecture for CMfg Services, and proposed construction method of manufacturing the cloud service pool, and we have achieved access management of cloud services, including unified interface definition and management, certification, etc. Besides, we have also achieved manufacturing cloud formation, aggregation, storage method, and the method that dynamically build and deploy, decompose, Optimal allocation of resources services cooperative scheduling in CMfg (Lin et al. 2012a, b; Ren et al. 2012; Yang et al. 2012a; Tao et al. 2011b).

The life cycle of CMfg involved the model-based activities. Under the environment of Internet-oriented cloud, a large number of black box, heterogeneous models were accumulated, and the traditional retrieval and match methods could not satisfy the demand of massive model management. Thus, the model composition should be taken into account in the process and the correctness of the time dynamic characteristics after composition should be ensured. Based on the composition approach of top-level modeling, semantic descriptions to each event of interactive scenarios among models were added, and existing redundant and mutually exclusive in the composition process was eliminated by planning method. Thus, the combination of dynamic behavior was realized. The application example showed that the characteristics of highly scalable and on-demand aggregation could be achieved in the model resources management of CMfg by introducing models automatic composition technology (Lin et al. 2012b).

Aiming at the requirements such as rapid deployment and construction of large-scale simulation system and getting stable simulation service according to need in complex application, a co-simulation supporting framework was proposed, especially the 16 core services such as modeling, system development and construction, simulation execution, and result analysis as well as evaluation were researched. The key technology of cloud simulation resource management was also conquered. The proposed framework was integrated into FEDEP and local adjustment was made. To promote the automation of simulation application process, the sharing and agile using of simulation resource, the sustained and stable simulation service getting on demand, and the high efficient collaboration for two types, an Application Process of Co-Simulation (APCS) model was proposed. Typical application case was given to verify the feasibility and effectiveness of co-simulation framework and APCS model (Yang et al. 2012a; Laili at el. 2012).

When establishing a virtual enterprise (VE), how some enterprises with different resources and advantages form the alliances by providing respective services to compose the VE for certain profits, it is actually a process of services composition. There always exist a lot of web services with similar functions but different nonfunctionality properties in the process of web services composition. In order to enhance the quality of service composition, nonfunctionality properties, i.e., quality of service (QoS) are usually considered. However, most existing works about QoS-based service composition treat the services involved in composition as independent from each other, and their correlations are usually ignored. In fact, most services have some correlations with each other, and the correlations can affect the whole quality of service composition badly. In our research, three kinds of correlations in services composition are investigated, and a correlation-aware QoS model is studied. The impact of each kind of correlation on the whole QoS of services composition is investigated. A case study indicates that higher quality of services composition can be achieved when considering the correlations among composite services (Guo et al. 2010; Tao et al. 2012a).

3.5 Operation Technology for Virtual Manufacturing Services Environment

In this technology, we have researched, developed, and implemented collaborative operation technology and CMfg resource service operation and fault-tolerant processing technology for CMfg systems. And we have achieved a high performance, intelligent search and matching technology for CMfg services, and this technology initially realized dynamic combination technology for cloud services (Xiao et al. 2012; Tao et al. 2012c).

Existing works on service composition are primarily based on the requirements of service composition, such as describing language supporting service composition, service composition framework, mechanism and method for service composition, and service composition validation. Few works have been carried out from the perspective of combinable relationship among services and service network (Guo et al. 2012; Tao et al. 2011b, 2012b).

Based on the analysis of the conditions of shuffled frog leaping (SFL) algorithm, a modified SFL (MSFL) was proposed to deal with the premature and slow convergence. In MSFL, convergence was analyzed based on the theory of geometrical sequence, and two premature judgments were defined, including the dispersion and the fitness variance, to adaptively adjust the coefficients, for balancing convergence accuracy and rate. Six benchmark functions were proposed to test the performances of MSFL, and experiment simulations show that MSFL has higher precision and good stability (Xiao et al. 2012; Yang et al. 2012b).

3.6 Evaluation Technology for Virtual Manufacturing Service Environment

After some researches, we have given four indicators decision model on comprehensive assessment of cloud services QoS. And proposed energy conservation assessment methods of complex products based on BOM, which can build a model suitable product-level system. And these methods developed a green assessment and analysis system for cloud services, and achieved multi-level, multi-stage visual assessments for products.

The problems such as energy consumption in the process of service data collection, manufacturing service access, service flexible combination, manufacturing service comprehensive evaluation and service application were ignored. Aiming at these problems, the energy consumption comprehensive evaluation framework and energy consumption computational model of cloud resource service based on Internet of things were proposed. All kinds of energy consumption computing matrixes were established under the condition of manufacturing servicers load, no load and additional energy consumption. The computing method of manufacturing service comprehensive energy consumption was given and the key technologies for various levels in energy consumption evaluation framework were analyzed. The manufacturing resources were taken as examples to describe the conceptual model and resource energy consumption information with physical markup language (PML). A manufacturing service unit's comprehensive energy consumption was calculated and a kind of CMfg resources access and information management system was preliminary designed (Xiang et al. 2012).

Energy saving and emission reduction (ESER) has caught attention around the world due to the severe environmental challenges. An Evaluation System to assess the consumption and emission of a product during its life cycle is proposed in our research. The system integrates with the enterprise information systems by adding the consumption and emission data into bill of material (BOM). Based on ESER database, the whole system takes product life cycle as the main line and improves the accuracy of evaluation on ESER. Besides, design and optimization of green manufacturing can also be accomplished by using this system. The pro-type and framework of designed ESER system is described in our research (Lv et al. 2012).

Due to the continuous growth in the application of networks in manufacturing, quality of service (QoS) has become an important issue. In our research, the concept of QoS for manufacturing networks is discussed. To provide overall performance assurance for manufacturing networks, a service framework integrating the QoS mechanisms of the networked resource service management function and the communication networks is proposed. The novel framework maps an application to resource services and then to communication networks, adopts an intelligent optimization algorithm for QoS management of resource services, and provides QoS schemes for data transfer across communication networks. A prototype implementation has been realized and a set of simulation experiments conducted to evaluate the validity of the framework. The results obtained demonstrate the ability of the framework to satisfy the various performance requirements posed by such applications and provide efficient overall performance assurance for manufacturing networks (Xu et al. 2012).

3.7 Trusted Virtual Manufacturing Services Technology

This technology is to design CMfg security and trustiness system and to propose evaluation model for CMfg resource service, and design-related evaluation algorithm. Besides, based on these thoughts, we have developed simulation resource virtualization security technology for cloud simulation.

The security system of virtualization simulation module and access control method of virtualization simulation resource based on privilege security control region has been realized.

CMfg security system including:

(1) The secure self-organizing service of virtualization simulation system
(2) Gateway security agent technology of virtualization simulation system
(3) Access control method of virtualization simulation resource based on privilege security control region: Inserting privilege security control region between simulation resource layer and virtualization simulation resource pool, intercepting and capturing the calls for operation system by VMM, and estimate the security of these calls based on security knowledge base. Control network invade, malicious software, unauthorized access, and so on. On the basis of the research of access control method of virtualization simulation resource based on privilege security control region, research, and design the security module of virtualization simulation system. Security knowledge base in the privilege region Domain 0 is the corn support component of the security risk-assessment system, including security strategy set, authorized access interoperate matrix, security regular set, knowledge base of invade

characteristic, and so on, used to assist security risk estimate module to distinguish all kinds of network invade, malicious codes, unauthorized access, and so on. Access control module communicates with physical resources via computer kernel on the basis of classification of the security risk estimate module.

Technologies above all will be used in other parts of CMfg.

To solve the trustiness evaluation problem between multi-agents in CMfg, the blurring direct trust quantitative model was studied. A direct trust prediction model based on trading experience and weighted moving average method was proposed, and the dynamic updating algorithm of model was given. The impact factors of weight in recommended trust value were analyzed and a weight calculation algorithm was proposed based on the similarity and the direct trust value between the Agents. By combining with direct transaction threshold value, the dynamic updating algorithm for the recommended trust weight was put forward (Gan and Duan 2011).

In our research, we proposed a new video watermarking algorithm based on shot segmentation and block classification to enhance the robustness, imperceptibility, and real-time performance based on the H.264/AVC codec. A method of selecting host frames is proposed based on shot segmentation to avoid embedding watermark frame by frame, so as to improve the robustness and the real-time performance. The watermark signal is cropped into small watermarks according to the number of shots in the host video, and small watermarks are, respectively embedded into different shots. The watermarking capability and the perceptual quality are greatly improved by this way. A method of selecting host coefficients is proposed based on block classification in the Discrete Cosine Transformation (DCT) compressed domain. The texture characteristics of host blocks are considered in the classification and the places of host coefficients can change adaptively according to the content of the video. The imperceptibility of the watermarked video is greatly improved by this way. The simplified quantization index modulation (QIM) is applied to embed watermark. It brings fewer artifacts to the host signal than the current main watermarking method, such as spread spectrum (SS), differential energy watermarking (DEW), and so on. The experiment results show that the proposed scheme has a good performance in maintaining real-time performance and resisting Gaussian noising, frame swapping, MPEG compression, etc. (Jiang et al. 2011).

CMfg provides a high-efficiency model to realize manufacturing resources sharing and on-demand using. Though private cloud service platform which has advantages of safety and efficiency is applicable for group enterprise, there is limit of network isolated. In order to increase profits and reduce costs simultaneously, some manufacturing enterprises will combine their private cloud service platforms with public cloud platform to release excess manufacturing resources and abilities and obtain more services. After analyzing the architecture of CMfg service platform, a hybrid cloud infrastructure applied for manufacturing enterprises is proposed, and some functional features of the hybrid cloud platform are described.

Furthermore, a mechanism for private cloud combining with public cloud is presented. Finally, some challenges that enterprise might face while operating in hybrid cloud environment are discussed (Mai et al. 2012).

3.8 Pervasive Human–Machine Interaction Technology

In this technology, we have researched, developed and implemented: user intent perception and fusion technology based on draft context information, visualization technology of mass services, cloud services characterization techniques based on ontology and personalized interface customization technology on demand.

Traditional adaptive interface lacks the ability of self-learning user interaction history and cannot predict user's intent effectively with their experiences. Based on the theory of cognitive psychology, this chapter proposes an adaptive user interface model based on experience awareness. First, the model is described by static interface elements, dynamic interactive actions as well as adaptive strategies. Second, we study the implementation architectures, key techniques, and modeling method of adaptive interface model. We built a personal virtual furniture customizing system and verified it by experiments. The experiments demonstrate that the model can obtain the user's intent fairly well, adjust interface layout and interactive actions in real-time to adapt to them (Ma et al. 2011).

In order to enable user's interface satisfy various individualized requirements in CMfg environment, characteristics of user's interface were first analyzed, such as ubiquitous, natural, intelligent and mobile, virtualized and loose coupled, individualized, and be customized in entire product life cycle. Then the next generation Human–Computer Interaction (HCI) technologies were reviewed including reality-based HCI, natural user's interface, pen-based user interface, and context perception. Further, a research framework of ubiquitous HCI technology in CMfg was presented and the key technologies were analyzed in particular, among which dynamic requirement configuration of virtual resources in user's inter face, ubiquitous interface customization, natural interaction-oriented to ubiquitous equipment, visual analysis-oriented to ubiquitous information, and pen-based operation platform. Finally, the future of HCI technology research in CMfg was discussed (Fan et al. 2011).

This chapter presents an online synchronization input method based on paper form and electronic one. It is designed to address the problem that while a paper form is filled out with the digital pen, the target location of corresponding electronic form is not accurate, and the recognition rate is not high. Furthermore, the users tend to switch so frequently between paper and display device in such circumstance that their interaction experience would be reduced dramatically. This method consists of error compensation on pen information and intelligent locating in the target cell with context and knowledge base. Based on statistical language model and dictionary, we design the recognition algorithm which can improve the recognizer and increase its recognition efficiency. Moreover, the results also will be added into the dictionary and the error correction module to improve the system's adaptability. The method proposed in our research is based on online synchronization input frame—FSIF, which combines paper form with electronic form during input process. It contains four parts: pen information correction converters, intelligent locator, ink splitter, and adaptive handwriting recognizer. According to recognition and procedures, the handwriting information is carried out on collection, correction, change, locating, segmentation, classification and identification, etc. We verify the method in experiments. The results indicate that the method is able to improve the pen-based synchronization input on paper forms and electronic ones (Xiong et al. 2012).

3.9 Application Technology of CMfg Service Platform

In this technology, we have achieved transaction settlement model that supported multi-users collaborative, CMfg architecture of control practices, management, monitoring reports architecture and operation of monitoring reports and its operation, and have achieved management and integration of interdisciplinary field of knowledge, as well as examples validation and promotion for CMfg services.

Pirating various forms of intellectual property (IP) causes great economic loss to intellectual property rights (IPR) holders. IPR protection is becoming a key issue in our highly networked world. In order to further deepen our understanding of how to protect IPR and enhance information interchange and knowledge sharing among related entities, ontologies for IPR protection are proposed. This study contains three parts, which are developed to deal with different perspectives in this domain. The first part presents a static ontology, i.e., a hierarchy framework for the domain language, including primarily classes of participants, classes of IP works, classes of activities, and relations between these classes. In the second part, a dynamic ontology is shown to illustrate the IPR protection process. Third, a causal map is used to demonstrate how classes of IPR protection methodologies are causally related with classes of IP piracy methodologies. Finally, the case of Tomato Garden is offered to demonstrate how the proposed ontologies are used in the real-world. In respect of the ontology, it is first helpful to gain a comprehensive understanding of domain knowledge of IPR protection; second, IPR protection systems' design and development in this domain are facilitated and supported by these ontologies; third, the proposed ontologies are united in the Ontology Web Language (OWL) and the OWL rules languages framework, both of which are machine readable (Zhang et al. 2012b).

In order to realize the unified optimal resource service management (RSM) in CMfg, the resource services characteristics were studied as well as the role of knowledge in the whole resource services lifecycle. ARSM mechanism based on knowledge was put forward. A knowledge base system construction method was designed for the RSM in CMfg. The RSM process was analyzed. Finally, a prototype system was developed to validate the proposed method (Hu et al. 2012).

3.10 Informatized Manufacturing Technology

According to the technical features of CMfg system, we improve and upgrade the traditional manufacturing information system in order to integrate with CMfg services platform. This platform includes: integration framework, integration platform, and computer-aided collaborative work platform, multidisciplinary virtual prototype design, Web design, intelligent design, project management, quality management, complex system modeling, and simulation technology.

In paper (Li et al. 2011b), the definition, importance, and research issues of High-efficiency simulation for Complex system are proposed first. The application demands for the development of high-efficiency simulation technology for complex system are discussed from two aspects, which are the demand from Complex system modeling and simulation for high-end users and that of Providing "cloud simulation services" for massive users. Eight important technologies in current high-efficiency simulation for complex system from three aspects, which are high-efficiency simulation modeling technology, high-efficiency simulation system and infrastructure technology, and high-efficiency simulation application technology, are introduced in detail, including:

(1) High-efficiency simulation language for complex system
(2) High-efficiency simulation algorithm for complex system
(3) Hardware Optimization of high-efficiency simulation
(4) High-efficiency simulation software support technology
(5) High-efficiency cloud simulation technology
(6) Verification, Validation and Authentication (VV&A) technology on high-efficiency simulation system for complex system
(7) Massive data management technology
(8) Simulation experiment results analysis and evaluation technology for complex system.

Capability and speed of mold products development have been considered as one of the decisive factors for the survival of mold manufacturing enterprises in such a severe market competition environment. Networked manufacturing mode for mold products development can enhance the response speed of customer and market demands, promote the information level of enterprise, product development cycle and production cost can also be greatly reduced. Based on the analysis of mold manufacturing task, information composing and manufacturing capability, the information models of task and manufacturing capability of enterprise are built in UML language in our research. And the object-relation database model is established as well as the transformation from task and resource information models to relational database model are realized (Zhang et al. 2012c).

Aiming at cloud encapsulation of massive, dynamic, isomerous parts resource in the service platform of parts CMfg for small- and medium-sized enterprises, the related technique of parts resource cloud encapsulation is researched. Regard cloud integration and share of parts resource-oriented to small- and medium-sized enterprises as the starting point, the parts resource description model oriented to whole life cycle of parts is constructed, and five layers tree structure classification system and coding system and coding system are proposed in order to easy to networking, structured organization and management of parts, and then the operation criteria of 3D parts parametric model and cloud encapsulation model of parts are established. Thus, the cloud encapsulation of massive, dynamic, isomerous parts resource is realized, which has a positive significance for improving the sharing and reuse of parts resource, and promoting the rapid development of small- and medium-sized enterprises (Wang et al. 2012).

3.11 Features of CMfg in China

The aim is to improve enterprise's market competitiveness.

Based on the previous work fruits of manufacturing informatization, it is important to stand out the following points:

(1) Highlighting the cyclic and spiral ascending development path of application demand leading system constructing to promote synergetic development of person and organization, equipment and technology, and operating management in CMfg.
(2) Highlighting the deeply fuse of the new information technology, informatized manufacturing technology, and product-related specific technology.
(3) Highlighting the core status of establishing a new paradigm and a new approach in manufacturing informatization, which means establish a new paradigm with the feature of "network, collaboration, agility, green, service and intelligence" and a new manufacturing approach with the smart technical features of "digitalization, things networking, virtualization, service-oriented, collaboration and intelligence".
(4) Highlighting to service for two type users (manufacturing enterprises and product customers) to integrate and optimize three elements/five flows in full life cycle of product manufacturing.
(5) Highlighting the synchronous development of industrialization, informatization, and urbanization.
(6) Highlighting the team strength of government, industry, university and research institute, and application unit.

4 Applications of CMfg

With the supports from the national foundations like 863 projects, our teams have developed two typical application demonstrations for complex products in group enterprise and normal products in small- and medium-sized enterprise (SME), which effectively verified the research works of CMfg and improved the manufacturing abilities in certain enterprises. Furthermore, based on the research fruits of previous works and applications, our team builds the Internet-based CMfg platform named "Tianzhi Net" and provides abundant cloud manufacturing services for massive enterprises in various cities, which helps the overall progress of the smart manufacturing industry in those cities.

4.1 CMfg for Aerospace Complex Products in Group Enterprise

4.1.1 Platforms and Systems

We have developed a CMfg service platform and built a CMfg service system, which includes: a 100 trillion times high performance computing resources and 320TB storage resources, and more than 10 kinds, 300 sets of machinery, electronics, control, and other large-scale multidisciplinary design and analysis software and its license resources. And the high-end CNC machining equipment and enterprise unit manufacturing systems have been built in 3 factories, Manufacturing process professional capabilities on various stages such as Multidisciplinary design optimization capabilities, Multidisciplinary, systems and system simulation analysis capabilities, High-end semi-physical simulation capabilities, etc.

The architecture diagram is also important. First, in application layer, combined with major equipment development, the platform supports the overall department, profession units, factories, and other units to collaboratively research online in the demonstration, design, simulation, production stage. Second, in platform layer, the platform forms a "aerospace group manufacturing services cloud platform," and provides a set of CMfg service support engine and management tools with complete independent intellectual property rights for aerospace conglomerates. Finally, in resource layer, the IT infrastructure resources, digital production line and other hard manufacturing resources, design and analysis software, enterprise information systems ,and other soft manufacturing resources can all be accessed to the CMfg service platform. It is shown in Fig. 11.

4.1.2 Technology System of CMfg in Aerospace Group Enterprises

The technology system of CMfg in aerospace group enterprises makes up of several important key technologies of CMfg, including:

(1) Perception, virtualization, servitization technologies of resource and capability
(2) Unified management and on-demand composition technologies of cloud services
(3) High efficient collaboration and fault-tolerance technologies of cloud services
(4) Trading management and efficiency evaluation technologies of cloud services.

It is shown in Fig. 12.

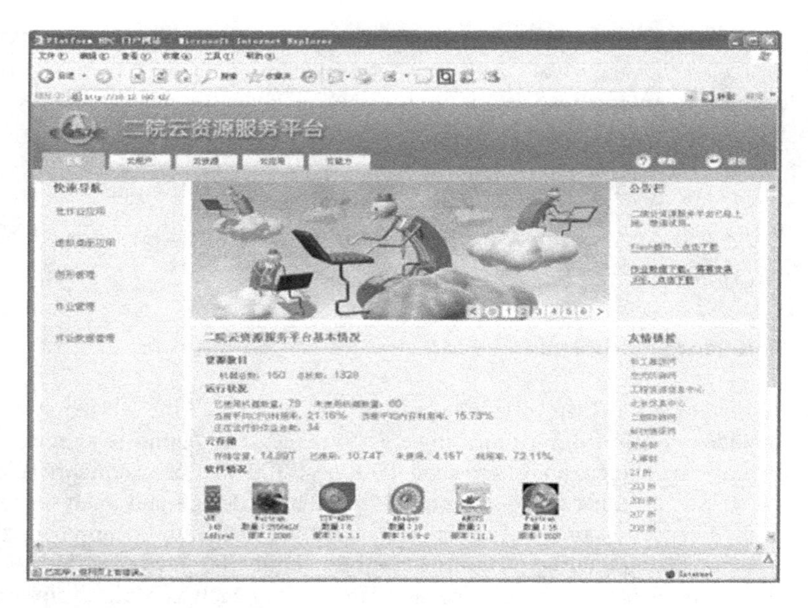

Fig. 11 CMfg service platform

4.1.3 New R&D Modes and Means of Aerospace Complex Products

First, the issue technology indexes provide an issue. Second, the structural, pneumatic, and control systems are designed and analyzed for the issue. Then multidisciplinary virtual prototype simulation is down. And the next step is process design, production. The last step is to assemble products and debug. New R&D modes and means of aerospace complex products are shown in Fig. 13.

4.1.4 Primary Application Effects of CMfg

The first effect is agility. The share of manufacturing resource and professional capability in the institution has been realized. Aerodynamics, aerodynamic heat, and aerodynamic load design and analysis of some aerospace products, which accelerate work progress on aerodynamics have been supported, and the time taken in an aerodynamic design has been shorten from 1 month to 1 week than before.

The second effect is servitization. The CMfg service platform has provided cloud services used in verification, design, simulation, and production stage, which enabled designers and craftsmen to access resource and capability services on demand through network, and focus on their business.

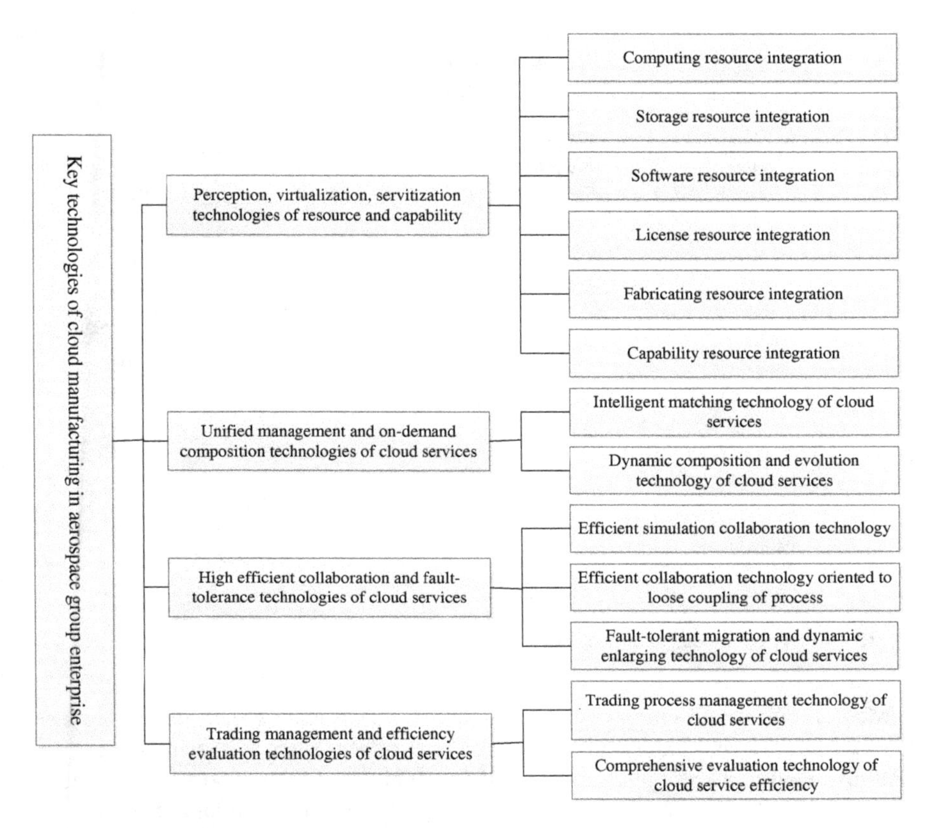

Fig. 12 Key technologies of CMfg in aerospace group enterprises

The third effect is green. The CMfg service platform has improved utility efficiency of resource in cloud by 5 %, avoided resource waste, and saved enterprise cost.

The last effect is intelligent. Based on knowledge and intelligent science, the CMfg service platform efficiently supported two-dimensional life cycle. They are the full life cycle of service and manufacturing process.

4.2 CMfg for Small- and Medium-Sized Enterprise

4.2.1 Application Demand

Application demand about close cooperation is requiring CMP to provide collaboration management capability, such as the fine ability to make a craft plan collaboratively, to allocate tasks and to trace the progress.

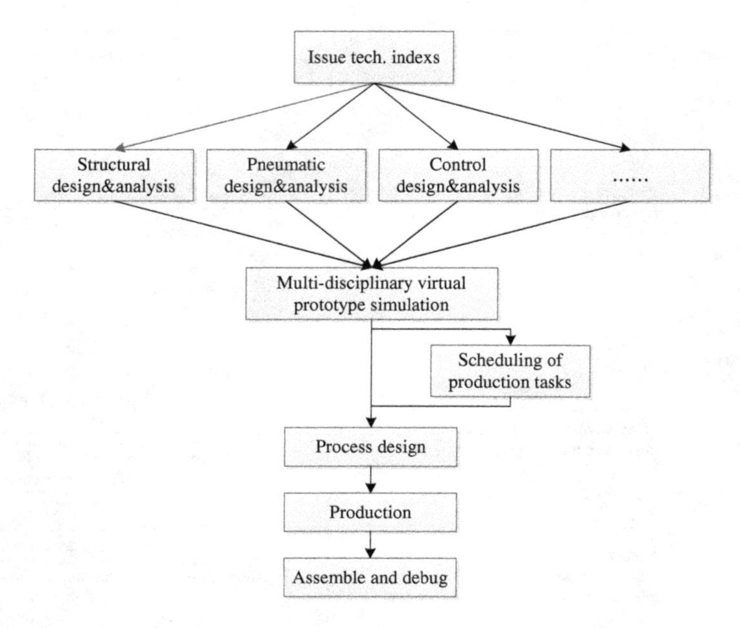

Fig. 13 New R&D modes and means of aerospace complex products

Medium and small enterprises also require CMP to realize the interconnection between interior and exterior business for enterprises, such as the integration of ERP inside with process plans outside, and they also need to support customizable management of exterior resources and to support customization and execution of exterior resources, in order to suit different manufacturing modes in enterprises.

4.2.2 Platform and Systems

A public service platform has been developed, which can support both individual management system for multi-enterprise and interconnection of data and business cooperation among enterprises.

4.2.3 Trading and Collaboration Mode

One multi-user resource trading and collaboration mode based on social network model has been proposed, which supports business circle building, business chance capture, and trading processing. So transaction among enterprises exists in a relatively stable commercial environment. Only when commercial needing extends to the commercial partners, enterprises will seek new partners in the public market through platform. It is shown in Fig. 14.

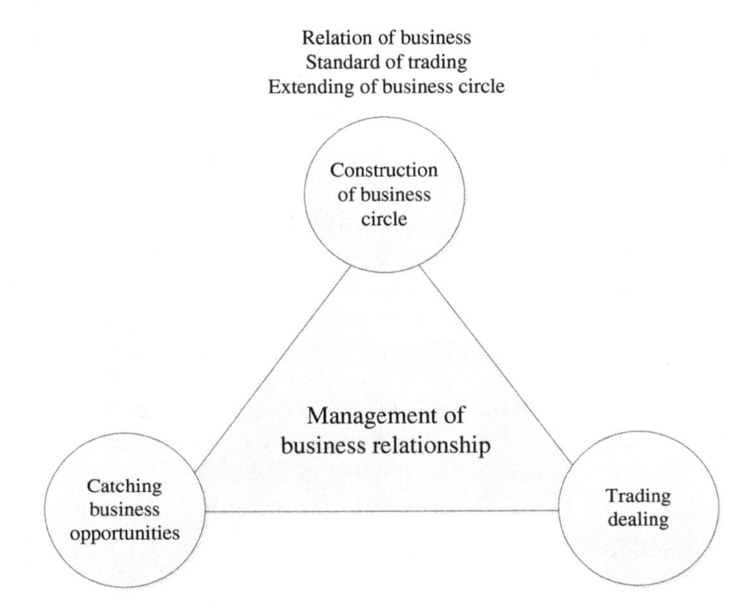

Fig. 14 Trading and collaboration mode

4.2.4 Business Mode

One business promotion mode which is realized by introducing the leading enterprise with partners on its supply chain (1 + N) to the CM platform has been explored. Local governments or industry societies can cooperate to build local or industrial service platforms. Leading enterprises are attracted to build supply chain management systems on CM platform, while pulling their upstream and downstream enterprises to CM platform. Related pricing scheme is customized and users can quickly publish applications by themselves, and acquire customizable value-added services.

4.2.5 Application

Shangteng platform was released in Nov. 4th. 2012. The platform has collected and classified 3,286 kinds of manufacturing knowledge bases from 32 different businesses, including equipment manufacturing and clothing industry. A complete system of operation organization and management, including market, implementation, customer service, and technology development has already been built. Promotion and cooperation activities with local societies and governments in Foshan, Sichuan province, Shanxi, Guangdong province have been launched. Application enterprises cover equipment, clothing, lighting, ceramics, appliance, and so on. The number of registered users has achieved 1,647 until Dec. 3th. 2013. Sales avenue of 11.89 million RMB was realized in 2012.

4.2.6 Primary Effects

In integration and virtualization aspect, BISWIT has integrated ERP-related resource and capability of different enterprises into CMP. In servitization aspect, platform and provide management service, such as service matching, evaluation, trading, account settlement has been released. In collaboration aspect, tight cooperation on process level, and realize the close integration with ERP has been supported. In intelligence aspect, support intelligent customization of CMP use. In aspect of supporting, a new mode of economic growth, including increasing share of resource and capability, lower manufacturing, and operation cost.

4.3 CMfg for Manufacturing in Smart Cities

Our team has contracted with local governments such as Foshan, Xiangyang, and Wuhan in China to implement CM in those cities where huge amounts of manufacturing enterprises are located. Some cities have put great importance in the implementation and application of CM to support resource share, industrial chain bridging, transaction, and business cooperation. Our aim is to improve enterprise competitiveness, including efficient configuration of resource and capability, promoting chain integration and cooperation capability, and providing comprehensive innovation means, and to speed up the manner change of city economic development, including building new industrial environment, changing the way of industry development and innovating investment ways. It is shown in Fig. 15.

Aiming at the manufacturing in Smart City, our team has established an Internet-based CMfg system which is named as Tianzhi Net in Chinese (www.cosimcloud.com). The design of Tianzhi Net is oriented to local industry chains and industry clusters, as shown in the following Fig. 16. Currently, more than 800 manufacturing enterprises have log on Tianzhi Net to acquire manufacturing resources services provided by the CMfg system, as well as implement the collaborative CMfg services to support their procurement, R&D, production, marketing, and after-sale services among different companies. With Tianzhi Net and its abundant CMfg services, those enterprises can easily establish their individual business social network and cooperates with enterprises online, which significantly reduce the cost and enrich the commercial opportunities. As for cities, Tianzhi Net helps to build a promising and smart environment for the manufacturing industry to collaborate and develop.

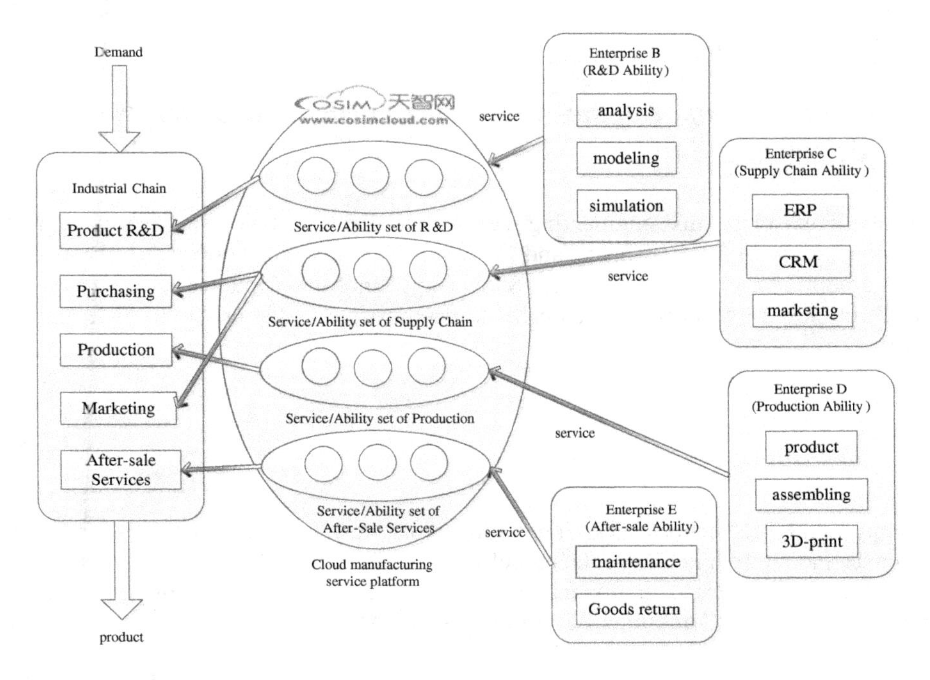

Fig. 15 Manufacturing in Smart City

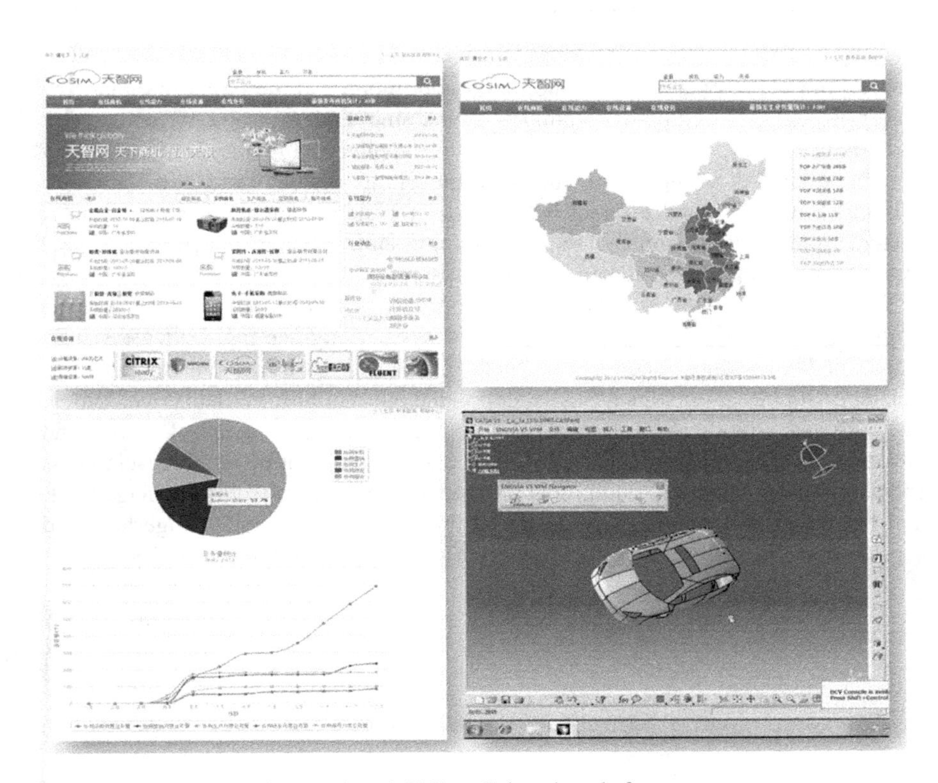

Fig. 16 Tianzhi Net, an internet based CMfg collaborative platform

5 Future Works

5.1 *Achievements' Engineering Industrialization and Deeply Applying*

Toolkits and platforms' engineering and industrialization; Promoting application demonstration in a few industry and including both the minor enterprise clusters and group enterprises in smart city; Supporting "Industry Cloud Innovation Project" proposed by ministry of industry to demonstrate CMfg in 10 province and city.

5.2 *Technology Extension*

Deeply applying technologies; Integrating more new information technologies, such as: mobile Internet, big data, model-based engineering, HPC, complex system simulation technology, intelligent analysis, and evaluation technology, e-commerce, and so on; Combining 3D printing with CMfg.

5.3 *Implementation Advices*

Around the transformation of economic growth mode and the target of enhancing enterprise market competition ability, developing CMfg in the virtuous cycle route is necessary, which means starting from application demands of enhancing enterprise market competition ability, establishing system based on demands, driving technology, and product R&D through system establishment, promoting system further improvement via technology and product development, and evolving new application depending with system.

CMfg system with connotation of complex system engineering must be implemented. CMfg system is complicated system engineering, and we should insist on the guiding ideology of "benefit-driving, top-down planning, distinguishing emphasis, implementing by step," and we should formulate development program and periodical implement scheme with enterprise's actual demands and situation. We should pay attention to integrated optimization of three elements (person and organization, equipment and technology and operating management) and five flows (information flow, material flow, capital flow, knowledge flow and service flow) in system-wide full life cycle product development activities.

References

Bao Y, Zhang L, Luo Y (2012) Massive data management in CMfg based on Hadoop. IEEE international conference on industrial informatics, pp 397–401

Cao J, Zhang W, Tan W (2012) Dynamic control of data streaming and processing in a virtualized environment. IEEE Trans Autom Sci Eng 9(2):365–376

Fan Y, Teng D, Yang H, Ma C, Dai G, Wang H (2011) An adaptive user interface model based on experience awareness. Chin J Comput 34(11):2211–2223

Gan J, Duan GJ (2011) Method of CMfg service trust evaluation. Comput Integr Manuf Syst 17(3):495–503

Guo H, Tao F, Zhang L, Su S, Si N (2010) Correlation-aware web services composition and QoS computation model in virtual enterprise. Int J Advanced manufacturing Technology 51(5–8): 817–827

Guo H, Tao F, Zhang L, Laili Y, Liu D (2012) Research on measurement method of resource service composition flexibility in service-oriented manufacturing system. Int J Comput Integr Manuf 25(2):113–135

Hu AR, Zhang L, Tao F, Luo YL (2012) Resource service management of CMfg based on knowledge. J Tongji Univ 40(7):1092–1101

Jiang X, Liu Q, Wu Q (2011) A new video watermarking algorithm based on shot segmentation and block classification. Multimedia Tools Appl doi: 10.1007/s11042-011-0857-3

Laili YJ, Tao F, Zhang L, Sarker BR (2012) A study of optimal allocation of computing resources in CMfg systems. Int J Adv Manuf Technol 63(5):671–690

Li BH, Zhang L, Wang SL, Tao F, Cao JW, Jiang XD, Song X, Chai XD (2010) CMfg: a new service-oriented networked manufacturing model. Comput Integr Manuf Syst 16(1):1–8

Li BH, Zhang L, Lei R, Chai XD, Tao F, Luo YL (2011a) Further discussion on CMfg. Comput Integr Manuf Syst 17(3):449–457

Li BH, Qin D, Chai X et al (2011b) Research on high efficiency simulation technology for complex system. AsiaSim 2011, Shanghai (keynote speech)

Li BH, Zhang L, Lei R, Chai XD, Tao F, Wang YZ, Yin C, Huang P, Zhao XP, Zhou ZD (2012a) Typical characteristics, technologies and applications of CMfg. Comput Integr Manuf Syst 18(7):1345–1356

Li BH, Quan C, Zhou X (2012b) Research and practice of Wisdom City, Peak BBS of Wisdom City. 2012-9, Ningbo

Lin TY, Chai X, Li BH (2012a) Top-level modeling theory of the multidiscipline virtual prototype. J Syst Eng Electron 25(3):267–290

Lin TY, Li BH, Chai X, Li T (2012b) CMfg oriented automatic composition technology of models. Comput Integr Manuf Syst 18(7):1379–1386

Lou P, Liu Q, Zhou Z, Wang H, Sun SX (2012) Multi-agent-based proactive-reactive scheduling for a job shop. Int J Adv Manuf Technol 59(1–4):311–324

Lv L, Tao F, Zhang HP, Huang W, Zhang L (2012) Framework of evaluation system for energy-saving and emission-reduction based on BOM. In: The IEEE 10th international conference on industrial informatics, pp 386–390

Ma CX, Ren L, Teng DX, Wang HA, Dai GZ (2011) Ubiquitous human-computer interaction in CMfg. Comput Integr Manuf Syst 17(03):504–510

Mai J, Zhang L, Tao F, Ren L (2012) Architecture of hybrid cloud for manufacturing enterprise. In: Asia simulation conference and the international conference on system simulation and scientific computing, pp 365–372

Ren L, Zhang L, Tao F, Zhang YB, Luo YL (2012) A methodology toward virtualization-based high performance simulation platform supporting multidisciplinary design of complex products. Enterp Inf Syst 25(3):267–290

Tao F, Zhang L, Venkatesh VC, Luo YL, Cheng Y (2011a) Cloud manufacturing: a computing and service-oriented manufacturing model. Proc Inst Mech Eng Part B J Eng Manuf 225(10):1969–1976

Tao F, Zhang L, Lu K, Zhao D (2011b) Research on manufacturing grid resource service optimal-selection and composition framework. Enterp Inf Syst 6(2):237–264

Tao F, Cheng Y, Zhang L, Zhao D (2012a) Utility modeling, equilibrium and collaboration of resource service transaction in service-oriented manufacturing. Proc Inst Mech Eng Part B J Eng Manuf 226(6):1099–1117

Tao F, Guo H, Zhang L, Cheng Y (2012b) Modeling of combinable relationship-based composition service network and theoretical proof of its scale-free characteristics. Enterp Inf Syst 6(4):373–404

Tao F, Cheng Y, Zhang L (2012c) A resource service optimized configuration of the fault tolerant management system. 201210335609

Wang W, Zhang YH, Xu XD, Chen C (2012) Research on technique of parts resource cloud encapsulation for the small and medium-sized enterprises. Mach Des Manuf 8:260–262

Xiang F, Hu YF, Tao F, Zhang L (2012) Energy consumption evaluation and application of CMfg resource service. Comput Integr Manuf Syst 18(9):2109–2166

Xiao YY, Chai XD, Li BH, Wang QS (2012) Convergence analysis of shuffled frog leaping algorithm and its modified algorithm. J Huazhong Univ Sci Technol (Nature Science Edition)

Xiong J, Fan Y, Teng D, Ma C, Dai G, Wang H (2012) An online synchronization input method based on paper forms and electronic ones. J Comput-Aided Des Comput Graph 24(09):125–1133

Xu W, Zhou Z, Pham DT, Liu Q, Ji C, Meng W (2012) Quality of service in manufacturing networks: a service framework and its implementation. Int J Adv Manuf Technol doi: 10.1007/s00170-012-3965-y

Yang C, Li BH, Chai XD (2012a) CMfg oriented cloud simulation supporting framework and its application process model. Comput Integr Manuf Syst 18(7):1444–1452

Yang C, Chai X, Zhang F (2012b) Research on co-simulation task scheduling based on virtualization technology under cloud simulation. In: Asia simulation conference and the international conference on system simulation and scientific computing 2012, Part II, pp 421–430

Zhang L, Luo YL, Fan WH, Tao F, Ren L (2011) Analyses of cloud manufacturing and related advanced manufacturing models. Comput Integr Manuf Syst 17(3):0458–0468

Zhang L, Luo YL, Tao F, Li BH, Ren L, Zhang XS, Guo H, Cheng Y, Hu AR, Liu YK (2012a) Cloud manufacturing: a new manufacturing paradigm. Enterp Inf Syst. doi: 10.1080/17517575.2012.683812

Zhang X, Liu Q, Wang H (2012b) Ontologies for intellectual property rights protection. Expert Syst Appl 39(1):1340–1388

Zhang YH, Wang W, Huo M (2012c) Information modeling for networked mold manufacturing task based on UML. Mach Des Manuf 8:240–242

Zhang L, Luo YL, Tao F, Zhang XS, Ren L (2013) A description method of cloud manufacturing capability supporting on-demand use and sharing of cloud manufacturing capability. 103020722A. 201210473792

Zhou J (2013) Research of manufacturing power strategy. Inner report of the Chinese academy of engineering

Zhou Z, Liu Q, Ai Q, Xu C (2011) Intelligent monitoring and diagnosis for modern mechanical equipment based on the integration of embedded technology and FBGS technology. Measurement 44(9):1499–1511

Future Manufacturing Industry with Cloud Manufacturing

Lin Zhang, Jingeng Mai, Bo Hu Li, Fei Tao, Chun Zhao, Lei Ren and Ralph C. Huntsinger

Abstract Cloud manufacturing (CMfg) is a new way to promote the further development of manufacturing industries. In this article, an envisioned picture of future manufacturing with CMfg which supports different advanced manufacturing modes, including intelligent manufacturing, sustainable manufacturing, agile manufacturing, and personalized social production mode was described. Some important technologies for building resources/capabilities and knowledge services platforms of CMfg were introduced. Several typical applications of CMfg, such as manufacturing communities, virtual industry clusters, 3D printing, cloud service evaluation, and hybrid cloud manufacturing were discussed.

Keywords Cloud manufacturing · Knowledge · Service · Big data · 3D printing

1 Introduction

Since the 1970s, along with the rapid development of computer technology, intelligent technology and communication technology, being intersected and integrated, all kinds of advanced information means have been applied in the manufacturing industry; meanwhile, digital manufacturing, networked manufacturing and intelligent manufacturing have gone deep into every stage of the manufacturing process. Moreover, a number of methods and modes have been put

L. Zhang (✉) · J. Mai · B. H. Li · F. Tao · C. Zhao · L. Ren
Beihang University (BUAA), Beijing, People's Republic of China
e-mail: zhanglin@buaa.edu.cn

B. H. Li
Beijing Simulation Center, Beijing, People's Republic of China

R. C. Huntsinger
The California State University—Chico Campus, Chico, CA, USA

D. Schaefer (ed.), *Cloud-Based Design and Manufacturing (CBDM)*,
DOI: 10.1007/978-3-319-07398-9_5, © Springer International Publishing Switzerland 2014

forward to improve the production efficiency, product quality, and enterprise's innovation ability, such as Lean Production (Womack et al. 2007), Concurrent Engineering (Sohlenius 1992), Agile Manufacturing (Ben Naylor et al. 1999), and Virtual Manufacturing (McLean et al. 1982), etc.

However, the further development of the global manufacturing industry still faces many challenges, such as the inefficient utilization and waste of manufacturing resources and capabilities, the lack of efficient coordination and innovation of manufacturing enterprises, the upgrade difficulties of the regional industry, etc. Through policy support and spending, the developed countries are trying to preempt the high-end manufacturing market, as well as promote the "re-industrialization" (Castells and Hall 1994) and the return of the manufacturing industry so as to sustain the recovery and growth of their manufacturing industry; on the contrary, facing the double pressure that the high-end market is preempted by the developed countries while the low-end industry is shifted to the emerging countries, the developing countries are adjusting their development strategies so as to seek a comprehensive upgrade for the traditional industry and realize a leap-forward development for the new industrialization (Hai-jian 2003). Actually, the manufacturing industry is in an urgent need to advance the informatization process so as to improve the enterprise's core competitiveness, accelerate the transformation and upgrading, and achieve the sustainable development.

In recent years, the Internet of Things (IoTs) has obtained rapid development under the support of RFID, sensor, intelligent technology, nanotechnology, etc., which are expected to promote the interconnection among all kinds of things. Therefore, the introduction of the Internet of Things definitely will be helpful to construct a shared platform, which is able to interconnect all kinds of manufacturing resources. In addition, the application of high-performance computer and the development of high-performance computing technology have provided possibilities to solve many more complicated manufacturing problems and to carry out large-scale collaborative manufacturing.

With the development and application of the new material, 3D printing, Big Data and other technologies, the personalized and customized production mode is affecting every link of the whole life cycle of manufacturing, including demand, design, production and processing, use, maintenance, scrapping, etc.

In the exploration for a new paradigm and means of the manufacturing informatization, in January 2010, the concept of cloud manufacturing (CMfg) was formally put forward (Li et al. 2010a, b). Cloud manufacturing is the extension and development upon infrastructure as a service (IaaS), platform as a service (PaaS), and software as a service (SaaS) provided by cloud computing. It has enriched and expanded the resource sharing content and service mode and technology of cloud computing (Li et al. 2011a, b). In a cloud manufacturing system, any manufacturing resource (e.g., equipment, software, database, etc.) can be accessed in anywhere and anytime, manufacturing resources and capabilities in the real world are mapped into the cyber space and encapsulated into services. Compared with the existing information-based manufacturing modes and technologies, cloud manufacturing has more outstanding typical technical features, for example, CMfg has the ability of self-adaptability and self-organization, has a flexible and open structure, can support massive resources and capability gathering and sharing, can provide on-demand services for the whole lifecycle of manufacturing (Zhang et al. 2011; Li et al. 2012a, b, c, d, e; Zhang et al. 2012a, b).

CMfg will exert a strong thrust on the sustainable development of the global manufacturing industry, and will create a new industry innovation mode, for which Lipson and Kurman have conducted a vivid description that cloud manufacturing will promote innovation by lowering down the market threshold in the book *Fabricated the New World of 3D Printing* (Lipson and Kurman 2013). They expressed that, the power of CMfg might be insignificant; however, just like the billions of mobile phones or ant-plants, the overall combined effectiveness will be greater than the sum of each part. D. Schaefer and D Wu, etc. studied about how to apply 3D printing as a service in cloud-based design and manufacturing (Wu et al. 2012, 2013; Thames et al. 2013), and some companies provide online 3D printing service recently. Combining with 3D printing, CMfg will gradually change the traditional manufacturing mode: with the manufacturing industry development involved with the wide participation and promotion of the public, the resulting personalized social production mode will inspire a group-based innovation.

The EU's Seventh Framework launched the "Manufacturing Cloud Project" (ManuCloud, Project-ID: 260142) in August 2010, which aimed to provide the users with the configurable manufacturing resource services under the support of SaaS. The CMfg mentioned in the project "Strategic Research over Construction and Promotion of China's Intelligent City" launched by the Chinese Academy of Engineering in 2012 was an important part of the intelligent manufacturing and design scheme. The CAPP-4-SMES project launched by the EU in December 2012 focused on the application of knowledge and service applications, as well as pointed out that cloud manufacturing will be an integrated solution to facilitate modular and configurable process planning services to increase the robustness of existing processes. Such pay-as-you-go services and options can be picked from the Cloud when necessary or applicable. In 2013, Germany's Federal Ministry of Teaching and Research and Germany's Federal Ministry of Economics and Technology jointly sponsored to launch the "Industry 4.0" project, which was designed to comprehensively promote the implement the Smart Plant, Smart Production, Green Production and Urban Production. In the production system of the "Industry 4.0" project, the product components will directly communicate with the production system and release instructions for the production processes required by the next step. The industry in the 4.0 era will thoroughly change the use of the production technology and the whole system will become more intelligent and more closely connected: specifically, different components can communicate with each other so that the efficiency will become faster while the response times will be made more quickly. In the U.S., Defense Advanced Research Projects Agency (DARPA) Manufacturing Experimentation and Outreach (MENTOR) program funded a project at Georgia Tech. This program focuses on engaging high school-age students in a series of collaborative design and distributed manufacturing experiments. It is planning to deploy up to a thousand computer-numerically controlled (CNC) manufacturing machines, such as 3D printers, to high schools nationwide.

With the in-depth study over the platform construction, operation, maintenance and other technologies of cloud manufacturing, the construction and applications are being carried out for 2 group enterprise private manufacturing clouds (Fan and Xiao 2011; Zhan et al. 2011; Lin et al. 2012; Yang et al. 2012a, b; Li et al. 2012a, b, c, d, e) and 2 public manufacturing clouds suitable for small and medium-sized enterprises (Yin et al. 2011a, b; Li et al. 2011a, b, 2012a, b, c, d, e; Yin et al. 2011a, b; Cao and Jiang 2012; Wang et al. 2012; Zhang et al. 2012a, b; Ji et al. 2012; Huang et al. 2013; Song et al. 2013). Under the support of cloud manufacturing technology, taking advantage of its original resources, the manufacturing enterprise, upon the "digitization, integration and collaboration", can realize the transformation and upgrading toward the "agility, service applications, greenization and intellectualization" so as to improve an enterprise's core competitiveness and sustainable development ability.

This article is to carry out discussions over the future manufacturing scenario based on CMfg, several typical applications and some related technologies.

2 Future Manufacturing Scenarios Based on CMfg

2.1 On-Demand Use and Intelligent Manufacturing

As shown in Fig. 1, there is no need for the user to directly deal with each resource node, as well as no need to grasp the specific location and condition of each resource node. When the user puts forward a demand on the terminal, the cloud manufacturing service platform will automatically structure a "virtual manufacturing environment" from the virtual manufacturing cloud pool for the user so that the user can use the required manufacturing resources and capabilities just like the use of water, electricity, coal, and gas. Actually, through the formation of the operation mode integrated with provider of manufacturing resources, operator of manufacturing cloud and user of manufacturing resources, it is to realize the on-demand use for manufacturing resources.

Taking advantage of the existing resources, CMfg will provide manufacturing services that are various in function, type, and granularity as needed to the user so as to meet the demand of the user maximally. On the one hand, through the sharing of manufacturing capabilities, many resources hard to be shared independently and directly can be provided to the user in the form of services, such as the design model, simulation data and all kinds of experiential knowledge. On the other hand, when providing the user with the original resources, CMfg can provide the user with a complete solution and additional services at the same time; meanwhile, through the servitization technology of the manufacturing capability, CMfg can

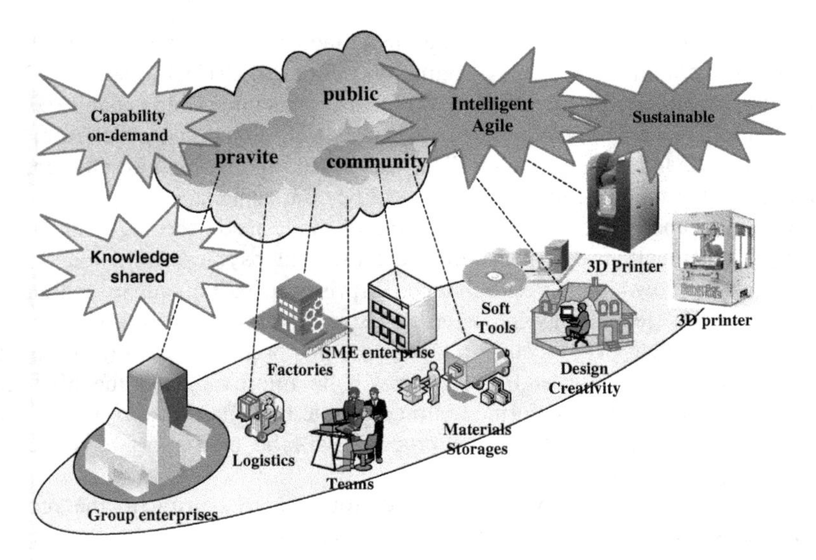

Fig. 1 Future manufacturing scenario based on cloud manufacturing

further realize intelligent and flexible combination and decomposition of manufacturing capabilities.

A CMfg platform contains vast amounts of resources, users, transactions, and other data, and due to big data technology (Lynch 2008; Manyika et al. 2011), more information can be processed in a deep way so that the platform-delivered demand and service information obtained by the user will be the contents that the user is most interested in. Taking advantage of the efficient intelligent retrieval tools based on knowledge (Hu et al. 2012) and semantics (Zeng et al. 2009), the user can obtain optimized and recommended services and service combination (Cheng et al. 2010; Tao et al. 2012a, b, c, 2013; Guo et al. 2011); therefore, even remaining within doors, the user can obtain great business opportunities, optimize low-cost resource services, and complete a manufacturing business efficiently.

2.2 Sustainable Manufacturing Ecology

CMfg platform supports a wide range of resources integration and access, such as the enterprise's traditional information-based resources/capabilities, human resources, high-end software resources, personalized design capabilities, processing and manufacturing capabilities, logistics resources, 3D printing equipment, etc., which will be provided to the user in the form of service, such as argument as a service (AaaS), design as a service (DaaS), fabrication as a service (FaaS), experiment as a service (EaaS), simulation as a service (SIMaaS), management as a service (MaaS), integration as a service (INaaS), etc. (Li et al. 2011a, b).

The gathered vast amounts of distributed manufacturing resources can be shared repeatedly and combined to be used optimally (Laili et al. 2012; Tao et al. 2012a, b, c): actually, the utilization rate of the manufacturing resources and capabilities will be substantially increased (Zhou et al. 2011), as well as the energy consumption will be reduced, and the emission will be decreased, upon which an efficient and green sustainable manufacturing ecology will be constructed.

The sharing and reuse of high-end resources have made the high-end manufacturing open to all users; meanwhile, the use and payment on-demand have greatly lowered down the threshold for enterprises to participate in a high-end industry, and the technology-intensive industry holding a high scientific and technological content will occupy a dominant position; at the same time, the utilization rate of the high-end resources will be improved, and the high-tech content of which will be transformed into economic benefits. Moreover, as a high value-added product, the intellectual property will be an important source of profits for high-tech enterprises.

The product value chain will be shifted from manufacturing to the product design and knowledge service so that the continuous value-added utility of knowledge, capability, and high-end product in the manufacturing industry will get highlighted (Luo et al. 2012a, b); actually, taking advantage of the knowledge service provided by the CMfg platform to improve the knowledge value, technology value, and service value of the product itself is an important approach to realize sustainable manufacturing.

In a cloud manufacturing platform, the knowledge innovation mode of the "double helix" (Etzkowitz and Leydesdorff 2000; Etzkowitz and Zhou 2006) can be fully embodied, and the enterprise can quickly get all kinds of knowledge services to advance science and technology, as well as can conveniently transform the application of results into value and profits, upon which the knowledge innovation has become the core power for the sustainable growth of a manufacturing enterprise.

2.3 Agile Manufacturing

A CMfg platform can support the user to conduct unified and centralized intelligent management and operation over resources and services. The user can participate in a variety of manufacturing communities, as well as can construct virtual factories, virtual enterprises, and other new forms of "virtual manufacturing environments" (as shown in Fig. 2), which means that the user not only can realize short-term collaboration, but also can achieve the long-term operation and management (Guo et al. 2011; Zhang et al. 2010a, b). The manufacturing cloud set up in the virtual environments (Guo et al. 2010) will not be bound by physical constraints, which holds the ability for infinite extension. The gathered resources can be further managed, planned, and dispatched uniformly so as to realize the optimal utilization of resources in a sufficiently large scope. The existence of

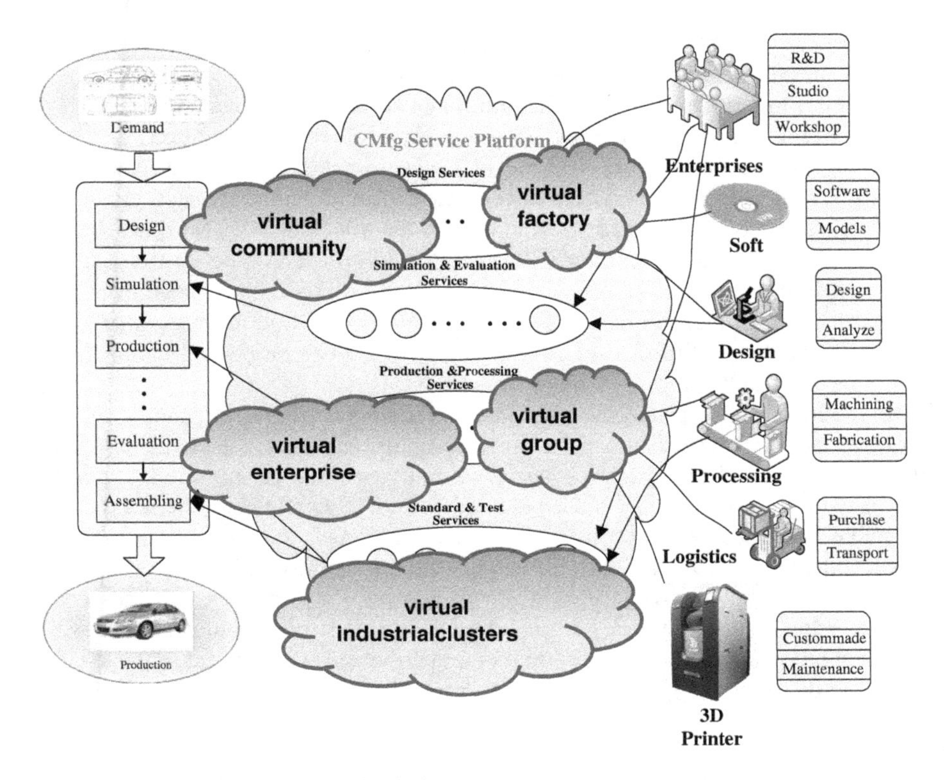

Fig. 2 Operation mode of the CMfg platform

intelligent, flexible, and efficient sharing and collaboration in every link of the whole product life cycle, including product demonstration, design, production and processing, experiment, simulation, operation and management, etc., can complete loosely coupled tasks collaboration like Witkey (Liu et al. 2007), Crowd-Sourcing (Howe 2006), etc.; meanwhile, tightly coupled tasks of complex products (Ulrich 1995) can also obtain dynamic, flexible (Johnstone and Kurtzhaltz 1984), intelligent collaborative services, which can facilitate the realization of collaborative manufacturing, agile manufacturing, and virtual manufacturing.

2.4 Personalized and Socialized Production Mode

In the subjective report "Manufacturing: A Third Industrial Revolution" (Markillie 2012), Paul Markillie expressed that, the application of the new materials, 3D printing, intelligent robot, and a series of network-based collaborative manufacturing services will form a resultant force, which will generate a great power large enough to change the process of economic society. Actually, small batch production will become more cost-effective and production will become more flexible while the required labor input will also be reduced. Meanwhile, the production mode has no longer highlighted mass manufacturing but has become more oriented toward personalization and socialization.

To print objects at will often appear in a science fiction movie as a scene. The appearance of 3D printers and the development of new materials have gradually made this scene become a reality. According to the principle of a 3D printer, based on the model or data, the printer will be able to print the product, which can be designed by a computer, or 3D scanning over the original object. Moreover, the physical model, which originally could only be obtained after several days, now can be fabricated and formed in several hours, which can be printed out again at any time if a modification is required.

In the future, the types of the 3D printing equipment will be extremely rich; the large printer will be used to print large complex parts, for which the use on-demand can be realized through a way of leasing; the small printer will be flexible and easy to use, which can be placed at home for use at any time. At the same time, printable materials will also be all-encompassing, such as nanomaterials, high molecular materials, metals, as well as even food materials and tissue cells.

3D printing is a catalyst of CMfg. CMfg will become a distributed system for the super large-scale network composed of the small-sized manufacturing enterprises, and it is a scheme to replace mass production and is composed of the small-scale and distributed nodes: to be specific, the individual manufacturing nodes are independent of each other but are interconnected at the same time so that the manufacturer can construct or reconstruct a temporary collection according to a special requirement of the project (Lipson and Kurman 2013). The small-scale and socialized production mode has made the new products spring up constantly and inspired a lot of creations and innovations, which, thus, has led to the group-based innovation and promoted sustainable development of the manufacturing industry.

Cloud manufacturing has embodied that there are two typical application scenarios for the personalized and socialized production.

First, under the premise that you do not have a factory of your own, if you need to complete the production for tens of thousands of customized water glasses in a short period of time, you can submit your design document and order to a cloud manufacturing platform, and then thousands of small businesses or individuals will respond to your need in a timely manner: through the 3D printer, these businesses and individuals will complete the products you need in a day or even a shorter period of time as well as will deliver the products to you quickly through logistics.

Second, to customize a car, if you are clear about the demands but you are incapable of design and do not have a factory of your own, then a manufacturing cloud, according to your demands, can recommend and match the optimal combination of service providers for you, including design, processing, assembly, testing, etc.: the schedule for all links will be under your control, which can be adjusted at any time; meanwhile, most of the parts will be processed and completed through customization by 3D printing. Within the scheduled time, a week or even shorter, you will be able to obtain this customized car.

It can be predicted that with the construction and operation of the CMfg platform, as well as the wide access and application of the numerous software and hardware resources and vast amounts of 3D printers, in the future:

(1) When we have a product idea, by taking advantage of the personal computer or mobile terminal, it will be convenient for us to locate the required software and tools in a cloud manufacturing platform to complete the whole process of manufacturing, including design, simulation, experiment, test, processing, transaction, etc.
(2) The leading factor of the profit model for the manufacturing industry will be transformed from the "product" into a "product + service"; meanwhile, high-end manufacturing will become the economic growth core of the industry.
(3) It will be very convenient for us to build a virtual factory, virtual enterprise, virtual group, and even a virtual industry. Meanwhile, the greenization, low-cost operation, and maintenance and convenience for transformation and upgrade will constitute the new enterprise ecology.
(4) Personalization, customization (Gardner and Piller 2009) and socialization will become the new mode of production. The useful 3D printing will be widely applied so that it will be very easy for everyone to produce their own products. Meanwhile, the social intercourse of manufacturing will be used in worldwide communications over manufacturing, for which creativity, innovation, and business opportunities will spring up constantly.

3 Construction and Typical Applications of a Cloud Manufacturing Platform

As shown in Fig. 3, through the construction of a cloud manufacturing resource/capability service platform, cloud manufacturing technology can gather/distribute the manufacturing resources to realize efficient sharing and on-demand use; through a knowledge service platform, cloud manufacturing technology can gather, share, and circulate high-end innovation resources; through the application of the cloud manufacturing community, the cloud manufacturing technology can support social manufacturing and personalized and socialized manufacturing to implement the application of innovation and group-based innovation.

3.1 Resource/Capability Service Platform

A resource/capability service platform will be used for the gathering, circulation, transaction, etc., of manufacturing capabilities. This construction, management, operation, and maintenance are involved with a series of technologies, such as the perception and access technology of the resource ability, the virtualization and

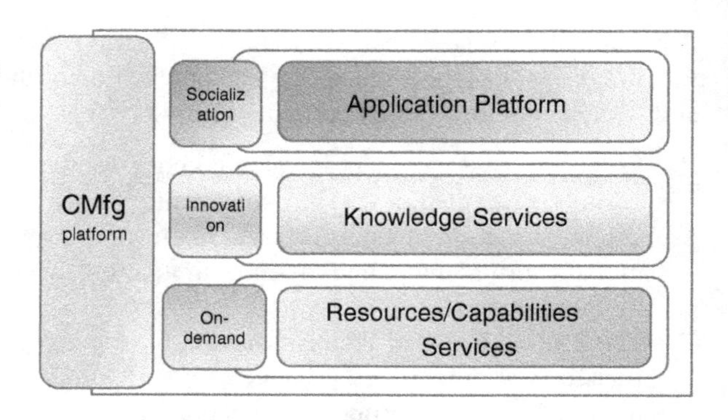

Fig. 3 Applications of a cloud manufacturing platform

servitization technology of the resource capability, the construction and management technology of the virtualized manufacturing service environment, the operation technology of the virtualized manufacturing service environment, the evaluation technology of the cloud service environment, service technology of trust and security manufacturing, the universal man–machine interaction technology, the application technology of the service platform, information-based manufacturing technology, etc., for which the team to which the Authors belong has carried out research in multiple areas. Due to limited space, hereby in this section, we will focus on the servitization technology of the resource/capability.

The servitization process of the manufacturing resource/capability: combining the virtualization access (Ren et al. 2011) (specifically, the perceptual access (Li et al. 2012a, b, c, d, e) of the physical manufacturing resources is involved with the real-time, online and distributed dynamic perception principles and technology of a manufacturing equipment's operating state, the adapter access mechanism and method of the manufacturing equipment resources in the heterogeneous network environment, as well as the intelligent analysis and processing technology of the manufacturing equipment resources' perceptual information) with all kinds of manufacturing resources with a manufacturing capability model and description mechanism to carry out the formalization description over the manufacturing capability so as to eventually release the manufacturing capability on to a cloud manufacturing service platform in the form of service and further to take advantage of service management tools to carry out intelligent management, such as the service retrieval, matching (Laili et al. 2012), combination (Tao et al. 2012a, b, c), transaction, execution, scheduling (Lou et al. 2012), settlement, evaluation (Guo et al. 2010), optimization (Tao et al. 2012a, b, c; Cao et al. 2013), etc.

3.1.1 Manufacturing Capability Model

To implement the use on-demand and free circulation of manufacturing capability under a cloud manufacturing mode, it will be of great importance to constitute an ideal manufacturing capability description model. The manufacturing capability under a cloud manufacturing mode will adopt an ontology-based "top-down"

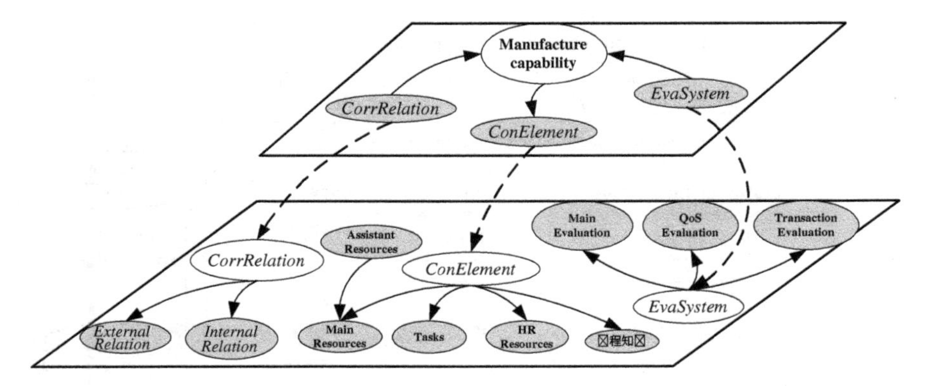

Fig. 4 Top-layer ontology model of the manufacturing capability (Zhang et al. 2010)

hierarchical description method to conduct the modeling, including the top-layer conceptual model (TopMeta_Model) and the bottom-layer case model (BottomIns_Model). The top-layer conceptual model can be represented by the ontology (Fig. 4) and be refined layer by layer: the upper layer is the ontology of the manufacturing capability, which mainly includes the three parts of the manufacturing capability, that is, incidence relation, element and evaluation system; the lower layer is the refinement over the upper ontology: for example, the ontology of the element includes the main resources, product tasks, human resources, process knowledge, and other attribute information. At the same time, there is a certain logic relationship among the three elements in the top-layer model, for example: the *InternalRelation* in the *CorrRelation* will reflect the collaborative relation for all kinds of resources kinds of resources the *ConElement* while the *EvaSystem* shall be established relying on the *ConElement*: for instance, different manufacturing capabilities or resources accordingly will have different evaluation systems and standards.

The bottom-layer model is the detailed analysis and refinement carried out for all kinds of information in the top-layer model: in this field, Luo and other scholars have conducted a great deal of research (Luo et al. 2012a, b, 2013).

3.1.2 Service-Oriented Description Mechanism of the Manufacturing Capability

The service-oriented description (SOD) of the manufacturing capability is the formalized description and representation for the manufacturing capability so that the manufacturing capability model can be embodied and accessed in a cloud manufacturing service platform through the form described, which can be understood by the computer. As shown in Fig. 5, the manufacturing capability is featured by uncertainty, incompleteness, dynamic, knowledge orientation, and other multiple characteristics so that different theoretical approaches are required to deal with the different characteristics of the manufacturing capability, for

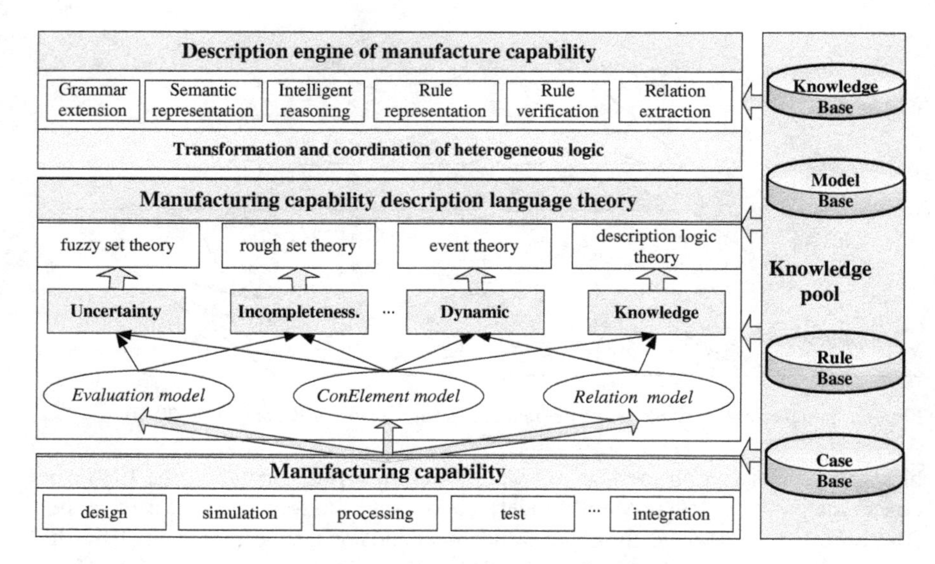

Fig. 5 Description framework of the manufacturing capability (Luo et al. 2012a, b)

example: the theory related to fuzzy sets can be used to analyze the uncertainty of the manufacturing capability while the dynamic theory, description logic theory, etc., can be used to describe the dynamic and knowledge orientation of the manufacturing capability.

The process of the SOD for the manufacturing capability is: first, to carry out extraction and classification over all kinds of characteristic attributes in the manufacturing capability information model. Second, as for the different characteristic information, different theoretical approaches shall be adopted for the solution: for example, targeted at the attributes of static state and certainty, the description logic (DL) can be adopted directly for the description; targeted at the fuzzy concept and relations, the fuzzy description logic (Fuzzy DL) can be adopted for processing; and targeted at the attribute information of the dynamic behavior for the manufacturing capability, the dynamic description logic (DDL) can be adopted for the formalized description. Finally, syntactic, semantics, deduction and rule representation and processing, and other technologies can solve the problems of logic compatibility and syntactic and semantic consistency for the related basic theoretical approaches in the same model so as to implement the unified description for the services of the manufacturing capability.

3.2 Cloud Manufacturing Knowledge Service Platform

When gathering all kinds of manufacturing resources and capabilities, the manufacturing cloud will gather all kinds of data, models, experience, and knowledge

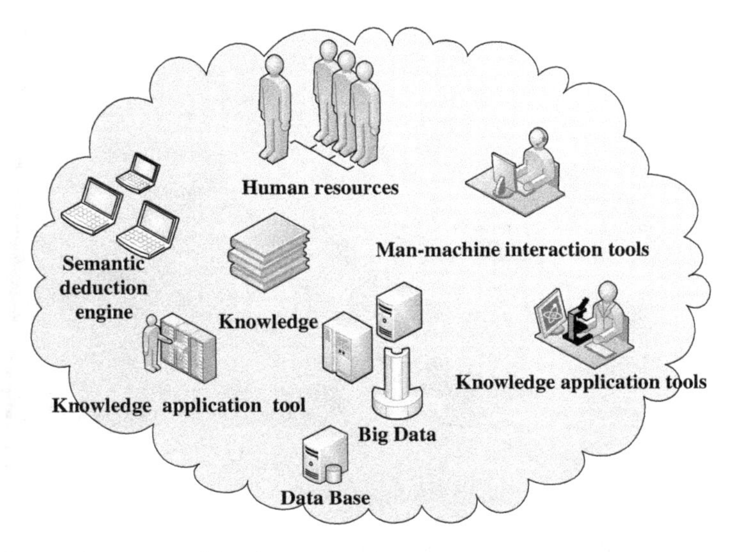

Fig. 6 Knowledge service of cloud manufacturing

(Fig. 6) at the same time; meanwhile, with the continuous evolution of the manufacturing cloud, the knowledge scale accumulated in the cloud will be expanded accordingly. The knowledge and knowledge service play a very important role in the cloud manufacturing system: first, the gathering of all kinds of data and knowledge can provide direct services for the whole life cycle of manufacturing so as to promote the innovation ability of the manufacturing industry; second, the support by the knowledge on the cloud manufacturing service platform itself makes the platform hold a high degree of intelligence, which, thus, can support the intellectualization of the manufacturing process; moreover, there are very large amounts of data and knowledge existing in a cloud manufacturing system, which, through the analysis and excavation by big data technology can provide personalized services. This section will introduce the knowledge servitization, knowledge cloud management, big data, and other technologies involved in the construction of the knowledge service platform.

3.2.1 Knowledge Servitization

The servitization encapsulation of knowledge is the foundation for knowledge sharing, including knowledge acquisition, language description, knowledge representation, servitization distribution, etc., which is as shown in Fig. 7. Taking advantage of an intelligent expert system, the knowledge acquisition will obtain the distribution, heterogeneousness, and dynamic knowledge existing in the whole life cycle of manufacturing, including domain knowledge, design knowledge, task knowledge, case knowledge, etc. In this process, it includes the knowledge obtained

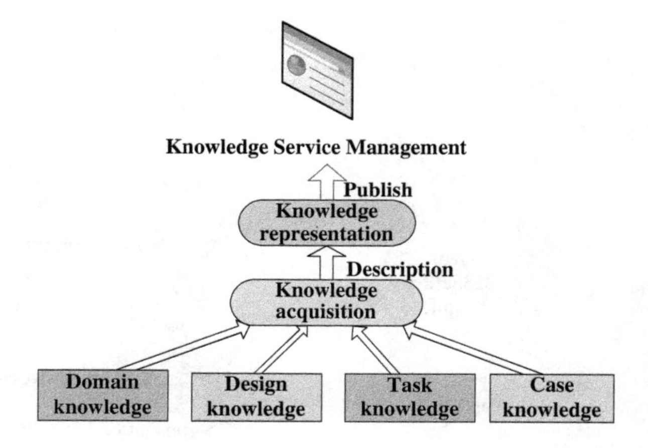

Fig. 7 Knowledge servitization

through the semi-automatic acquisition system by the knowledge uploader, the knowledge automatically obtained due to the various data generated in the operation of an automatic recording system, the new knowledge generated through knowledge deduction and excavation, etc. Through the analysis over the characteristics of all kinds of knowledge, the knowledge representation is to select an appropriate knowledge representation model to carry out the servitization description for the knowledge and release it onto a cloud platform.

3.2.2 Knowledge Cloud Management

In a cloud manufacturing platform, knowledge cloud management is a kind of knowledge management model to implement the intelligent collaboration among the very large amounts of domain knowledge (Hu et al. 2012). The knowledge cloud management system mainly includes four parts: unified retrieval, knowledge evaluation, knowledge submission, and knowledge base management (Hu et al. 2013), which is as shown in Fig. 8.

Unified retrieval: after the user has input the query request, through the Demand analysis rules and Q&A template, the system will carry out a demand analysis and then enter this into a semantic retrieval model: the retrieval model based on the vector space model will adopt the ontology computation requirements and the semantic similarity of the knowledge information to carry out the logic deduction so as to locate the knowledge required by the user in the knowledge information and feed it back to the user.

Knowledge evaluation: the user shall carry out evaluation over the knowledge fed back by the system, which, as historical information data, will be stored in a knowledge information pool.

Knowledge submission: through the knowledge submission module, the user can submit new knowledge to the system; through the fusion deduction module, the new knowledge will wait for the verification by the domain expert; the verified

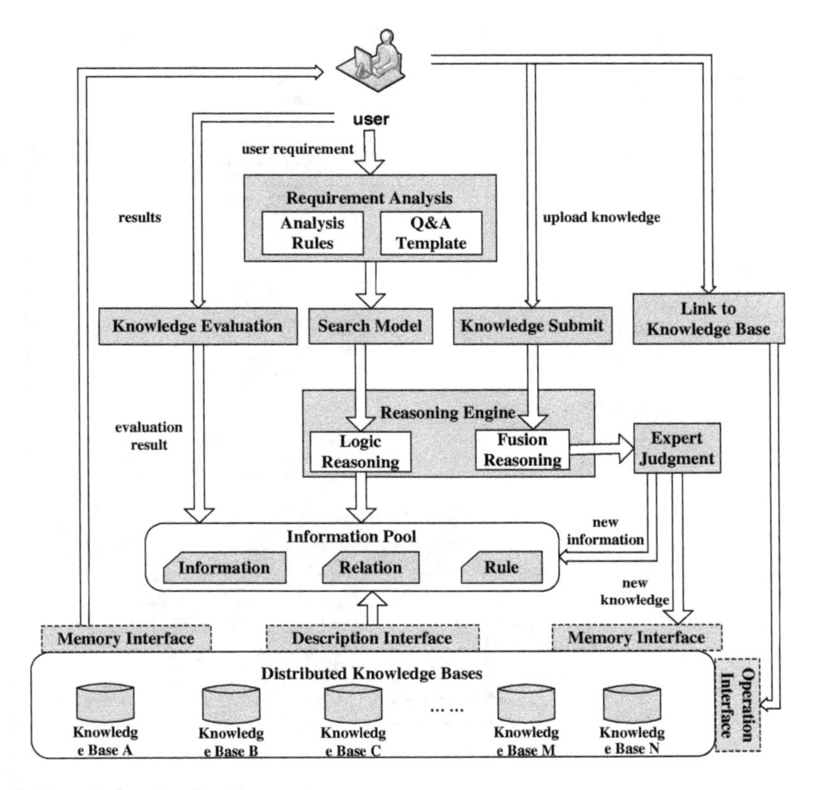

Fig. 8 Knowledge cloud management

knowledge will be stored in the knowledge base for classification, the description information of which will be stored into the knowledge information pool.

Knowledge base management: takes advantage of the knowledge base interface to directly carry out the checks and operations over the distributed knowledge within the allowed permission range.

Usually, the user needs to obtain the knowledge involved from many fields from a knowledge cloud; to implement the cross-database matching and search of knowledge, it is necessary to carry out a mapping description over the knowledge presented by different methods in accordance with a certain template and standard so as to provide information support for the establishment of the matching algorithm model (Hu et al. 2013). Targeted at the huge amounts of knowledge stored in a knowledge cloud, the traditional search based on keyword or theme has been unable to satisfy the demands of the user for accuracy, quickness, and comprehensiveness of search results; therefore, it is necessary to establish an efficient search pattern in a knowledge cloud, which can correctly understand the user's intent and can realize an intelligent retrieval.

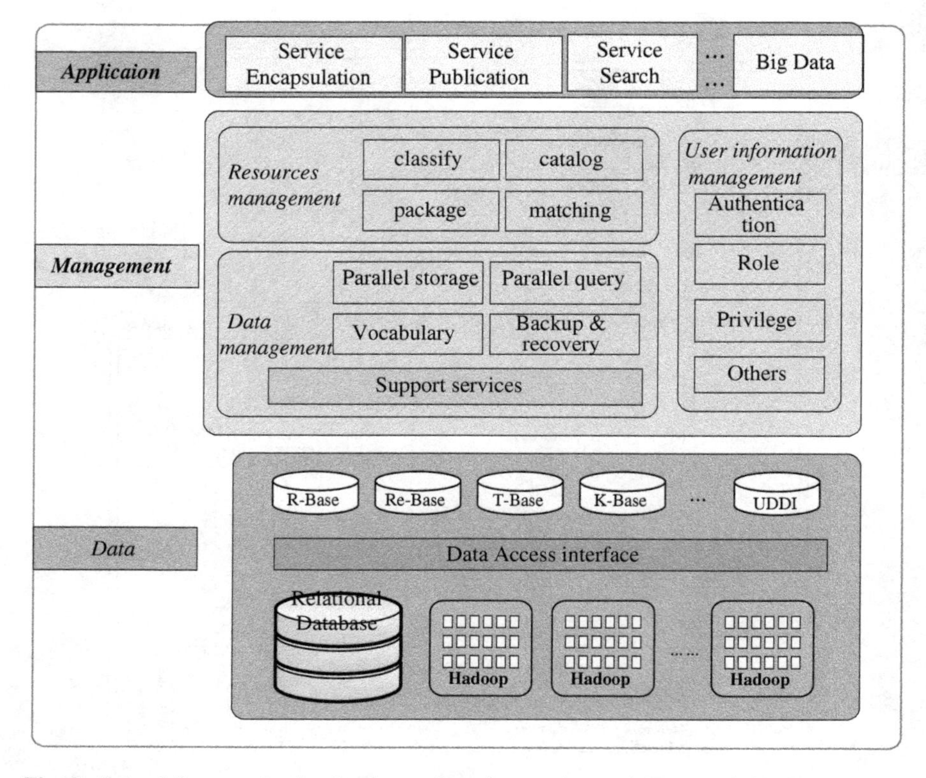

Fig. 9 General framework of a cloud manufacturing service pool (Bao et al. 2012)

3.2.3 Big Data Technology

In the cloud manufacturing platform, all kinds of resource/capability, model, knowledge and other services will exist in the service pool in miscellaneous data forms (Fig. 9). These data are characterized with varied structures, including structured data, semi-structured data, and unstructured data; all these data will present a trend of explosive growth so that the data volume is tremendous; quick and timely analysis is often required in an application; meanwhile, the authenticity and reliability of the processing results must be guaranteed. It is just like the 4 V characteristics of the big data: Volume, Velocity, Variety, and Veracity; the traditional infrastructure has been unable to cope with the situation so that the distributed system framework (such as Hadoop) shall be used for the storage and processing of the large amounts of heterogeneous data in a cloud manufacturing service pool (Bao et al. 2012).

The Hadoop framework transparently provides applications both reliability and data motion. Hadoop implements a computational paradigm named Map/Reduce, where the application is divided into many small fragments of work, each of which may be executed or re-executed on any node in the cluster. In addition, it provides a distributed file system (HDFS) that stores data on compute nodes, providing very high aggregate bandwidth across the cluster. Both Map/Reduce and the distributed file system are designed so that node failures are automatically handled by this framework (http://wiki.apache.org/hadoop/).

Through the analysis and knowledge excavation over huge amounts of data, a cloud manufacturing platform provides intelligent and personalized services, such as intelligent search services and cloud recommendations. A cloud manufacturing platform contains vast amounts of resources and service data, especially dynamic perception data, masses of user information, log, evaluation and social intercourse data, as well as the space, time, weather, and other potential relevant data, from which a cloud manufacturing platform will dig out the information rich in commercial value, such as personal preferences, interests, quantitative indicators of the product life cycle, hotspots and development trend of the industry, etc. The more data one has, the more accurate the understanding toward the user will be while the shorter the processing time will be, and the higher the information credibility will also be.

3.3 Typical Applications of Cloud Manufacturing

3.3.1 Cloud Manufacturing Community

The cloud manufacturing community provides the user with an integrated application environment, which can support the instant communication, discussion, business management, customized processing, transaction payment, and other various functions at the same time; meanwhile, the cloud manufacturing community supports the user's participation in or establishment of the manufacturing social intercourse, virtual enterprise, and virtual organization; moreover, it supports Crowd-Sourcing, Witkey and all kinds of new manufacturing modes.

As shown in Fig. 10, a basic resource/capability service platform can provide huge amounts of hardware and software resources and knowledge services to construct a manufacturing community; on-demand use technology, on the other hand, is used to realize the on-demand leasing and elastic pricing of services; moreover, the transaction evaluation, security authentication, and other technologies are used to provide the user with a credible and safe transaction environment. The manufacturing community contains huge amounts of dynamic operation information; the big data technology is used for the excavation of historical data, which it is able to quickly and immediately grasp the user's use habit and search intention so as to implement intelligent pushing, personalized customized search and other intelligent community services.

The virtual community has lowered down the threshold for the public to get involved in the manufacturing and obtain resources, as well as has made it easier to produce the personalized products than the traditional way, which, thus, has

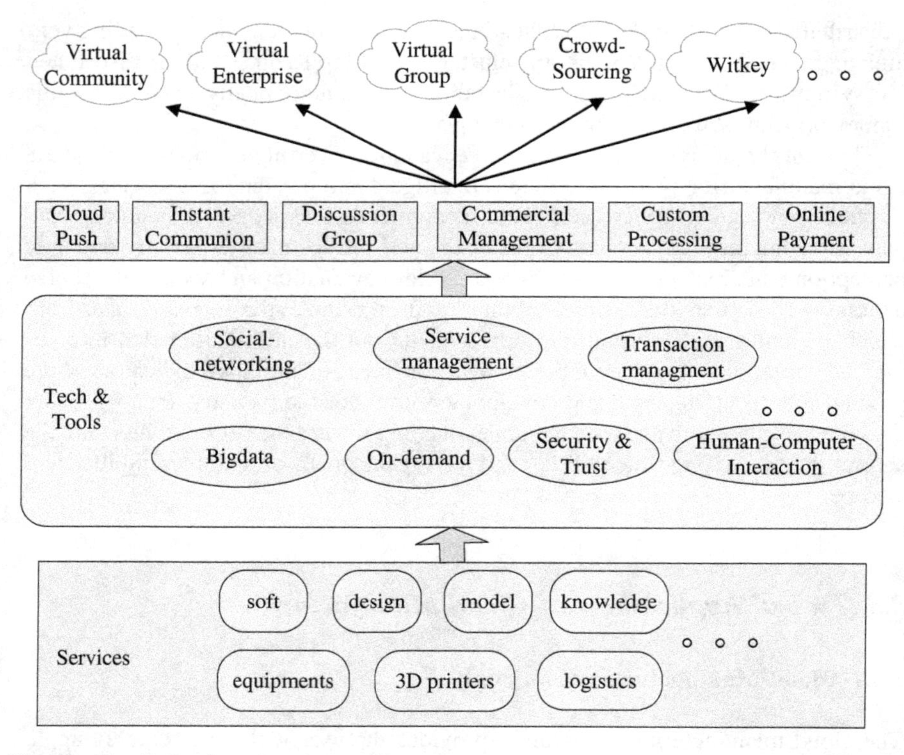

Fig. 10 Cloud manufacturing community

inspired the group-based innovation and led to the formation of various manufacturing business relations.

3.3.2 Virtual Industrial Cluster

Relying on shared infrastructure within the region, the traditional industrial cluster has formed a certain industry scale to maintain its economic benefits. However, after a period of rapid development, the further improvement of the industrial cluster is constrained by the physical space, service ability, service level, information communication and vision, on-demand use and collaboration ability and other factors.

In the manufacturing cloud, through the construction of the virtual unit, virtual factory, virtual enterprise and virtual group, the virtual space industry cluster will be forged (as shown in Fig. 11). Through the change over the development model of the traditional industrial cluster, the virtual industrial cluster will realize the sustainable development for various industries. First, the virtual industrial cluster will take advantage of the highly shared resources and services as required in a manufacturing cloud to carry out the efficient collaborative production so as to

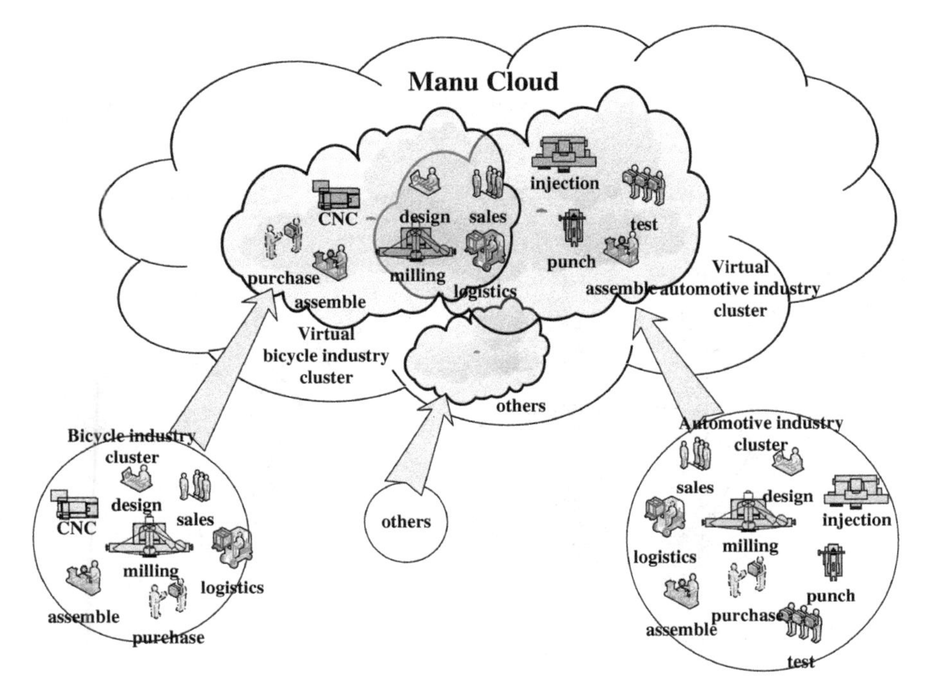

Fig. 11 Industrial cluster of the virtual space

lower down production costs and improve industrial benefits; second, holding the features for being scalable, reducible, sustainable, etc., a cloud platform can support the transformation, upgrade, sustainability, etc., of a virtual industrial cluster; moreover, intelligent optimization tools and technologies provided by a cloud manufacturing platform can be used for intelligent management of an industrial cluster.

3.3.3 3D Printing

As one of the rapid proto-typing technologies, based on digital model file, 3D printing will adopt the powder metal or plastic and other adhesive materials to construct the object through the layered printing (Sachs et al. 1993; Kruth et al. 1998; Gibson et al. 2010). Compared with other processing equipment requiring complex manual operation, 3D printing equipment holds a good digitalized interface, which can receive a model file and implement full-automatic cascading process.

At present, the application of the 3D printer is still in the initial stages: to carry out and realize its intelligent application in a cloud manufacturing platform still needs to solve the adapter access, printing management, equipment maintenance and other problems, which is as shown in Fig. 12. Adapter access: the standard

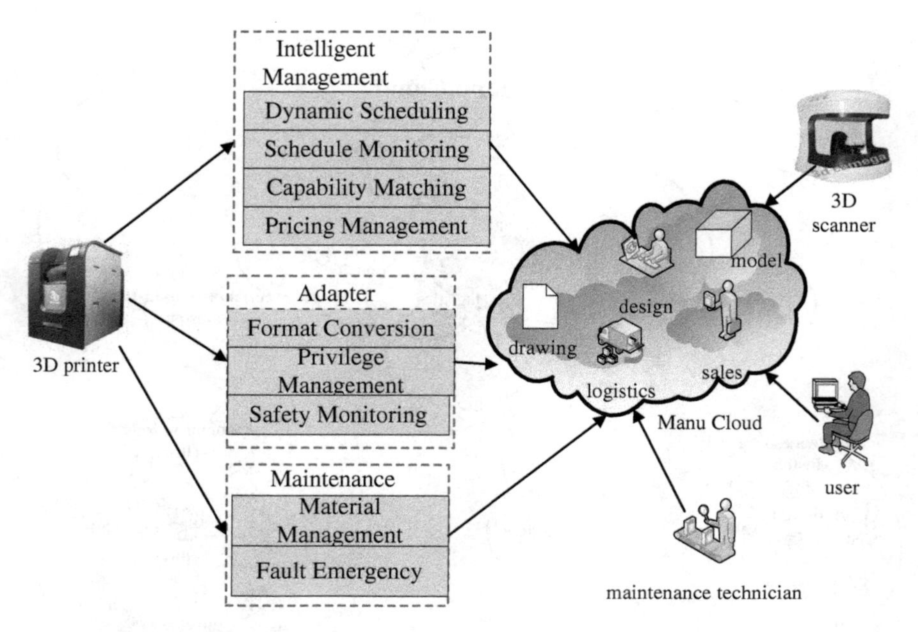

Fig. 12 3D printing for cloud manufacturing application

drawing files can be received and printed by all kinds of printers; the permission management and security monitoring technology will be used to ensure the safe usage of 3D printer. Intelligent management: it includes the dynamic production scheduling management, processing, processing capability management, pricing management, etc., which is used to realize the on-demand leasing and pricing. Equipment maintenance: the automatic batching management and intelligent fault emergency measure are used to implement all-weather automatic printing.

As shown in Fig. 13, in cloud manufacturing, the advantages for the user to use 3D printing to process a product lie in the following aspects: obtain vast amounts of design resources and renting many printing devices as required, which can lower down the production threshold and cost; produce personalized products quickly, which can shorten the cycle ranging from the design to the trial production and to small batch production; it is suitable to produce optimal customized components: the customized products generally can embody creative innovation, which, in light of the higher content of technology and knowledge, is kept at the upper end of the value chain.

3.3.4 Green Evaluation of the Cloud Service

The green evaluation of a cloud service is the real-time monitoring and quantitative evaluation conducted for the composition, source, and production mode of

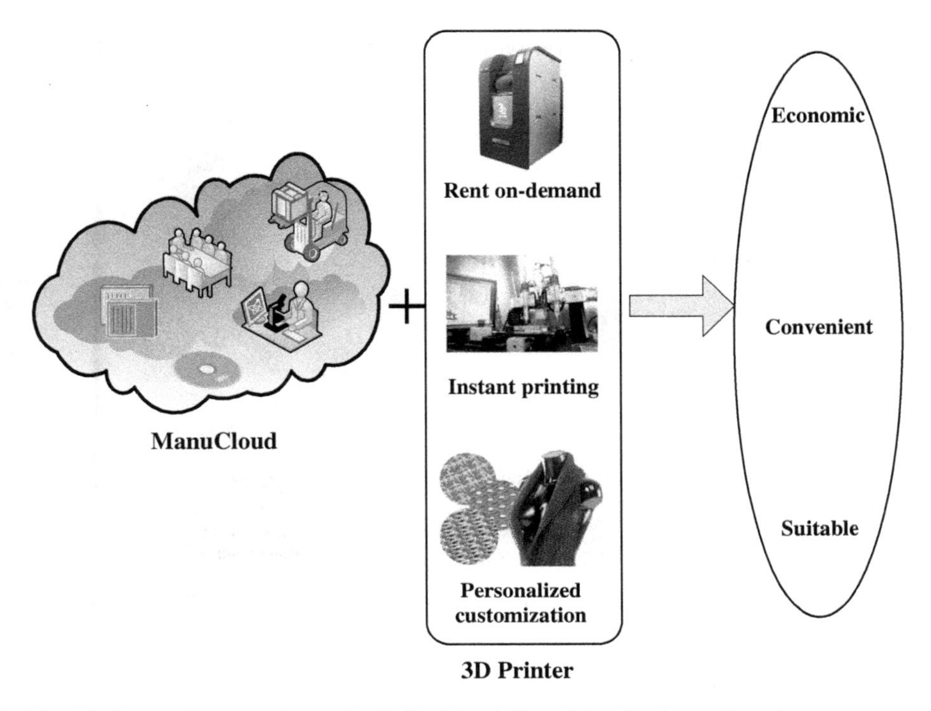

Fig. 13 Economic, convenient and suitable 3D printing of the cloud manufacturing

the raw materials and energy involved in the manufacturing process, which is used to quantify the energy consumption management and optimization. The team, to which the Authors belongs, has conducted research and development over the green evaluation application for the cloud service based on BOM, the principle of which is to take advantage of the various list data generated in the links of product design, process design, manufacturing assembly, etc., including an engineering bill of materials (EBOM), planning bill of materials (PBOM), manufacturing bill of materials (MBOM), etc., which is as shown in Fig. 14, to calculate all kinds of characteristic indicators, such as the energy consumption indicator and emission indicator, so as to provide basis for the optimized choice of the service resources. Its key technologies include: establishment of the evaluation model for the product life cycle; construction of the basic database for the energy conservation and emission reduction evaluation of the cloud service, such as energy database, emission database, standard database related to energy conservation and emissions reduction, raw material database, product database, process database, etc.

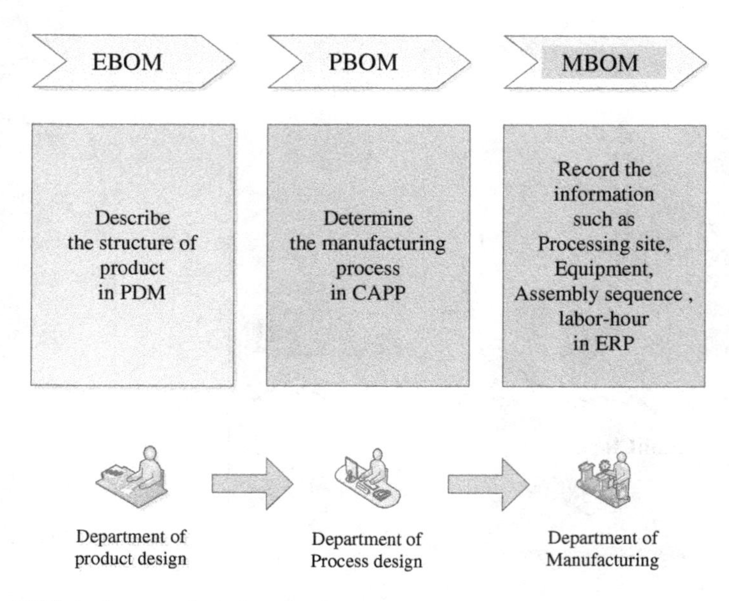

Fig. 14 BOMs in the manufacturing cloud

3.3.5 Hybrid Cloud

From the perspective of this implementation subject, a cloud manufacturing platform can be divided into three types, that is, an enterprise cloud, an industry (region) cloud, and an hybrid cloud (Fig. 15). (1).

The enterprise cloud (also called as a private cloud) is constructed based on the network inside the enterprise or group, which is to highlight the integration and service of the manufacturing resources and capabilities inside the enterprise or group, optimize the utilization rate of the resources and capabilities of the enterprise or group, and reduce the repeated construction of the repeated resources and capabilities so as to lower down the cost and improve competitiveness. (2) The industry (region) cloud (also called as the public cloud) is constructed based on the "public network" (such as the Internet and the Internet of Things), which is to highlight the integration of manufacturing resources and capabilities among the enterprises, improve the utilization rate of the whole society's manufacturing resources and capabilities, and realize the transaction of the manufacturing resources and capabilities; based on the third party enterprises, the corresponding public cloud manufacturing service platform will be constructed; all enterprises can provide the surplus or idle manufacturing resources and capabilities to the platform to make profits; meanwhile, all enterprises can purchase and use the services of resources and capabilities provided by platform as required. (3) The hybrid cloud, based on the basis of the existing public and private cloud platforms, mainly refers to the realization of public cloud integration among the regions/

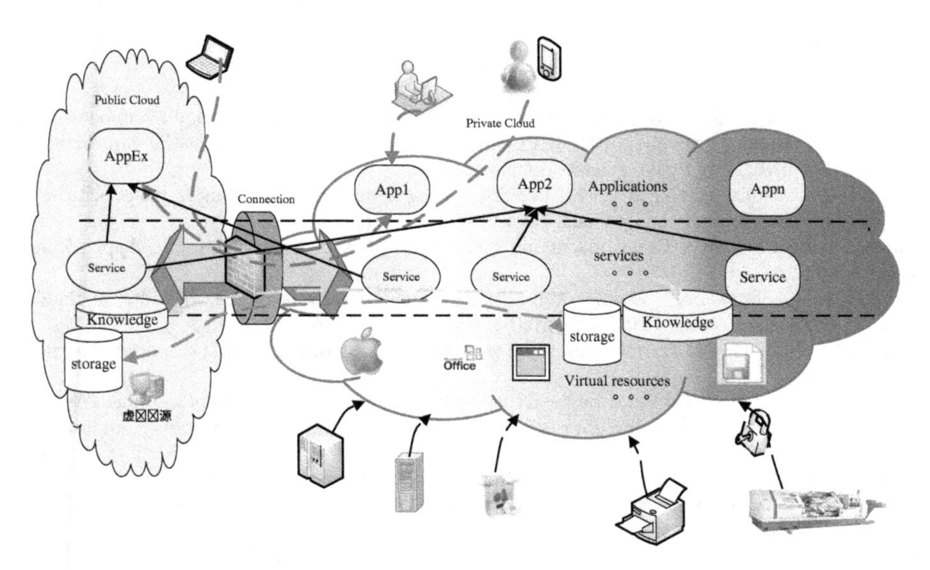

Fig. 15 Application of a hybrid manufacturing cloud (Mai et al. 2012)

industries, the integration of the public cloud and private cloud (Mai et al. 2012), the integration of the private clouds and the integration of the cloud platform and the existing information system (Li et al. 2012a, b, c, d, e).

4 Conclusion

At present, the global manufacturing industry is presenting a high socialization and personalization based on the network, as well as the development trend toward the highly intellectualization, greenization, and agility taking service as the core.

As a kind of new model and method, cloud manufacturing can support on-demand use, intelligent manufacturing, and sustainable manufacturing; by virtue of its powerful resource and knowledge gathering capabilities, cloud manufacturing can support manufacturing collaboration, group-based collaboration and agile manufacturing; moreover, cloud manufacturing can support the customization of personalized products and a socialized production so as to embody group-based innovation.

Acknowledgments This work is partially supported by the National Nature Science Foundation of China (61374199), and Beijing Natural Science Foundation (4142031). Thanks for the help from A. R. Hu, Y. F. Mai, Y. Zuo, L. Lv at the Laboratory of Manufacturing Integration and Simulation Technology (MIST) at Beihang University.

References

Bao Y, Ren L et al (2012) Massive sensor data management framework in cloud manufacturing based on Hadoop. In: 2012 10th IEEE international conference on industrial informatics (INDIN), IEEE. http://wiki.apache.org/hadoop/

Ben Naylor J, Naim MM et al (1999) Leagility: integrating the lean and agile manufacturing paradigms in the total supply chain. Int J Prod Econ 62(1):107–118

Cao W, Jiang PY (2012) Cloud machining community for social manufacturing. Appl Mech Mater 220:61–64

Cao J, Hwang K et al (2013) Optimal multiserver configuration for profit maximization in cloud computing. IEEE Trans Parallel Distrib Syst 24(6):1087–1096

Castells M, Hall P (1994).Technopoles of the world: the making of twenty-first-century industrial complexes

Cheng Y, Tao F et al (2010) Study on the utility model and utility equilibrium of resource service transaction in cloud manufacturing. In: 2010 IEEE international conference on, industrial engineering and engineering management (IEEM), IEEE

Etzkowitz H, Leydesdorff L (2000) The dynamics of innovation: from national systems and "Mode 2" to a Triple Helix of university–industry–government relations. Res Policy 29(2):109–123

Etzkowitz H, Zhou C (2006) Triple Helix twins: innovation and sustainability. Sci Public Policy 33(1):77–83

Fan W-H, Xiao T-Y (2011) Integrated architecture of cloud manufacturing based on federation mode. Comput Integr Manuf Syst 17(3):469–476 (in Chinese)

Gardner DJ, Piller F (2009) Mass customization: an enterprise-wide business strategy; how build to order, assemble to order, configure to order, make to order, and engineer to order manufacturers; increase profits and better satisfy customers, Happy About

Gibson I, Rosen DW et al (2010) Additive manufacturing technologies: rapid prototyping to direct digital manufacturing. Springer, New York

Guo H, Tao F et al (2010) Correlation-aware web services composition and QoS computation model in virtual enterprise. Int J Adv Manuf Technol 51(5–8):817–827

Guo H, Zhang L et al (2011) A framework for correlation relationship mining of cloud service in cloud manufacturing system. Adv Mater Res 314:2259–2262

Hai-jian CJ-HL (2003) On the way of new type of industrialization. China Ind Econ 1:006

Howe J (2006) The rise of crowdsourcing. Wired Mag 14(6):1–4

Hu A, Zhang L et al (2012) Resource service management of cloud manufacturing based on knowledge. J Tongji Univ Nat Sci 40(7):1092–1101

Hu A, Zhang L, Tao F, Hu X (2013) Lifecycle management of knowledge in a cloud manufacturing system. Proceedings of the ASME 2013 international manufacturing science and engineering conference MSEC2013, Madison, Wisconsin, USA, 10–14 June 2013

Hu X, Zhang L et al (2013) Knowledge semantic search in cloud manufacturing. The 25th European modeling and simulation symposium, Athens, Greece, 25–27 September

Huang B, Li C et al (2013) Cloud manufacturing service platform for small-and medium-sized enterprises. Int J Adv Manuf Technol 56:1–12

Ji H, Yang QJ et al (2012) Enterprise product platform services in cloud manufacturing. Dev Innov Mach Electr Prod 3:016 (in Chinese)

Johnstone R, Kurtzhaltz JE (1984) Flexible manufacturing system, Google Patents

Kruth J-P, Leu M et al (1998) Progress in additive manufacturing and rapid prototyping. CIRP Ann Manuf Technol 47(2):525–540

Laili Y, Tao F et al (2012) A study of optimal allocation of computing resources in cloud manufacturing systems. Int J Adv Manuf Technol 63(5–8):671–690

Li B-H, Zhang L et al (2010a) Cloud manufacturing: a new service-oriented networked manufacturing model. Comput Integr Manuf Syst 16(1):1–7 (in Chinese)

Li B-H, Zhang L, Chai XD (2010b) Introduction to cloud manufacturing. ZTE Commun 8(4):6–9 (in Chinese)

Li B-H, Zhang L et al (2011a) Further discussion on cloud manufacturing. Comput Integr Manuf Syst 17(3):449–457 (in Chinese)

Li Q, Qin B et al (2011b) Exploration on mould coordinated manufacturing based on cloud manufacturing. Forging Stamping Technol 3:140–143 (in Chinese)

Li B-H, Zhang L et al (2012a) Typical characteristics, technologies and applications of cloud manufacturing. Comput Integr Manuf Syst 18(7):1345–1356 (in Chinese)

Li X-Q, Yang H-C et al (2012b) Knowledge service modeling approach for group enterprise cloud manufacturing. Comput Integr Manuf Syst 18(8):1869–1880 (in Chinese)

Li B, Zhang G-J et al (2012c) Mould industry cloud manufacturing platform supporting cooperation and its key technologies. Comput Integr Manuf Syst 18(7):1620–1626 (in Chinese)

Li R-F, Liu Q et al (2012d) Perception and access adaptation of equipment resources in cloud manufacturing. Comput Integr Manuf Syst 18(7):1547–1553 (in Chinese)

Li Q, Wang Z-Y et al (2012e) Applications integration in a hybrid cloud computing environment: modelling and platform. Enterp Inf Syst (ahead-of-print):1–35

Lin T, Chai X et al (2012) Top-level modeling theory of multi-discipline virtual prototype. J Syst Eng Electron 23(3):425–437

Lipson H, Kurman M (2013) Fabricated: the new world of 3d printing. Wiley, Indianapolis

Liu F, Zhang L et al (2007) The application of knowledge management in the internet—witkey mode in China. Int J Knowl Syst Sci 4(4):32–41

Lou P, Liu Q et al (2012) Multi-agent-based proactive–reactive scheduling for a job shop. Int J Adv Manuf Technol 59(1–4):311–324

Luo Y-L, Zhang L et al (2012a) Key technologies of manufacturing capability modeling in cloud manufacturing mode. Comput Integr Manuf Syst 18(7):1357–1367 (in Chinese)

Luo Y-L, Zhang L et al (2012) Research on the knowledge-based multi-dimensional information model of manufacturing capability in CMfg. Adv Mater Res 472:2592–2595

Luo Y, Zhang L et al (2013) A modeling and description method of multidimensional information for manufacturing capability in cloud manufacturing system. Int J Adv Manuf Technol 1–15

Lynch C (2008) Big data: how do your data grow? Nature 455(7209):28–29

Mai J, Zhang L et al (2012). Architecture of hybrid cloud for manufacturing enterprise. System simulation and scientific computing, Springer, pp 365–372

Manyika J, Chui M et al (2011) Big data: the next frontier for innovation, competition, and productivity

Markillie P (2012). A third industrial revolution. The Economist 1–2

McLean C, Bloom H et al (1982) The virtual manufacturing cell. Proceedings of Fourth IFAC/IFIP conference on information control problems in manufacturing technology

Ren L, Zhang L et al (2011) Resource virtualization in cloud manufacturing. Comput Integr Manuf Syst 17(3):511–518 (in Chinese)

Sachs E, Cima M et al (1993) Three-dimensional printing: the physics and implications of additive manufacturing. CIRP Ann Manuf Technol 42(1):257–260

Sohlenius G (1992) Concurrent engineering. CIRP Ann Manuf Technol 41(2):645–655

Song T-X, Cheng-lei Z et al (2013) Cloud manufacturing service platform for small and medium enterprises. Comput Int Manuf Syst 19(5):1147–1154 (in Chinese)

Tao F, Laili Y et al (2013) FC-PACO-RM: a parallel method for service composition optimal-selection in cloud manufacturing system. IEEE Trans Ind Inform 9(4):2023–2033

Tao F, Cheng Y et al (2012a) Utility modelling, equilibrium, and coordination of resource service transaction in service-oriented manufacturing system. Proc Inst Mech Eng Part B J Eng Manufact 226(6):1099–1117

Tao F, Zhang L et al (2012b) Research on manufacturing grid resource service optimal-selection and composition framework. Enterp Inf Syst 6(2):237–264

Tao F, Guo H et al (2012c) Modelling of combinable relationship-based composition service network and the theoretical proof of its scale-free characteristics. Enterp Inf Syst 6(4):373–404

Thames JL, Rosen DW et al (2013) Enhancing the product realization process with cloud-based design and manufacturing systems. Innovation 17:18

Ulrich K (1995) The role of product architecture in the manufacturing firm. Res Policy 24(3):419–440

Wang W, Zhang Y-H et al (2012) Research on technique of parts resource cloud encapsulation for the small and medium-sized enterprises-sized enterprises. Mach Des Manuf 8:260–262 (in Chinese)

Womack JP, Jones DT et al (2007) The machine that changed the world: the story of lean production—Toyota's secret weapon in the global car wars that is now revolutionizing world industry, SimonandSchuster. com

Wu D, Thames JL et al (2012) Towards a cloud-based design and manufacturing paradigm: looking backward, looking forward. Innovation 17:18

Wu D, Greer MJ et al (2013) Cloud manufacturing: drivers, current status, and future trends. Proceedings of the ASME 2013 international manufacturing science and engineering conference (MSEC13). Paper Number: MSEC2013-1106, Madison, Wisconsin, US

Yang C, Chai X et al (2012a) Research on co-simulation task scheduling based on virtualization technology under cloud simulation. In: AsiaSim 2012, Springer, pp 421–430

Yang C, Li B-H et al (2012b) Cloud manufacturing oriented cloud simulation supporting framework and its application process model. Comput Integr Manufact Syst 18(7):1444–1452 (in Chinese)

Yin C, Huang B-Q et al (2011a) Common key technology system of cloud manufacturing service platform for small and medium enterprises. Comput Integr Manuf Syst 17(3):495–503 (in Chinese)

Yin S, Yin C et al (2011b) Outsourcing resources integration service mode and semantic description in cloud manufacturing environment. Comput Integr Manuf Syst 17(3): 525–532 (in Chinese)

Zeng C, Guo X et al (2009) Cloud computing service composition and search based on semantic. In: Cloud computing, Springer, pp 290–300

Zhan D-C, Zhao X-B et al (2011) Cloud manufacturing service platform for group enterprises oriented to manufacturing and management. Comput Integr Manuf Syst 17(3):487–494 (in Chinese)

Zhang L, Guo H et al (2010a) Flexible management of resource service composition in cloud manufacturing. In: 2010 IEEE international conference on, industrial engineering and engineering management (IEEM), IEEE

Zhang L, Luo Y-L et al (2010). Key technologies for the construction of manufacturing cloud. Comput Integr Manufact Syst 16(11): 2510–2520 (in Chinese)

Zhang L, Luo Y-L et al (2011) Analyses of cloud manufacturing and related advanced manufacturing models. Comput Integr Manuf Syst 17(3):458–468 (in Chinese)

Zhang L, Luo Y et al (2012a) Cloud manufacturing: a new manufacturing paradigm. Enterp Inf Syst (ahead-of-print):1–21

Zhang F, Jiang P et al (2012b) Modeling and analyzing of an enterprise collaboration network supported by service-oriented manufacturing. Proc Inst Mech Eng Part B J Eng Manufact 226(9):1579–1593

Zhou Z, Liu Q et al (2011) From digital manufacturing to cloud manufacturing. Int J Eng Innov Manag 1(1):1–14

Enabling Product Customisation in Manufacturing Clouds

Arthur L. K. Yip, Ursula Rauschecker, Jonathan Corney, Yi Qin and Ananda Jagadeesan

Abstract Cloud manufacturing has emerged as new manufacturing paradigm providing a service-oriented approach to integrate distributed manufacturing resources and to utilise available manufacturing capabilities for collaborative and networked production. With increasing demand for complex customer-oriented products, cloud manufacturing presents provides a promising solution to address the challenges involved in customised specification and production across the supply chain. In this chapter, a concept and architecture is proposed to enable the dynamic customisation of products based on the availabilities of the production network from the cloud manufacturing concept of Manufacturing-as-a-Service (MaaS). An overview of the MaaS concept and architecture is described, which include the core components for product configuration, manufacturing service management and the integration of factory-IT systems. In addition, the proposed manufacturing service description used to enable the provision of customised options is presented. Finally, a case study is presented which demonstrates the feasibility of MaaS concept for customised products. This is followed by an evaluation of the implemented cloud manufacturing platform provided from an industrial perspective.

Keywords Manufacturing service · Manufacturing capability · Architecture · Service description · Customisation

A. L. K. Yip (✉) · J. Corney · Y. Qin · A. Jagadeesan
Department of Design, Manufacture and Engineering Management,
University of Strathclyde, Glasgow, UK
e-mail: arthur.yip@strath.ac.uk

U. Rauschecker
Fraunhofer IPA, Nobelstraße 12, 70569 Stuttgart, Germany
e-mail: ursula.rauschecker@ipa.fraunhofer.de

D. Schaefer (ed.), *Cloud-Based Design and Manufacturing (CBDM)*,
DOI: 10.1007/978-3-319-07398-9_6,

1 Introduction

The vision of mass customisation (i.e. bespoke products at mass-produced prices) emerged in the 1980s (Davis 1987) and has progressed as an evolution, rather than a revolution, to the point where today the ability to tailor products to one's own specification is not uncommon in many industries (Salvador et al. 2009). From advancements in the areas of manufacturing automation and information technologies, many forms of mass customisation have become well established for products such as cars, running shoes and computers. In these cases, online customisation of products is provided by software systems known as 'product configurators' which support the creation of a product specification interactively based on the individual customer's requirements. Typically, the product options available for customisation are restricted to a fixed selection of 'standard' catalogue choices that are communicated between the configuration and the manufacturing system (Jiao and Helander 2006; Ma et al. 2008). In many of today's implementations, product configurators have only limited interaction with the manufacturing capabilities of the factories that will fulfil the customer's order.

As customer demand shifts towards complex customer-oriented products, enabling customisation requires the coordination of the entire supply chain. A production network for the production of complex products may involve different suppliers and manufacturers working in collaboration within a distributed manufacturing environment. This presents challenges across different tiers of the supply chain such as the management of customised product specifications and to coordinate customised production across the different production network participants. Consequently to support product customisation within production networks, manufacturing systems need to be responsive to customer demand and dynamically adjust the entire production network.

The recent emergence of cloud manufacturing as a new paradigm has become an area of much interest for collaborative and networked production from the transfer of cloud computing principles to the manufacturing domain (Tao et al. 2011; Xu 2012). The basis of cloud manufacturing considers a shift from a production-oriented towards service-oriented manufacturing, in which manufacturing resource and capabilities can be provided as manufacturing services that can be utilised to support different phases of the product lifecycle. In a wider context, the potential of cloud manufacturing and its applications can be extended towards a view of cloud-based design and manufacturing (CBDM) (Wu et al. 2012). With this mind, the overall aim of this research is to enable product customisation within a cloud manufacturing environment.

This chapter describes a concept and associated architecture to achieve a highly flexible production network that integrates distributed factories and their manufacturing capabilities in overall cloud manufacturing infrastructure to enable collaborative speciation and manufacture of customised products. The proposed approach here is to manage a product consisting of various sub-products and the related manufacturing by means of *manufacturing services* and associated *service*

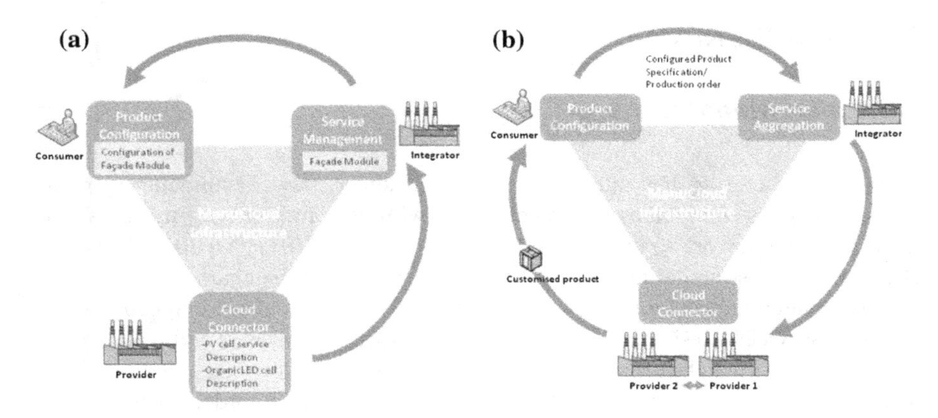

Fig. 1 **a** Provision of real manufacturing capabilities and availabilities, i.e. customisation options, **b** Product configuration and associated manufacturing and delivery

descriptions. This methodology adapts concepts from cloud computing (X-as-a-Service) (Mell and Grance 2011) to the manufacturing domain with the aim of establishing a network to execute MaaS. As a consequence, the MaaS concept enables the provision of customisation options based on available manufacturing capabilities and resources of the production network. In other words, the approach offers end-users access to dynamic product customisation that is *limited only by the capability of the manufacturing facilities* rather than a predefined range from a configuration catalogue in Fig. 1a.

Furthermore, this approach also enables the production networks to be dynamically configured in response to customer demand. This is supported by the provision of planning and control features on the production network level, which today are more commonly associated with individual (i.e. isolated) factories. From the manufacturer's perspective, individual product configurations are transformed into orders and sub-orders, and then assigned to the corresponding production network participants, illustrated in Fig. 1b. Overall, the MaaS architecture described here could not only provide customisable products from a production network but also generate a process plan for physical manufacture of the good throughout the supply chain and support the exchange of production data such as product tracking in the across distributed production sites.

This chapter is structured as follows: Sect. 2 discusses the challenges and approaches involved in the specification and manufacture of customised products through a state-of-the-art review. Section 3 introduces the concept of MaaS with the principles and components of a cloud manufacturing infrastructure which comprises of a user front-end, manufacturing service management and integration with factory IT systems. This is followed by a case study based on the customisation of an integrated solar and lighting facade element reported in Sect. 4. Finally, some conclusions and future work are presented at the end of this chapter.

2 Related Research

2.1 Cloud Manufacturing

Cloud manufacturing has attracted much attention as new manufacturing paradigm, which has emerged over in recent years with its definition varying across the focus of research work to where it is discussed (Tao et al. 2011; Xu 2012; Wu et al. 2013). From existing definitions, cloud manufacturing can be characterised by the provision and management of distributed manufacturing resources and capabilities based on services. Besides this, cloud manufacturing organises services according to users' needs, which means the ability to not only request or use manufacturing services, but also to combine and even configure them in order to best fulfil customers' demands. As a result, much research effort has been dedicated in the areas of manufacturing resource, capabilities and services for cloud manufacturing. Liu and Li (2012) described an ontology-based method for modelling manufacturing resources and capabilities to be virtualised as manufacturing services. In another study, Luo et al. (2013) proposed an approach to define a model of the manufacturing capability in cloud manufacturing in order to capture multidimensional information.

In towards the realisation of cloud manufacturing, a number of working implementations have been proposed (Huang et al. 2012; Wang and Xu 2013; Valilai and Houshmand 2013). Such platforms and prototype implementations have demonstrated the setup and functioning of collaborative and distributed manufacturing environments based on manufacturing services. These provide a useful conceptual framework in which solutions to product customisation can be understood.

For enabling product customisation, it is well understood that the customisation of products throughout production networks is a complex issue because of the many aspects that need to be considered from product design and customisation to the management of production and supply chain activities (Blecker and Friedrich 2007). In the case of production networks, it is essential that the integration of these activities are necessary to ensure a consistent and integrated information flow from customers to all concerned suppliers in order to achieve a synchronised product flow from production sites throughout the network to the customer.

While customisation strategies nowadays mainly are based on and restricted to pre-defined options, cloud manufacturing would enable customisation based on available manufacturing capabilities. In other words, this means that cloud manufacturing could support build-to-order strategies and the management of virtual organisations by means of providing the necessary infrastructure.

The following sections provide a review of current concepts, approaches and applications that have been proposed and are relevant to enable product customisation within the context of production networks.

2.2 Product Configuration

Product configuration systems or configurators have become well established as a key enabler for implementing mass customisation and supporting the specification of customised products (Fogliatto et al. 2012). During this time, different configuration methods have been proposed, from constraint satisfaction problem (CSP), case-based reasoning (CBR) to ontology-based approaches in the development of configuration systems (Xie et al. 2005; Tseng et al. 2005; Yang et al. 2008). However, much research attention has been focused in the related area of configuration knowledge representation and modelling to support the configuration task (Aldanado et al. 2003; Zhang et al. 2005; Felfernig 2007). The challenge faced in this technical area is to define a generic product model that is applicable across different product domains and applications.

Within the last decade, the growth of web technologies enabled connectivity and consumer access to product configurators, driving the trend in the development of web-based configuration systems (Ong et al. 2006; Zhang et al. 2007). This has also enabled product configuration systems to become an important interface for the flow of communication (i.e. customer, retailer, manufacturers) between all stakeholders in the supply chain for the delivery of customised products. A study by Jiao and Helander (2006) developed an online configure-to-order platform to support configuration design, production planning and enterprise systems integration between customers and companies.

However in recent years, further advancements in web and IT technologies have impacted the development of configuration research. Recent studies have adopted the application of service-oriented principles to connect customers and suppliers across a customised product supply chain. Ma et al. (2008) proposed a web-service oriented system to better support customisation and dynamic update in electronic product catalogues. The feasibility of a service-oriented architecture (SOA) approach was investigated and implemented for product customisation in the shoe industry (Dietrich at al. 2007).

Today, many of the current research on configuration are motivated by a wider context to address the lifecycle of customised products and to achieve the seamless integration of order fulfilment activities (Zhang et al. 2010a, b; Helo et al. 2010). The ability to match manufacturing capabilities with customer needs; management of dynamic product data and integration with production and supply chains represent some of the present research challenges that should be addressed in today's product customisation environment (Yip et al. 2011). With the advancement of IT technologies, these play an increasingly important role in the development of next generation product configurators.

2.3 Build-to-Order Strategy

The most important supply chain management method with regard to product customisation is to classify the manufacturing of products with regard to the position of the order penetration point. In a spectrum that ranges from make-to-stock (i.e. mass production), assembly-to-order (i.e. limited customisation based on pre-defined components) and build-to-order (i.e. manufacturing a product completely after receiving the order which enables it's build it according to a customer's specification) (Schönsleben 2007). Gunasekaran and Ngai (2005) state that the application of advanced IT technologies to integrate supply chain participants and support operations management across them is an essential pre-condition in order to achieve successful built-to-order strategies. However, there is no common IT concept or architecture yet proposed to enable this.

2.4 Virtual Enterprises

However, defining a customisation strategy for a supply chain is in most cases not sufficient. Especially when it comes to flexible composition of supply chains based on the fact that suppliers have to be integrated depending on product component selection, additional concepts like virtual enterprises become relevant.

Virtual enterprises, also called virtual organisations are according to (Schönsleben 2007), supply chains that are established from a production network for the manufacturing of a singular product/fulfilment of one order and which are dispersed after the delivery of this product. For the next order which will be received from a customer, a new virtual enterprise (with possibly different contributors) is established in the production network. During customer's interaction, the virtual enterprise acts towards the end-customer as one organisation. The main reasons for organisations to join a virtual enterprise are an increased flexibility and adaptability to environmental changes (including customers' needs), access to a pool of competencies and resources (in order to concentrate on the own core competencies) and the achievement of a critical organisation size (in order to meet market constraints) (Martinez et al. 2001).

As virtual enterprises aim at maximum flexibility with regard to the fulfilment of customer orders by integrating independent organisations on-demand, various interests have to be considered common priorities to be found during their management. The complexity of harmonising distributed production processes, aligning financial interests and demands, fulfilling intellectual property requirements, etc. all at the same time is daunting. For this reason, the application of decision support systems or decision-making systems frequently arises in the field of virtual organisations. Those decision support systems depend on the type of the applied control concept (hierarchical or distributed). Especially for distributed, non-hierarchical controlled systems, the approach of finding solutions which are appropriate for all participants can be challenging. Due to this fact, decision support systems are usually implemented as complex IT systems which, e.g. apply multi-agent system approaches, self-learning algorithms like neural networks, etc. (Tuma 1998).

2.5 Information Exchange in Production Networks

There are already many ideas, approaches and concepts which describe how to apply and manage services and related information within manufacturing networks. For example, Yan et al. (2010) and Lilan et al. (2007) describe the synergies which could result from the transfer of Service Oriented Architectures (SOA) to the manufacturing domain due to the fact that manufacturing service concepts focus on resource sharing and SOA provides concepts for distributed architectures, which are to be considered as enablers for interoperability and extendability, easy integration and simplicity.

Current research efforts are focusing on the development of appropriate descriptions for manufacturing services. Many of them have chosen the approach of extending web services and related software technologies such as Web Service Description Language (WSDL), Universal Description, Discovery and Integration (UDDI), Darpa Agent Markup Language (DAML) and Web Ontology Language for Web Services (OWL-S) as these technologies in their pure form are not sufficient for the implementation of manufacturing service descriptions (Jang et al. 2008). Consequently, the enrichment of service descriptions by means of ontologies is an approach which many researchers in this field have chosen. For example, various ontologies have been proposed in order to represent knowledge about manufacturing services (Lemaignan et al. 2006; Hu et al. 2009; Ameri and Patil 2010). A further example is by Cai et al. (2011), who added ontologies to web services via annotations. These approaches also provide the basis for the management of services through service search, match-making, dependency management, discovery, selection and composition, service granularity and aggregation, and network structures and performance evaluation (Huang et al. 2011).

However, for the management and execution of manufacturing services within a flexible production network, this is not sufficient as e.g. costs, capacities and logistics are not considered in detail. For this reason, generic service descriptions have been proposed such as Unified Service Description Language (USDL), which include the description of technical, economic, and operational aspects such as product and service categories, pricing models, business conditions, availability, etc. (Kursawe 2010). Due to USDL's manual usage and lack of integration with existing manufacturing systems, this approach requires further extension in order to achieve manufacturing service management and execution within flexible production networks.

Furthermore, there is also a wide range of commercial supply chain management tools (e.g. SAP-based applications) in use, which focus on acceleration and robustness of supply chain processes and supplier integration (Knolmayer et al. 2002). One essential precondition for their successful application is the electronic exchange of business documents like orders and invoices. Several standards exist such as EDI (Electronic Data Interchange), GS1-related standards (e.g. GDSN) and other standardisation efforts like Universal Business Language (UBL), Electronic Business using XML (ebXML) and Business Process Execution Language (BPEL).

All of the above supply chain management and manufacturing service description approaches have in common is that they address a subset of the general requirements for manufacturing services and their management (ranging from product- and production related content up to business level information). One exception is RosettaNet which goes beyond sharing of business-level data as it includes Partner Interface Processes (PIPs) not only for order, inventory, and marketing information and management but also for manufacturing (e.g. for exchange of design, configuration, process, quality) (Chong and Ooi 2008). However, this standard is complex to implement and does not provide the ability to build up production networks flexibly by means of manufacturing service aggregation.

3 Manufacturing-as-a-Service Overview

The preceding section has shown that there are various technical challenges to be addressed to enable product customisation within a production network environment. A significant challenge exists in managing the information flow and coordinating the interactions between the customer and the different manufacturers and suppliers of a supply chain. This demands a flexible infrastructure that can connect customisation and production IT systems, in order to address the complete manufacturing lifecycle from product specification to production and transportation.

In adapting cloud computing principles to the manufacturing domain, the new paradigm of MaaS is proposed to better support the on-demand manufacturing scenarios (Meier et al. 2010; Rauschecker et al. 2011). The motivation is to achieve a highly flexible production network that integrates distributed factories and their manufacturing capabilities in an overall cloud manufacturing infrastructure to enable the collaborative specification and manufacture of customised products. This is supported by the application of virtual manufacturing services and their corresponding descriptions, which provide the basis to utilise the available manufacturing capabilities to manage configurable production networks for enabling product customisation.

Based on the paradigm of MaaS, the manufacturing resources and capabilities of a production facility are transformed from physical resources and encapsulated as virtualised manufacturing services by means of a mapping mechanism. In this process, the manufacturing capabilities can be abstracted at different hierarchical levels, from the equipment/tools and station/lines to the factory level, as manufacturing service descriptions and then published to a MaaS infrastructure. Within the MaaS environment, users have centralised access to search, manage and execute manufacturing services that are directly provided from distributed locations. A manufacturing service can represent the provision of a final product, sub-components or the capability of a production process.

A key principle of MaaS is the aggregation of individual manufacturing services to establish configurable virtual production networks for complex products. In other words, integrating individual manufacturing services from different suppliers (i.e. sub-products) allows further complex and value-added manufacturing service (i.e. assembled end-products) to be provided to end-users of the MaaS infrastructure. The result is a parent manufacturing service description composed of interlinked child service descriptions, which represent the structure of a configurable production network for a specific product. As product characteristics, configuration options and dependencies at the sub-product level are aggregated into the parent service description, the respective production network is able to provide end-products with available customisation options to end-users. This is also supported by the definition of 'product templates', which provide a starting point to configure available products.

As a product is configured and submitted to the MaaS infrastructure, the resulting product configuration is linked to the aggregated manufacturing service and transformed as a customer order to be processed by the production network. The customer order is checked for availability with confirmation of pricing and lead times to ensure that it can be produced. With no constraints violated, the order is forwarded to the respective organisation of the production network for manufacturing service execution to enable customer order fulfilment. Overall, the MaaS infrastructure supports the execution and control of processes at both the business and production level.

3.1 MaaS Stakeholders

The proposed concept of the MaaS infrastructure for supporting customised products is shown in Fig. 2 and can be illustrated by the description of the following target user groups:

- *Service Providers* are parties with manufacturing facilities who would like to deliver products or production services (processes) to the manufacturing cloud by defining manufacturing services and their descriptions that they can offer within the cloud manufacturing environment. They represent manufacturers or suppliers who are motivated in the benefits of scaling their production load or to strengthen their market position providing additional information and flexibility to their customers, such as in terms of product specification and scheduling of delivery dates.

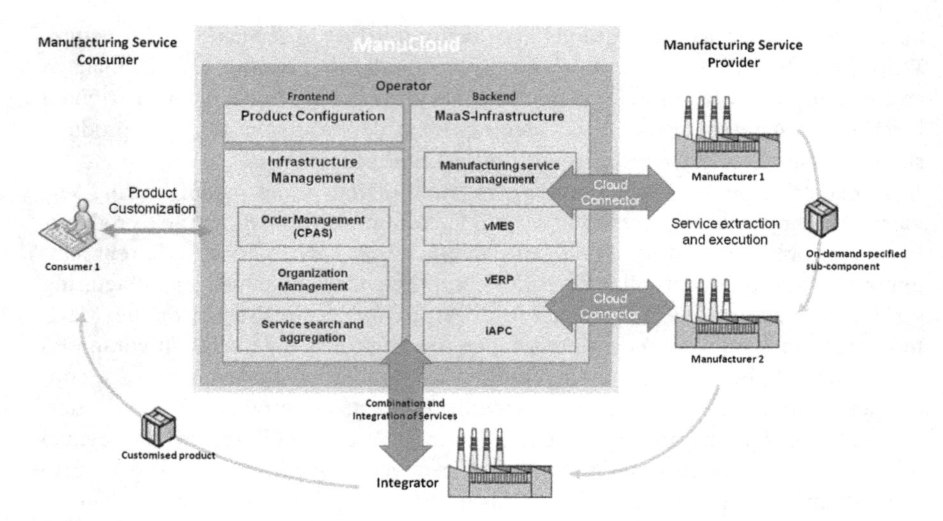

Fig. 2 Manufacturing-as-Service (MaaS architecture and components)

- *Cloud Consumers* represent end-users or customers who want to purchase a product or sub-component but are not able to manufacture the products by themselves due to lack of facilities, know-how or cost restrictions. Cloud Consumers are able to customise products and also services provided by available manufacturing services in the MaaS environment. Examples of this group are product designers or engineers who could bring new concepts and developments into industrial production without having direct access to manufacturing expertise, capabilities, or capacities. In addition, they would be able to verify the manufacturability of new designs from the manufacturing cloud.
- *Service Integrators* have a key role in the manufacturing cloud, involving the composition of available manufacturing services to define a certain type of product, which can be further specified by customers by determining the provided configuration options. By doing so, the related production network is automatically assembled. This enables them to effectively manage the supply chain and to adapt it, based on the options provided by the services. In effect, the Service Integrator is a combination of both roles as a Manufacturing Service Provider and Cloud Consumer.

3.2 MaaS Architecture

The architecture of an appropriate MaaS Infrastructure can be derived from functional and non-functional requirements gathered from several representative use cases and are described in the following (Rauschecker et al. 2013):

- *Register organisation.* A pre-requisite for publicising manufacturing services is to register the organisation via the manufacturing cloud portal by giving general business details and their verification.
- *Connect factory.* On successful registration/verification, a cloud-specific mapping and connection tool can be used to establish a link between the factory internal IT systems and the cloud platform.
- *Publish manufacturing service.* Factories then publicise their manufacturing services to the cloud as service providers. The description of the manufacturing services at factory level is transformed into the manufacturing service description applicable in the cloud by means of an appropriate mapping tool. The service description is made up of various aspects (product information such as product parameters and product visualisation, process parameters, financial aspects such as price conditions and cost models, logistic parameters and also descriptions of service levels and the contractual conditions to be observed), which are validated during the publication process in the cloud.
- *Aggregate manufacturing service.* The pool of manufacturing services can now be aggregated by the service integrators to form higher value manufacturing services. The aim is either to publicise them as manufacturing services or to offer an end-product directly via a marketplace. If a product or process idea exists, the cloud is searched for the corresponding manufacturing services required to realise it. The services chosen are then aggregated to form new manufacturing services. The parameters of the new manufacturing services may depend on the parameters of the manufacturing services selected from the cloud. These dependencies must be described by the service integrator in the form of so-called validation and calculation expressions. To complete the aggregation process and once the decision has been made to supply an end-product, if required the service provider can draw up a product website directly with a product configurator and link it to the marketplace.
- *Configure product.* End customers use the cloud marketplace to look for products and browse through catalogues and category trees. If a product is chosen which has a product website with a configurator, the customer is guided to that site before carrying out the order process. Product configurators show the adaptations made by customers directly in a 3-D model of the product. The price and delivery times of the products are determined in dependence upon the configuration based on simulations using filed logistics parameters and cost models and displayed provisional values. Configurations can be saved and made available to other users in the form of templates. Product suppliers may also provide templates for the product configuration.
- *Set up supply chain.* On the basis of their configuration, end customers can order products via the marketplace. Part of the supply/process chain is already defined by the configuration made. Based on further criteria, such as the various cost models of the manufacturing service alternatives, the system can now ascertain the remaining links in the chain. Using the resulting supply/process chain, quotes can be obtained from the associated suppliers of the manufacturing

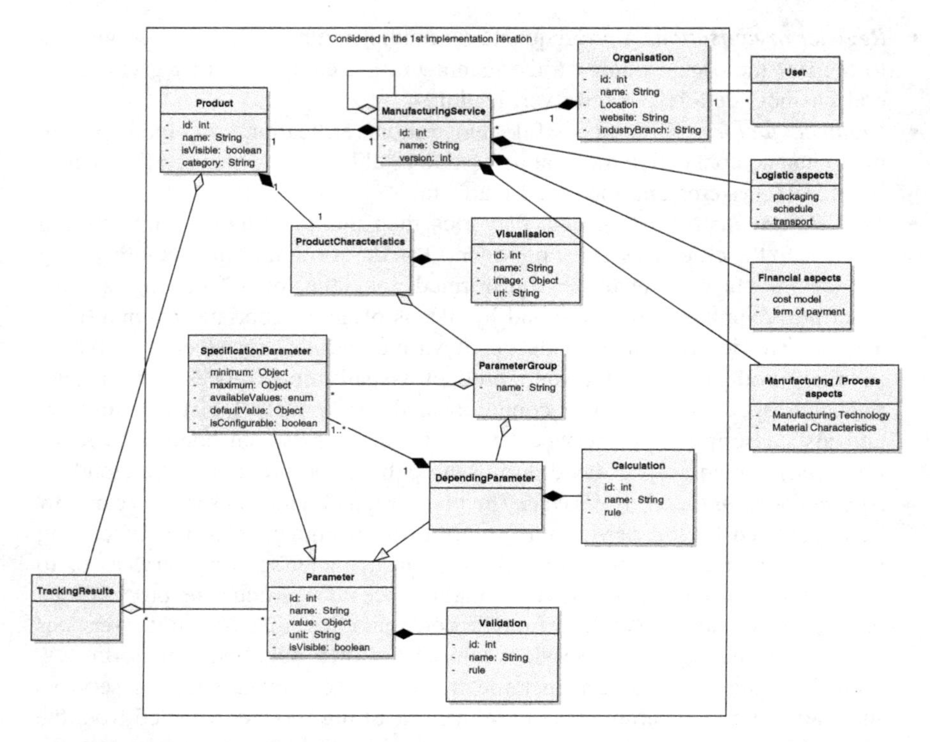

Fig. 3 Manufacturing service description structure

services concerned (including specific delivery times, prices, conditions of payment, etc.) before confirming a certain order.

- *Manufacture product*. The product is then manufactured along the defined supplier/process chain according to the specifications made during product configuration. Taking into account the sequence of the factories involved in the process, the manufacturing orders are distributed at cloud level among the various connected factories and their internal IT systems. Aggregated tracking data is used to monitor execution of the orders. If problems arise during the manufacturing process, these are also passed on to the cloud level of the factories. One feasible concept would be to exchange a manufacturing service during runtime for an alternative manufacturing service. In order to provide those features, components are needed in the manufacturing cloud, which implement a sub-function of a factory-level manufacturing execution system.

In order to fulfil the identified use cases, an architecture for the MaaS infrastructure has been implemented as shown in Fig. 2. Its main architectural components are as follows:

- A front-end comprises of a web-portal which provides interfaces for product customisation, order management, and utilisation and composition of manufacturing services. Additionally, it includes general features for the management of the infrastructure such as user and organisation administration. The core of this component is a web portal, to which specific user interfaces are integrated as portlets.
- A back-end provides middleware services for the management of manufacturing services and related functionalities like the management of product configurations, order management, multi-site manufacturing execution, and product tracking. This component consists of several sub-layers responsible, which include data consistency, business logic and web service provision.
- A cloud connector component provides the mapping interface for the integration of factory-level IT systems such as ERP (Enterprise Resource Planning) or MES (Manufacturing Execution System) to support manufacturing service extraction and execution tasks.

The manufacturing services and their descriptions which are handled by these architectural components are discussed in more detail in the following subsections.

3.3 Manufacturing Service Descriptions

The aim of the proposed MaaS approach is to provide a common formal description for manufacturing services which supports the management of supply chains, production networks, and virtual organisations on all levels. Based on the research described in (Rauschecker and Stöhr 2012), it has been concluded that the manufacturing services must be described so they can be:

- *Extended.* The structure of manufacture services should be extendable in order to not restrict their usage to the integration of pre-defined manufacturing facilities. This includes the representation of various aspects in their descriptions which range from product characteristics; quality constraints to manufacturing process specific details; organisational information; financial aspects; and logistics information.
- *Combined.* The aggregation of manufacturing services must be possible to establish and manage production networks that are represented by the structure of interlinked manufacturing services and also act as virtual organisations.
- *Configured.* Manufacturing services should be configurable and also provide configuration options based on the product options (e.g. parameters) that are made available by the respective manufacturing facilities.
- *Manufactured.* An important requirement is that the options provided by manufacturing services are manufacturable by the expressed resources and capabilities. Therefore, matching, validation, and calculation mechanisms must be included to represent the respective interrelationships among product parameters. Also, the interdependencies have also to be enabled across singular

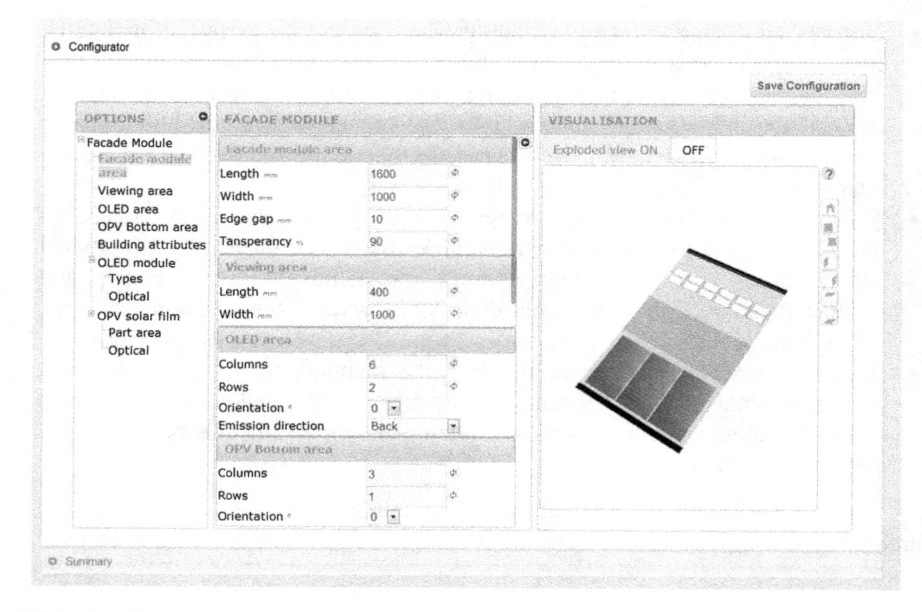

Fig. 4 Product configurator interface

services in order to support the assessment of producibility for aggregated services.

- *Owned*. Manufacturing services should include a static interlinkage to their sources in order to reason and respond to customised product queries such as delivery dates and costs.
- *Minimal*. Manufacturing service descriptions should include the minimum level of information required in order to avoid unnecessary administration efforts and therefore make the integration and aggregation of services applicable for a wide range of user groups.

In consideration of these requirements, the main structure and elements of the proposed manufacturing service description is represented in Fig. 3. At the top level, a manufacturing service is composed by the description of different service aspects. The definition of a product is described by the means of *ProductCharacteristics* with *Parameters* as their basic elements, and the inclusion of a *Visualisation* aspect such as a 2D/3D CAD file.

The product parameters can be summarised to *ParameterGroups* such as geometrical, electrical and optical characteristics. When providing customisable options for a product, this is supported by the introduction of *Specification* and *Depending* parameters. The former can be defined to a user-specific parameter value and the latter is calculated by other parameter values. Dependencies between parameters are represented by *Validation* and *Calculation* expressions that define the necessary rules and constraints to enable product configuration and service aggregation tasks.

With regard to the description of the *Manufacturing/Process* aspect, the service structure can be extended to support ontologies, which allow the representation of domain-specific manufacturing process knowledge and relationships. This is to allow the application of service descriptions in different industries such as automotive or consumer electronics

In addition to the representation of product and manufacturing-related information, the following service description aspects are also considered:

- *Organisational aspects*. Basic details of an organisation that is necessary for registration and participation within the manufacturing cloud.
- *Financial aspects*. Description of cost models, discount conditions and terms of payment that are associated with the provision of manufacturing services.
- *Logistics aspects*. Information of delivery dates, location of facilities, transport routes, packaging that are necessary to support production network configuration.

For information interchange within MaaS infrastructure, the manufacturing service structure is automatically transformed via a web service interface into XML structures. The web service provides all the necessary operations and uses SOAP for message encapsulation. All the service elements and web service operations are described in a WSDL file, which include XSD elements to represent the service and aspect entities.

The main architectural components within the MaaS infrastructure and their core functionalities to enable product customisation are described in further detail in the following sections.

3.4 Managing Manufacturing Clouds Based on Service Descriptions

This section describes the features required of cloud-based manufacturing systems based on service descriptions. Essentially the provision of products in the production network is based on manufacturing services, those services and their descriptions—as well as aggregated service structures. Consequently, these have to be managed within an overall cloud manufacturing infrastructure in order to make them available for production network participants and customers.

First of all, it has to be ensured that all manufacturing services are stored in a consistent and persistent way which is accessible from the whole network, as soon as access rights allow this. The data model has to correspond with the concept of aggregating manufacturing services e.g. to enable production of complex assembled products. Even more, it has to provide the ability to reuse a manufacturing service within various high-level services. For this reason, the data structure of manufacturing services has to provide linkage points to be used for integrating references e.g. to sub-services. However, complex service structures may also make service search and browsing quite complex or under-performing. For this reason, it is necessary to not only put effort into appropriate data models, but also into the selection of suitable IT technologies which can range from the application of cloud computing technologies to the usage of e.g. graph-based databases which provide advantages for interlinked data structures.

In addition to storing manufacturing services and their aggregated structures, a MaaS platform has also to provide the ability to update, disable, search, and aggregate them. One of the most challenging topics of manufacturing services is their aggregation to complex structures. To do this, it is necessary to define the dependencies among manufacturing services and the manner in which their characteristics are inherited or constrained during their aggregation (for example consider a window which must not be larger than the frame to which it is assembled). Furthermore, it has to be ensured that the services which are intended to be aggregated are compatible (e.g. it may not be possible to execute a certain coating process on an inappropriate material). Finally, consistent manufacturing service aggregation also requires features like a version management in order to ensure that a high-level manufacturing service does not refer to an obsolete sub-service. The above properties of aggregated manufacturing services should be visible to customers through a user interface (generated semi-automatically) that allows their configuration for specific products.

The configuration of the network for production of specific products must also support the planning of processes and product routes throughout the production network. Effectively this is equivalent to selecting a path through the network by determining where a product will be processed (in cases where the same or similar service are offered by more than one production site) based on decision criteria like transport durations and costs.

Additionally, this path may have to be updated online according to (unforeseen) events like delivery delays or changes of priority. However, it is not only scheduling in the supply chain. Orders which are received from customers for concrete products have to be forwarded in the form of sub-orders to the identified service contributors. During the execution phase, products and service progress have to be tracked and traced in order to monitor the current status of the overall manufacturing process by means of time stamps, product or service status information.

However, these functional features for manufacturing service management are not sufficient to achieve acceptance from users (end-customers and manufacturing service providers). The most important requirement is the implementation of security and privacy throughout various layers which range from message level security and secure access to user-, role-, and organisation-specific access to data.

3.5 Product Configuration Based on Configurable Service Descriptions

In the manufacturing cloud, the customisation of products is realised by the integration of a product configurator as part of the front-end to the MaaS environment. The development is a generic system that can automatically adapt to the available manufacturing capability of the production network to enable product configuration based on manufacturing services (Yip et al. 2013).

Based on the concept of MaaS, the definition of manufacturing services also includes the description of configuration options for the customisation of provided end-products. In utilising the underlying service descriptions, the proposed approach provides end-users with a level of product customisation that is directly integrated and supported by the available manufacturing capabilities of the production network. The result is a configurable product configurator whose structure can be dynamically generated by manufacturing services to automatically reflect new capabilities within the configuration options available in the configurator interface.

At the front-end, customers can browse and select from default product configurations which represent available customisable end-products offered by the manufacturing service integrator/provider in the manufacturing cloud. As an end-product is selected for configuration, a manufacturing service request is sent by the product configurator to back-end of the manufacturing cloud. In response, the corresponding product-related data (i.e. product aspects) of the manufacturing service description is loaded into the configurator to initiate the start of a new customer configuration. This provides the underlying configuration knowledge-base (i.e. product parameters, values, configuration rules and constraints) and additional product characteristics (e.g. visualisation aspect) to enable the system to be dynamically generated at run-time. One key advantage is that the configurator can be easily adapted to different end-products provided in the MaaS environment based on only the given manufacturing service description.

As a default configuration is loaded, the configurator graphical user interface is dynamically generated according to the selected end-product with the corresponding 2D/3D product visualisation, as shown in Fig. 4. In the user interface, the customer can navigate and specify the available configuration options to individual requirements using the tree view and input fields, respectively.

For solving the configuration task, the product data from the service description provides the underlying configuration model input for the implemented configuration engine. To extend support for different product configuration problems, the configurator can be implemented with different solving technologies such as rule-based or constraint-based solvers. This allows flexibility in selecting a specific solver that best suits the complexity of the product configuration and requirements of the product domain.

During the configuration process, the product visualisation can be dynamically updated and manipulated interactively according to the configured options selected by the customer. This is achieved by mapping of the configuration parameters with the visualisation model within the manufacturing service description. Furthermore, the system also supports the functionality to provide design assessment of the product during configuration. This includes the calculation of output product parameters and the generation of product feedback, for example, if certain product constraints are violated during the customer configuration.

Lastly, the output of the product configurator is the generation of a valid product specification by configuring the available customisation options supported by the system. The final completed customer configuration can then be reviewed and submitted to the MaaS back-end for the next stage of order processing of the configured product.

3.6 MaaS Integration to Factory-IT Systems

An important aspect of MaaS is the integration of distributed production sites and their shop floor capabilities to the manufacturing cloud by the connection of internal factory-IT systems. Depending on the level of integration and automation at each factory, manufacturing service description data can be published by several methods. In a fully integrated environment, the automatic extraction of service descriptions from factory-IT systems can be achieved. This is followed by a semi-automatic approach, which requires manual confirmation of the extracted service data. Without a direct connection to the factory-IT systems, manual generation of service descriptions is required using a provided web-interface. To support this, the proposed concept to extract and map parameters throughout all automation levels is illustrated in Fig. 5.

On the equipment level, it is intended to establish equipment self-descriptions, which are used to support easy integration to factory level IT systems like MES (Manufacturing Execution System). This includes parameters and events that can be used to communicate with the tools; information about services (low-level processes), which are supported by them and also the provided parameter range for their configuration.

On the MES/Process level, the equipment descriptions are aggregated to whole process flows, again specified by services they provide and the respective parameter ranges. From these processing services, manufacturing services (product level) are extracted via the cloud connector interface with the mapping of parameters in order to achieve an appropriate granularity. As a result, manufacturing services will not only provide information regarding the capability to manufacture a certain product but include configurability for this product, as far as the respective production facilities allow it.

After the manufacturing services have been published, aggregated and configured by a user, an order is generated for the manufacture of the required product at the production site via the cloud connector. Using this interface, tracking and tracing information is reported back to the cloud manufacturing system in order to monitor the overall manufacturing status of all products in the production network. For measurement and status of production, the level of granularity for the reporting is dependent on how the parameters are mapped among the different layers, from the manufacturing cloud down to equipment level.

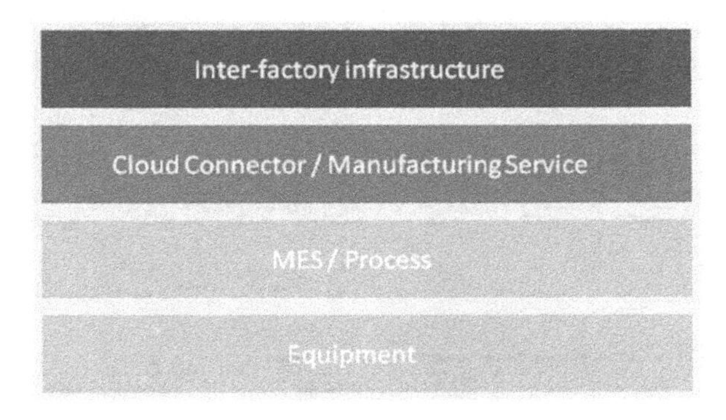

Fig. 5 Extraction and mapping of manufacturing service descriptions

4 Case Study

To validate the proposed architecture and MaaS concept, a reference implementation for the customisation of products in a manufacturing cloud was set up for organic semiconductor products. The work is motivated by the increasing demand for high-value, low-volume customised products in the competitive European organic LED (OLED) and organic photovoltaic (OPV) industries. The example in this case is the growing market for customer-specific lighting solutions and façade elements for building integration.

Figure 6 gives an overview on the setup of a customisable solar and lighting façade module selected for the case study. The façade module consists of OLED and OPV elements, which can be integrated together to provide new degrees of customisation freedom for architects, system designers and façade constructors.

Based on the proposed MaaS architecture, a manufacturing cloud environment was setup with prototype systems of the front-end product configuration, manufacturing service management and cloud connector interfaces developed using Java and Web Service technologies. For the study of façade module customisation and production, the following implementation process was identified according to the use cases analysed in Sect. 3.2:

- *Product and production analysis.* Firstly, an analysis of production processes and product portfolio is conducted by the Manufacturing Service Providers (i.e. OLED and OPV manufacturer). The task is to select the manufacturing capabilities to be provided as services and determine the available customisation options to be offered when providing a product to customers. Examples for those options from a customer's view are product parameters relating to shape, length, and width of OLED and OPV elements as well as their colour and transparency.
- *Service extraction and aggregation.* Afterwards, the OLED and OPV service descriptions were extracted automatically from the internal factory IT systems by installed cloud connectors at each production site and published to the MaaS architecture. From composition of OLED and OPV services, an overall façade

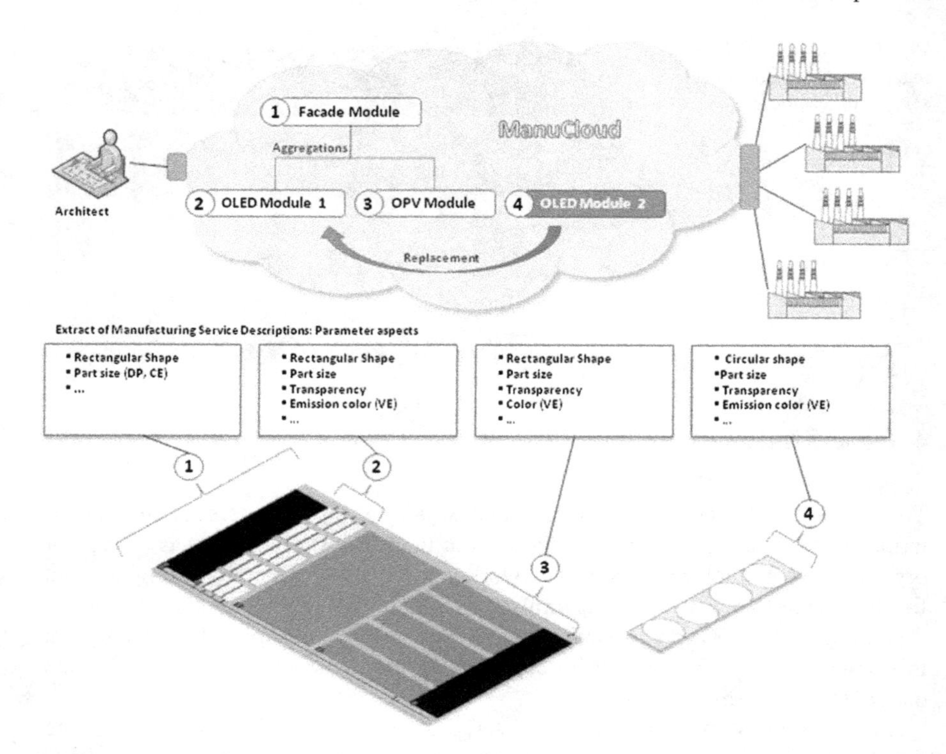

Fig. 6 Customisation of façade module overview

module service description is generated. The resulting aggregated manufacturing service contains all the necessary information to enable configuration and manage the production network of the façade module, shown in Fig. 7.

- *Configure façade module*. At the front-end web portal (refer also to Fig. 4), the product configurator is dynamically generated from the façade module service description. From the user interface, an architect is able to configure a range of available customisable product options such as façade module size, shape, colour of the OLED and OPV elements. During configuration, the architect can obtain feedback on the final product based configured parameters and their dependencies. For example, error messages are generated when configured size of OLED or OPV element exceeds the limits specified for the façade module. Similarly, it is also possible to calculate the energy consumption or efficiency of OLED or OPV elements from their size and transparency.
- *Route façade module through the production network*. After customisation, the configured product instance of the façade module is sent to the back-end of the MaaS architecture for process planning. An overall production plan is generated from information extracted from the service descriptions, which include production process, logistics and financial aspects. For instance, the delivery date is calculated from the capacity and lead time of the module lamination

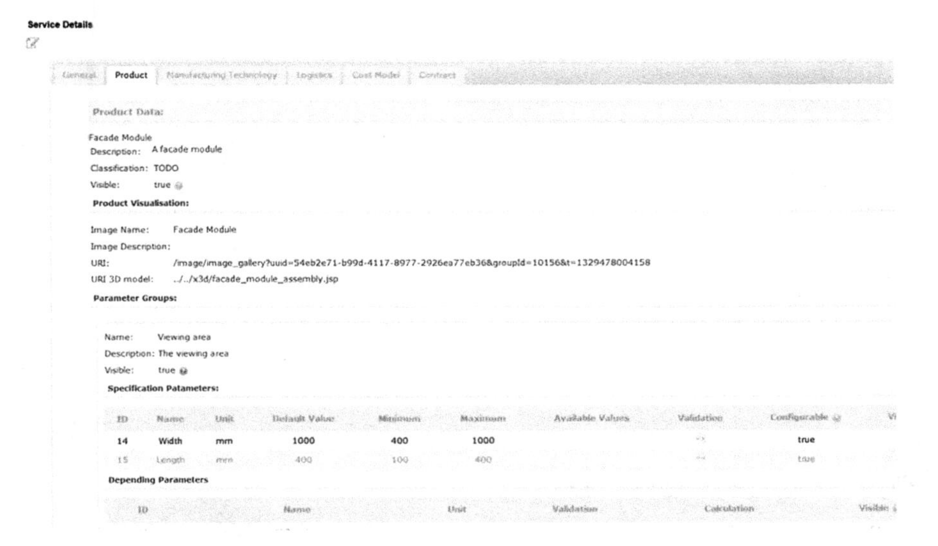

Fig. 7 Façade module service definition

process; the transportation time based on location and logistics information and lastly, the capacities and lead times for the OLED and OPV manufacturing (Fig. 8). Orders are dispatched to production network participants according to the generated plan.

- *Execute façade module production*. From the production plan, the configured product parameters are mapped back to process and equipment parameters using the cloud connector interface to drive production execution. For example, a certain colour specified for an OLED element is mapped to the individual organic material required during its manufacturing. At the same time, tracking data is collected and provided to the MaaS infrastructure to be used for quality assurance such as guarantee estimations or as additional input for subsequent process steps at the different OLED and OPV production sites.

An additional scenario evaluated in the case study was the replacement of subservices in the structure of the manufacturing service descriptions. This was successfully demonstrated with the substitution of OLED Module 2 subservice into the aggregated service description, shown in Fig. 6, to enable customisation and execute the production of the façade module in the manufacturing cloud.

4.1 Discussion and Industrial Feedback

From the overall demonstrator system and execution of the façade module production, the impact of product customisation in cloud manufacturing was validated by the industrial partners involved. This was evaluated by considering responses from the industrial partners with regards to:

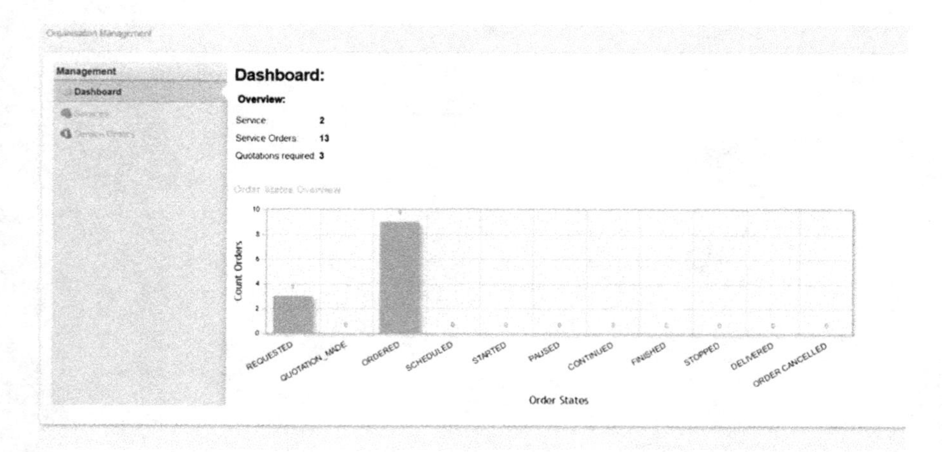

Fig. 8 Management of product orders

- Company activities and industry branch
- Degree of customisation of products
- Relevance of cloud manufacturing concepts for the company
- Potential roles the respective company could cover
- Expectation and opportunities, the companies see when considering a participation in the ManuCloud platform

The overall feedback regarding the expected impact highlighted the point that the measurement of profitability of the ManuCloud application cannot easily be done by means of financial figures. In fact, non-financial indicators are most relevant for the estimation of the benefits of the platform. The most relevant indicators and examples for potential applications which were identified are:

- Access to a larger group of potential customers and also to new customer groups by providing customised products.
- Using the cloud platform as a collaboration tool for associations of SMEs which jointly developed new product ideas/provide them to customers.
- Using the platform (manufacturing service description repository) as a knowledge database for industry clusters, etc. in order to enable fast search for capabilities of potential collaboration partners.

Furthermore, feedback for potential improvements of the platform and functionalities were collected from the industrial users. A main point was the complexity of manufacturing service descriptions and the effort required to generate them. Whilst the platform has been optimised with regard to usability of the manufacturing service creation and editing, there remains the issue of dealing with the large numbers of parameters and configuration options due to the complexity of the product involved.

In summary, the platform was regarded to be beneficial for company associations which have no existing supply chain management and wish to jointly provide customisable products in a production network. They should be supported by the platform provider by integrating their manufacturing capabilities/describing their services. The financial benefit of the platform usage under these conditions cannot be estimated since it strongly depends on the kind of products which is sold via it and the business models related to the platform operations.

5 Conclusion and Outlook

This chapter has described an approach and architecture for cloud manufacturing based on the concept of developing a MaaS version of the XaaS approach (X-as-a-Service). The resulting model, generated by a broad consortium of industrial and academics stakeholders, has considered the requirements and challenges of every layer of the system and demonstrated how shop floor capabilities can be expressed and propagated through a production network to be utilised by customers. A case study has demonstrated the feasibility of the approach for one product domain and has established the viability of a MaaS system architected around Manufacturing Service Descriptions to enable product customisation.

Further work is required and will address the following challenges:

- Further development of the prototype systems to support detailed product tracking, versioning and security.
- Investigate the scalability of the architecture to support larger production networks involving integration of multiple manufacturers and suppliers and consequently.
- Evaluate the aggregation of more complex manufacturing services consisting of further parameters and dependencies.
- Simplification of manufacturing service descriptions, i.e. their creation, management, etc.
- Application of MaaS in different product domains (e.g. 3D mechanical products).

This work has offered a vision of a technology with the potential to make cloud manufacturing and the concept of MaaS as ubiquitous and established as cloud computing applications such as email. Beyond the technology of languages and protocols, there is the possibility of cloud computing enabling a manufacturing system many times more responsive than today's and so forming a cornerstone of a new industrial revolution.

Acknowledgments Parts of the work described in this section as well as the case study have been executed within the ManuCloud project, which is funded by the European Commission grant agreement FP7-260-142.

References

Aldanado M, Hadji-Hamou K, Moynard G, Lamothe J (2003) Mass customization and configuration: requirements analysis and constraint based modeling propositions. Integr Comput Aided Eng 10(2):177–189

Ameri F, Patil L (2010) Digital manufacturing market: a semantic web-based framework for agile supply chain deployment. J Intell Manuf doi:10.1007/s10845-010-0495-z

Blecker T, Friedrich G (2007) Guest editorial: mass customization manufacturing systems. IEEE Trans. Eng. Manage. 54(1):4–11. doi: 10.1109/TEM.2006.889063

Cai M, Zhang WY, Zhang K (2011) ManuHub: a semantic web system for ontology-based service management in distributed manufacturing environments. IEEE Trans Syst Man Cybern—Part A: Syst Hum 41(3):574–582. doi: 10.1109/TSMCA.2010.2076395

Chong AY, Ooi K (2008) Adoption of interorganizational system standards in supply chains: an empirical analysis of RosettaNet standards. Ind Manage Data Syst 108(4):529–547. doi: 10.1108/02635570810868371

Davis SM (1987) Future perfect. Addison-Wesley, Reading

Dietrich AJ, Kirn S, Sugumaran V (2007) A service-oriented architecture for mass—a shoe industry case study. IEEE Trans Eng Manage 54(1):190–204

Felfernig A (2007) Standardized configuration knowledge representations as technological foundation for mass customization. IEEE Trans Eng Manage 54(1):41–56

Fogliatto SF, Da Silveira G, Borenstein D (2012) The mass customization decade: an updated review of the literature. Int J Prod Econ 138(1):14–25

Gunasekaran A, Ngai E (2005) Build-to-order supply chain management: a literature review and framework for development. J Oper Manage 23(5):423–451. doi:10.1016/j.jom.2004.10.005

Helo PT, Xu QL, Kyllonen SJ, Jiao RJ (2010) Integrated vehicle configuration system—connecting the domains of mass customization. Comput Ind 61:44–52

Hu Y, Tao F, Zhao D, Zhou Z (2009) Manufacturing grid resource and resource service digital description. Int J Adv Manuf Technol 44(9–10):1024–1035. doi: 10.1007/s00170-008-1899-1

Huang S, Zeng S, Fan Y, Huang GQ (2011) Optimal service selection and composition for service-oriented manufacturing network. Int J Comput Integr Manuf 24(5):416–430. doi: 10.1080/0951192X.2010.511657

Huang B, Li C, Yin C, Zhao X (2012) Cloud manufacturing service platform for small-and medium-sized enterprises. Int J Adv Manuf Technol 65(9–12):1261–1272

Jang J, Jeong B, Kulvatunyou B, Chang J, Cho H (2008) Discovering and integrating distributed manufacturing services with semantic manufacturing capability profiles. Int J Comput Integr Manuf 21(6):631–646. doi: 10.1080/09511920701350920

Jiao J, Helander MG (2006) Development of an electronic configure-to-order platform for customized product development. Comput Ind 57(3):231–244. doi: 10.1016/j.compind.2005.12.001

Knolmayer G, Mertens P, Zeier A (2002) Supply chain management based on SAP systems: order management in manufacturing companies. Springer, Berlin

Kursawe R (2010) Evaluierung der modellierung realer dienstleistungen in der Dienstbeschreibungssprache USDL. http://texo.inf.tu-dresden.de/publications. Accessed 13 Feb 2012

Lemaignan S, Siadat A, Dantan J, Semenenko A (2006) MASON: a proposal for an ontology of manufacturing domain. In: IEEE Workshop on Distributed Intelligent Systems: Collective Intelligence and Its Applications (DIS'06), pp 195–200

Lilan L, Yuan H, Tao Y, Zonghui X (2007) Research on SOA-based manufacturing grid and service modes. In: Sixth International Conference on Grid and Cooperative Computing (Gcc 2007), pp 608–613

Liu N, Li X (2012) A resource virtualization mechanism for cloud manufacturing systems. In: Enterprise Interoperability: proceedings of the 4th International IFIP Working Conference (IWEI 2013), vol 122, pp 46–59

Luo Y, Zhang L, Tao F, Ren L, Liu Y, Zhang Z (2013) A modeling and description method of multidimensional information for manufacturing capability in cloud manufacturing system. Intern J Adv Manuf Technol 69(5–8):961–975

Ma Y, Jiao J, Deng Y (2008) Web service-oriented electronic catalogs for product customization. Concurrent Eng 16(4):263–270. doi: 10.1177/1063293X08100026

Martinez M, Fouletier P, Park K, Favrel J (2001) Virtual enterprise—organisation, evolution and control. Int J Prod Econom 74(1–3):225–238. doi: 10.1016/S0925-5273(01)00129-3

Meier M, Seidelmann J, Mezgár I (2010) ManuCloud: the next-generation manufacturing as a service environment. ERCIM News 83:33–34

Mell P, Grance T (2011) The NIST definition of cloud computing. NIST Spec Publ 800(145):1–7

Ong SK, Lin Q, Nee AYC (2006) Web-based configuration design system for product customization. Int J Prod Res 44(2):351–382

Rauschecker U, Meier M, Muckenhirn R, Yip ALK, Jagadeesan AP, Corney J (2011) Cloud-based manufacturing-as-a-service environment for customized products. In: Cunningham P, Cunningham M (eds) eChallenges e-2011 Conference Proceedings. IIMC International Information Management Corporation, Dublin, Ireland

Rauschecker U, Stöhr M (2012) Using manufacturing service descriptions for flexible integration of production facilities to manufacturing clouds. In: IEEE ICE2012 Conference Proceedings

Rauschecker U, Stöhr M., Schel D (2013) Requirements and concept for a manufacturing service management and execution platform for customizable products (2013) In: ASME 2013 International Manufacturing Science and Engineering Conference (MSEC 2013), Madison, Wisconsin, Paper MSEC2013-1021, 10–14 June 2013

Salvador F, de Holan PM, Piller F (2009) Cracking the code of mass customization. MIT Sloan Manage Rev 50(3):70–79

Schönsleben P (2007) Integrales logistik management: operations und supply chain management in umfassenden Wertschöpfungsnetzwerken, 5th edn. Springer, Heidelberg

Tao F, Zhang L, Venkatesh, VC, Luo Y, Cheng Y (2011) Cloud manufacturing: a computing and service-oriented model. Proc Inst Mech Eng Part B: J Eng Manuf 225(10):1969–1976

Tseng HE, Chang CC, Chang SH (2005) Applying case-based reasoning for product configuration in mass customization environments. Expert Syst Appl 29(4):913–925

Tuma A (1998) Configuration and coordination of virtual production networks. Int J Prod Econom 56–57:641–648. doi: 10.1016/S0925-5273(97)00146-1

Valilai OF, Houshamnd M (2013) A collaborative and integrated platform to support distributed manufacturing system using a service-oriented approach based on cloud computing paradigm. Robot Comput-Integr Manuf 29(1):110–127

Wang XV, Xu XW (2013) An interoperable solution for Cloud manufacturing. Robot Comput-Integr Manuf 29(4):232–247

Wu D, Thames JL, Rosen DW, Schaefer D (2012) Towards a cloud-based design and manufacturing paradigm: looking backward, looking forward. In: Proceedings of the ASME 2012 International Design Engineering Technical Conference & Computers and Information in Engineering Conference (IDETC/CIE12), Paper Number: DETC2012-70780, Chicago, U.S

Wu D, Greer MJ, Rosen DW, Schaefer D (2013) Cloud manufacturing: strategic vision and state-of-the-art. J Manuf Syst 32(4):564–579

Xie H, Henderson P, Kernahan M (2005) Modelling and solving engineering product configuration problems by constraint satisfaction. Int Prod Res 43(20):4455–4469

Xu X (2012) From cloud computing to cloud manufacturing. Robot Comput-Integr Manuf 28(1):75–86. doi: 10.1016/j.rcim.2011.07.002

Yang D, Dong M, Miao R (2008) Development of a product configuration system with an ontology-based approach. Comput Aided Des 40(8):864–878

Yip ALK, Jagadeesan AP, Corney, JR, Qin Y, Rauschecker U (2011) A front-end system to support cloud-based manufacturing of customised products. In Harrison DK, Wood BM, Evans D (eds) Advances in Manufacturing Technology XXV—Proceedings of the Ninth International Conference on Manufacturing Research (ICMR 2011), Glasgow, Scotland. pp 193–198, 6–8th Sept 2011

Yip, ALK, Corney JR, Jagadeesan AP, Qin Y (2013) A product configurator for cloud manufacturing (2013) In: ASME 2013 International Manufacturing Science and Engineering Conference (MSEC2013), Madison, Wisconsin, Paper MSEC2013-1250, 10–14 June 2013
Yan J, Ye K, Wang H, Hua Z (2010) Ontology of collaborative manufacturing: alignment of service-oriented framework with service-dominant logic. Expert Syst with Appl 37(3): 2222–2231. doi: 10.1016/j.eswa.2009.07.051
Zhang J, Wang Q, Wan L, Zhong Y (2005) Configuration-oriented product modelling and knowledge management for made-to-order manufacturing enterprises. Int J Adv Manuf Technol 25(1–2):41–52
Zhang X, Yang Y, Liu S, Liu F (2007) Realization of a development platform for web-based product customization systems. Int J Comput Integr Manuf 20(2–3):254–264
Zhang L, Lee C, Xu Q (2010a) Towards product customization: an integrated order fulfillment system. Comput Ind 61:213–222
Zhang L, Guo H, Tao F, Luo YL, Si N (2010) Flexible management of resource service composition in cloud manufacturing. In: 2010 IEEE International Conference on Industrial Engineering and Engineering Management, pp 2278–2282

A Manufacturing Ontology Model to Enable Data Integration Services in Cloud Manufacturing using Axiomatic Design Theory

Omid Fatahi Valilai and Mahmoud Houshmand

Abstract Today's global enterprises have faced with a growing increase in the competitiveness that forces them to adopt and develop new strategies and methods. Recent researches have proposed the Cloud manufacturing as the new paradigm for global manufacturing enterprises. The first challenge in cloud manufacturing paradigm is the manufacturing data integration concept. The integration of enterprise-level business systems with manufacturing systems is found to be one of the inevitable drivers for productivity. However, integrating heterogeneous and autonomous data sources through enterprises which are numerous in number and also in approaches is a significant challenge. Moreover, the data integration should involve the integration of autonomous, distributed, and heterogeneous database sources into a single data source associated with a global schema in Cloud manufacturing paradigm. In this chapter, the concepts of Cloud manufacturing data integration have been discussed comprehensively. Discussing the dominant researches for both manufacturing ontology models and solutions proposed as enabler approaches in global manufacturing data integration, the chapter illustrates two main characteristics of an efficient global Cloud manufacturing ontology model. The first characteristic emphasizes on a consistent manufacturing data model, while the second one looks for an efficient management structure to fulfill the improvements and developments in manufacturing discourses. The chapter uses the axiomatic design theory to propose an efficient idea for a global Cloud manufacturing ontology model. The proposed idea fulfills the first axiom of the axiomatic design theory which insures the capabilities of the proposed ontology model to fulfill the characteristics of an efficient global manufacturing ontology model. The capability of the proposed approach is discussed by proposing

O. F. Valilai (✉) · M. Houshmand
Advanced Manufacturing Laboratory, Department of Industrial Engineering,
Sharif University of Technology, Tehran, Iran
e-mail: FValilai@sharif.edu

M. Houshmand
e-mail: Hoshmand@sharif.edu

D. Schaefer (ed.), *Cloud-Based Design and Manufacturing (CBDM)*,
DOI: 10.1007/978-3-319-07398-9_7,

implementation structures for a manufacturing ontology model based on the international standard organization standards related to ISO 10303, ISO 15531, and ISO 18629. The proposed Cloud manufacturing ontology model can be considered as the basic step toward achieving the global Cloud manufacturing.

1 Introduction

Today's global enterprises have faced with a growing increase in the competitiveness that forces them to adopt and develop new strategies and methods with relevant challenges in integrating product, process, and enterprise dimensions and life cycles (Subrahmanian et al. 2005; Americon and Antonio 2011; AbdulKadir et al. 2011; Barbau et al. 2012). The Cloud manufacturing is the proposed paradigm for today's global manufacturing (Xu 2012; Shen et al. 2011). This new look considers a wide cloud of encapsulated manufacturing resources usually distributed over the globe with the capability of centralized management (Tao et al. 2010). This paradigm inherits the cloud computing paradigm (Lai et al. 2012; Bohm and Kanne 2011; Goscinski and Brock 2010) in which manufacturing agents distributed over the globe use the cloud services to fulfill their manufacturing operations requirements (Li et al. 2010). The first challenge to achieve the Cloud manufacturing paradigm is the global manufacturing data integration.

The integration of enterprise-level business systems with manufacturing systems is found to be one of the inevitable drivers for productivity and making businesses more responsive to supply chain demands (Zurawski 2007; Tu and Dean 2011; Koren 2010; Ahmadn and Cuenca 2012). Combination of systems and technologies designed to integrate the data and information of global enterprises is known to be one of the key competitive strategies for industries in the twenty-first century (Ang 1987, 1989; Hannam 1997; Shen et al. 2010; Rudberg and Olhager 2003). Considering that the product life cycle has been shorten, the constant communication and transmission of ideas between different sections throughout the product life cycle in different processes should be integrated to foster a fast and effective development process (Lu and Storch 2011; Abdelkafi et al. 2011; Tu and Dean 2011; Dangayach and Deshmukh 2006; Bloomfield et al. 2012).

However, integrating heterogeneous and autonomous data sources through enterprises' structures which are numerous in number and also in approaches for data management is a significant challenge (Kosanke 2005; Bellatreche et al. 2006; Cutting-Decelle et al. 2007). This challenge increases the requirements for proposing researches for development of tools and techniques to integrate the various enterprises' data sources (Young et al. 2005; Martin 2005). This integration should support the aforementioned sources for working based on their own data ontology design, structures, and procedures for data management and representation (Valilai and Houshmand 2012, 2013a, b). Moreover, the data integration should involve the integration of autonomous, distributed, and heterogeneous databases sources into a single data source associated with a global schema (Bellatreche et al. 2006; Wu et al. 2009; Paredes-Moreno et al. 2010).

The enterprise data integration is known to be a painful procedure due to problems like occurrence of different types of conflicts and not credible accumulation or storage of the information (Santarek 1998; Vassiliadis et al. 2001), challenges due to information sharing among the various management information systems (Zha and Du 2002; Wolfert et al. 2010), existence of a harmonized data model for appropriate representation of enterprise's manufacturing discourse information (Zauner et al. 1993; Zhao et al. 2006), enterprises' diverse requirements for information systems and IT rapid development which persuade information systems in different industry fields to becomes more complicated (Wu et al. 2012; Whitfield et al. 2012). Although different researches have been conducted to fulfill the enterprises' requirements for integration, but still there is an essential need for a solution for enterprises' data integration (Valilai and Houshmand 2010, 2011; Houshmand and Valilai 2012).

In the next section, the chapter studies the concepts of global manufacturing data integration. Discussing the manufacturing ontology model researches, the chapter studies the ISO-based standards approaches and reviews the related researches and solutions offered as an enabler approach in global manufacturing data integration. Talking over the axiomatic design theory, the chapter proposes an efficient solution for global manufacturing ontology model. Finally, the chapter proposes an implementation sample of the proposed idea for global manufacturing ontology model using ISO standard framework.

2 Global Manufacturing Data Integration in Cloud Manufacturing

Researches show that enterprises' information is considered as a highly valuable asset (Paredes-Moreno et al. 2010; Bachlaus et al. 2008; Zhou et al. 2007; Wiendahl et al. 2007). However, the manufacturing enterprises' data are often scattered over different areas, formats, and software packages (Zhang et al. 2011a, b; Scannapieco et al. 2004; Rudberg and Olhager 2003). This necessitates the proposal of solutions to manage the enterprises' information from its various sources updated and integrated. Such solutions must include structure and procedures in a centralized and sophisticated manner to exploit their information effectively and profitably (Wang et al. 2010; Feng and Wu 2009; Patel et al. 2006). However, this integration is usually limited by the current state of technology in terms of the structure and functionalities of current manufacturing information systems (Lebreton et al. 2010; Hwang and Katayama 2009). Also, the proper solution for integration of manufacturing enterprises' processes should cover a wide range of manufacturing discourses like integrating engineering design databases for product information (Jiao and Helander 2006), integration of business activities (Liu et al. 2011), supply chain integration (Bachlaus et al. 2008; Wang et al. 2010; Molina et al. 2007), and integration of information related to the manufacturing operation analysis (Hernandez-Matias et al. 2008).

The consistent data model for various manufacturing ontology scope is one of the important aspects of global manufacturing data integration (Verstichel et al. 2011; Runger et al. 2011). This data model is vital to enable the exchange of information through manufacturing enterprises' processes (Lampathaki et al. 2009). The concept of manufacturing ontology scope defines sets of data and their related structure which enables the experts to share information in a manufacturing domain of interest based on knowledge management and computer-supported cooperative work (Chang and Terpenny 2009; Curran et al. 2010). The manufacturing ontology can be modeled along with the rules and the relationships between the objects. This model enables the interoperability through the manufacturing domain (Blornqvist and Ohgren 2006; Alsafi and Vyatkin 2010).

2.1 Manufacturing Ontology Models

During the last decades, significant researches have been conducted to propose information model supporting decision-making processes throughout the manufacturing operations (Guerra-Zubiaga and Young 2006; Grubic and Fan 2010). Molina et al. (1995) proposed a consistent source of manufacturing information to support CAE users and software packages. This model enables the manufacturing capability of a manufacturing facility presentation. The proposed model considers three types of information, categorized as manufacturing resources, processes, and strategies. The proposed manufacturing data model can be implemented in four levels proposed as factory level, shop floor level, working cell level, and working station level. Uschold et al. (1998) proposed Enterprise Ontology named EO to improve the communication between humans through enterprise processes, provide a data model to support the end-user applications and support interoperability among them (Grubic and Fan 2010). The proposed model consists of five types of information, categorized as meta-Ontology and time, activity, plan, capability and resource, organization, strategy and marketing. The proposed ontology model lacks to cover the whole enterprise ontology data model; however, it has been used by the later researchers for further works.

AMICE. CIMOSA (1993), an organization consists of major companies, including users, vendors, consulting companies, and academia established a project in CIM concept. This project aimed to develop a solution for Computer Integrated Manufacturing Open Systems Architecture named CIMOSA. CIMOSA is an enterprise modeling framework which is able to support the enterprise integration of machines, computers, and people. The framework of CIMOSA is based on the system life cycle concept, and offers a modeling language, methodology, and supporting technology for enterprise integration (Berio and Vernadat 1999). AMICE consortium has promoted the application of CIMOSA for whole enterprise lifecycle support.

Steele et al. (2001) proposed an object-oriented approach to integrate engineering functions through enterprise processes like planning, analysis, and control of resources such as machine tools and fixtures. This object-oriented resource model links information from different enterprise ontology domains. The model establishes a one–to-one relation among the software packages and physical objects. Madni et al. (2001) proposed a data model foundation for designing, reinventing, managing, and controlling for collaborative and distributed enterprises. This model named IDEON, considers four types of information categorized as enterprise data model for environmental data exchange, organizational structure in the enterprises, enterprise processes, and resource and product. Feng and Song in (2003) proposed an object-oriented manufacturing process information model. The model is capable to elaborate the hierarchical structure of the information representing manufacturing processes by means of recursive definition. The proposed model consists of information concerning to manufacturing artifact, manufacturing activities, workpiece, manufacturing equipment, estimated cost and time, and manufacturing process sequences.

Guerra-Zubiaga and Young (2006) offered a manufacturing facility information and knowledge model named MFIKM. This model aimed to enable the management of information and knowledge related to the manufacturing facility. This model focused on process and resource knowledge in manufacturing enterprises. Lin and Harding (2007) proposed a knowledge representation scheme for general manufacturing system engineering called MSE ontology model. The model aims to facilitate the communication and information exchange through the enterprises. The proposed model focuses on information autonomy which enables the enterprise agents to maintain their own data structures. The model focuses on main types of information which can be named as process, resource, strategy, and projects.

Ye et al. (2008) proposed an ontology-based architecture to fulfill the problem of semantic integration through supply chain management concepts. This model, named supply chain ontology (SCO), considers the skeletal method to capture information concepts through the supply chain management processes. This model considers the main information types like supply chain structure, activity, resource, and performance. Tursi et al. (2009) proposed a novel model focusing on a product centric to represent its technical data. This model enables interoperability among applications which are involved in manufacturing enterprises. The researchers tried to include all technical data which may be needed through manufacturing operations. This model is based on a product ontology which formalizes all technical data and concepts contributing to the definition of a product.

2.1.1 ISO-based Data Model Standards for Manufacturing Integration

Considering the aforementioned researches to define a basis for data exchange and sharing, one of the best efforts is the work of the ISO committees during the last two decades (Valilai and Houshmand 2010; Cutting-Decelle et al. 2007; Carnahan et al. 2005; Kramer and Xu 2009). Through the committees' researches, subcommittee 4 (SC 4) of technical committee (TC 184) focused on integration of manufacturing operations in the field of product information model (Jiang et al. 2010; Valilai and Houshmand 2011; Zhou et al. 2007; Panetto and Molina 2008; Zhao et al. 2009; Ray and Jones 2006). The outcome of this committee was proposed as the STEP standard—ISO 10303—which is improving under auspices of the International Standard Organization and is believed to be one of the most successful solutions (Liang and O'Grady 1998; Wang et al. 2009; Zhao and Liu 2008; Xu and Newman 2006; Panetto et al. 2012). Considering the problems encountered where the STEP standard is applied (Gielingh 2008; Ball et al. 2008; Lee et al. 2007; Pratt et al. 2005), different researches have been conducted to propose an efficient approach to overcome these limitations (Houshmand and Valilai 2012; Valilai and Houshmand 2010; Ma et al. 2009; Nylund and Andersson 2010).

Of other related activities of TC184/SC4, the PLib Standard, officially named ISO 13584 Standard series, has defined a model and an exchange format for digital libraries of technical components. PLib is fully interoperable with STEP and in its libraries, component representations may be defined as STEP data and vice versa; in STEP product data, PLib components may be represented by simple references (Maropoulos and Ceglarek 2010; Pierra et al. 1998). MANDATE (ISO 15531) is of other International Standard proposed for the computer-interpretable representation and exchange of industrial manufacturing management data. This standard is suitable not only for neutral file exchange, but also acts as a basis for implementing and sharing manufacturing management databases and archiving (Steele et al. 2001; Cutting-Decelle et al. 2007; Tolio et al. 2010; Cerovsek 2011). The purpose is to facilitate the integration between the numerous industrial applications by means of common standardized software tools. PSL standard (ISO 18629) aims to establish a neutral, high-level language for specifying processes and the interaction related to multiple process-related applications considering the manufacturing life cycle (Dzeng and Tommelein 2004; Chituc et al. 2008; Cutting-Deeelle et al. 2006).

2.2 *Overview of Current Researches for Global Manufacturing Data Integration*

Different researches have been conducted to propose solutions for global manufacturing data integration. To discuss the dominant proposed approaches, this chapter studies the scope which these approaches cover in the world of manufacturing ontology. Moreover, the structures of the proposed manufacturing ontology model are also discussed based on the approaches in aforementioned section.

Oztemel and Tekez (2009) by a reference model for intelligent manufacturing system, Chen et al. (2009) by a product lifecycle ontology model, and Steele et al. (2001) by an object-oriented resource model tried to propose global manufacturing data integration solutions to cover a wide range of manufacturing operations. These operations are mainly focused on computer-aided design, computer-aided manufacturing and production planning, and quality control processes. The researchers have developed their own manufacturing ontology model.

Wiesner et al. (2011) using a formal ontology named OntoCAPE for the domain of Computer-Aided Process Engineering with its application for the integration of design data in chemical engineering, Madni et al. (2001) using a four-categorized data model for computer-aided design processes, Molina et al. (1995) using a three-categorized data model for the domain of production planning and resource control and quality control, Zauner et al. (1993) using a relational database within integrated manufacturing model for the domain of quality control and management, and Ye et al. (2008) using a supply chain management ontology model for supply chain management processes provided solutions each on a specific manufacturing domain.

Zha and Du (2002), Johansson et al. (2004), Vichare et al. (2009), Valente et al. (2010), Valilai and Houshmand (2010), Zhang et al. (2011a, b), Kretz et al. (2011), Ridwan and Xu (2012), Valilai and Houshmand (2013a, b) provided solutions for manufacturing data integration based on ISO 10303 standard. The researchers manufacturing domain were mainly concentrated on computer-aided design, computer-aided manufacturing and CNC machining operations, and in some cases production planning and resource control operations. Considering the ISO 10303 shortcomings and imitations, some solutions were proposed in researches to facilitate the manufacturing data integration (Valilai and Houshmand 2010, 2013a, b).

Naciri et al. (2011) using a generic product modeling languages called (GPM), Tursi et al. (2009) using a product definition ontology model, and Uschold et al. (1998) using a five-categorized data model provided solutions for whole manufacturing lifecycle data integration. These researchers provided ontology models which were designed to be capable of covering all information requirements in different manufacturing operations.

2.2.1 Discussion of Researches for Global Manufacturing Data Integration

Considering the current approaches for global manufacturing data integration as shown in Table 1, the approaches can be discussed in three main groups. The first group of the approaches has developed their own data model structures. These data model structures enable the integration of manufacturing data through specific manufacturing processes but these approaches cannot insure the necessary improvements and refinements due to their own developed methodologies. This group raises a risk of reliability for future manufacturing domain improvements and extensions. The second group of the approaches has used international

Table 1 Current researches of global manufacturing data integration solutions

Researchers	Scope of manufacturing discourses Integration[a]	Data model
Oztemel and Tekez (2009)	NPD/CAD/CAM/PPRC/QC-M	A reference model for intelligent integrated manufacturing system
Chen et al. (2009)		Establishing a product lifecycle ontology
Steele et al. (2001)	CAD/CAM/PPRC/QC-M/A-RMS	An object-oriented resource model that links information from different knowledge domains
Wiesner et al. (2011)	A-RMS	A formal ontology named OntoCAPE
Madni et al. (2001)	CAD	Four types of information categorized as 1. Enterprise data model for environmental data exchange, 2. Organizational structure in the enterprises, 3. Enterprise processes, and 4. Resource and product
Molina et al. (1995)	A-PL/PPRC/MHS	Three types of information, categorized as 1. Manufacturing resources, 2. Processes, and 3. Strategies
Zauner et al. (1993)	QC-M	A relational data base within a computer-integrated manufacturing data model
Ye et al. (2008)	SCM	The model considers the main information types like supply chain structure, activity, resource and performance
Zha and Du (2002)	CAD/A-PL	ISO 10303 (STEP) standard
Johansson et al. (2004)	CAM	
Vichare et al. (2009)	PPRC/CNC	
Valente et al. (2010)	A-PL	
Valilai and Houshmand (2010)	CAD/CAM/CNC	
Zhang et al. (2011a, b)	CAM/CNC	
Kretz et al. (2011)	CAM	
Ridwan and Xu (2012)	CNC	
Valilai and Houshmand (2013a, b)	CAD/CAM/CNC	

(continued)

Table 1 (continued)

Researchers	Scope of manufacturing discourses Integration[a]	Data model
Naciri et al. (2011)	WEL	Using a modeling language called Generic Product Model (GPM)
Tursi et al. (2009)		A product ontology which formalizes all technical data and concepts contributing to the definition of a product
Uschold et al. (1998)		Five types of information, categorized as 1. Meta-Ontology and time, 2. Activity, plan, capability and resource, 3. Organization, 4. Strategy, and 5. Marketing

[a] A-PL: Assembly line/Production line control
A-RMS: Computer systems for automated/robotic operation and control
CAD: Computer-Aided Design
CAM: Computer-Aided Manufacturing
CNC: CNC machining
MSM: Manufacturing Strategy Management/Planning
MHS: Material Handling System
MCM: Manufacturing cost/financial management
NPD: New Product Development
PMC: Preventive Maintenance/Control
PPRC: Production planning and resource control
QC-M: Quality control/Management
SCM: Supply Chain Management/Planning
WEL: Whole Enterprise Life cycle

standard. These approaches can insure continuous and ongoing processes for data model improvements but the solutions do not propose any approaches to overcome the aforementioned criticisms discussed for international standards. Furthermore, the researchers have concentrated on a limited scope of manufacturing operations in the area of design and manufacturing which is indeed a small scope of global manufacturing operations.

The third group of the approaches has proposed whole manufacturing lifecycle data integration solutions. However, like the first group, these approaches cannot insure the necessary improvements and refinements due to their own developed manufacturing otology model structures. So this group also raises a risk of reliability for future improvements and manufacturing domain extensions due to the improvements and developments in manufacturing discourses.

Considering the aforementioned studied researches, two major characteristics can be proposed for a global manufacturing data integration solution. A global manufacturing data integration solution should insure the capability to support whole manufacturing discourses data integration and also should support ongoing improvement procedures due to the improvements and developments in manufacturing discourses. Proceeding, the chapter discusses an efficient framework named axiomatic design theory. Using this theory, the chapter proposes a manufacturing ontology model to efficiently fulfill the major characteristics of the global manufacturing data integration.

2.3 Axiomatic Design Theory

Considering the complexity of the characteristics of a global data integration solution in the aforementioned section, the chapter focuses on an efficient approach to propose an idea for the design of a road map. This chapter considers the axiomatic design axioms as an efficient approach to propose a manufacturing ontology model. The axiomatic design framework provides the fundamental axioms for analysis and decision making. It also proposes a systematic approach to the design process, which has usually employed empirical methods (Suh 1990; Albano and Suh 1994; Bae et al. 2002; Shirwaiker and Okudan 2011). The axiomatic design framework consists of two axioms which are very effective in both the conceptual design stage and the detailed design stage (Brown 2005; Cebi and Kahraman 2010; Cebi et al. 2010). The axiomatic design theory insures an efficient approach for fulfilling "what we need" by "How we are going to satisfy the requirements".

In axiomatic design framework, the first axiom is called the Independence Axiom and the second axiom is called the Information Axiom. The first axiom, Independence Axiom, proposes that the independence of functional requirements (FRs) should always be maintained, where FRs are defined as the minimum set of independent requirements that characterizes the design goals through functional domain. The second axiom, Information Axiom, proposes that the design having

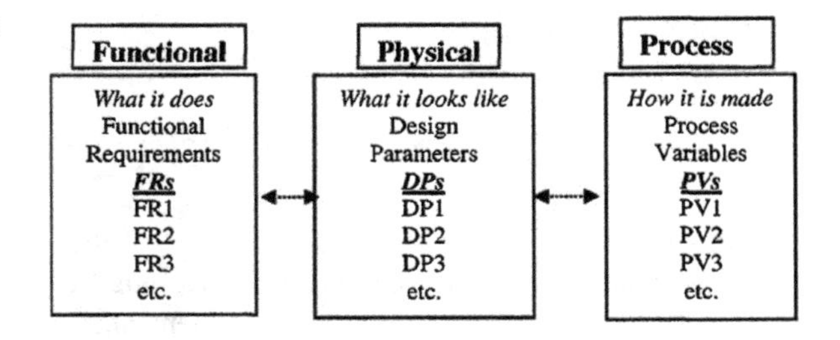

Fig. 1 The axiomatic design domains known (Brown 2005)

the smallest information content is the best design among those designs that satisfy the Independence Axiom (Suh 2001; Chen and Feng 2003; Togay et al. 2008; Ferrer et al. 2010; Cebi and Kahraman 2010). A formal representation of the Independence Axiom is possible (Gazdik 1996; Goncalves-Coelho and Mourao 2007; Heoa and Lee 2007; Peck et al. 2010), if the set of FRs is regarded as an m-component vector and the set of DPs—which represent the design parameters to satisfy the FRs—as an n-component vector in physical domain. Then the following relationship holds:

$$FR = [A]DP \tag{1}$$

Besides the FRs in functional domain and DPs in physical domain, the axiomatic design uses parameter values called PVs in the process domain as shown in Fig. 1. The Tis domain is used as a check to see if a reasonable manufacturing process and process variables exist which are beneficial in concurrent or simultaneous engineering (Brown 2005). Where analyzing the functional domain and physical domain, the "product design" is accomplished. Finally, moving from the physical to the process domain, "process design" is accomplished (Goncalves-Coelho and Mourao 2007).

In Eq. 1, [A] is a mn-design matrix. The structure of this matrix makes it possible to assess the quality of the design. The following alternatives are possible:

- Decoupled design:

$$\forall (m, n) \in \mathrm{N},\ m \neq n : A_{mn} = 0 \tag{2}$$

This design is called the best possible design. In this state, every functional requirement is contingent on one design parameter only. So, there is a one-to-one mapping between every member in the sets of DP and FR, through A. The parameters affecting the object designed are not coupled to one another through the design. Often, only simple mechanisms and structures can be shown to meet the criteria of decoupled design (Heoa and Lee 2007; Thielman et al. 2005).

- Quasi-coupled design:

$$(\forall(m,n) \in \mathrm{N},\ m > n : A_{mn} = 0)\ \text{or}\ (\forall(m,n) \in \mathrm{N},\ m<n : A_{mn} = 0) \quad (3)$$

In this design, the elements on the one side of the diagonal of [A] are equal to nil. Unfortunately, there is a one-to-many mapping between the sets of DP and FR, through A. This design is sequence dependent and can behave like a decoupled design if things are done in the correct order (Gazdik 1996; Kulak et al. 2005; Yi and Park 2005).

- Coupled design:

$$\exists(m_1,n_1),(m_2,n_2) \in \mathrm{N},\ (m_1 > n_1)\ \text{and}\ (m_2<n_2) : A_{m_1n_1} \neq 0\ \text{and}\ A_{m_2n_2} \neq 0 \quad (4)$$

This design is the least desirable alternative. There is a many-to-many mapping between the sets of DP and FR, through [A]. Any design parameter affects not only the design requirement it is supposed to affect, but also other functional requirements, and is itself affected by them (Hirani and Suh 2005; Janthong et al. 2010). As the second axiom, Information Axiom states, among the designs which satisfy the Independence Axiom, the design that has the smallest information content is the best design (Gazdik 1996; Pappalardo and Naddeo 2005; Ogot 2011). The axiomatic design framework proposes the decomposition of the high-level FRs to develop the design details by zigzag approach (Goncalves-Coelho and Mourao 2007; Stiassnie and Shpitalni 2007). This approach includes moving from FRs domain to DPs domain to illustrate the DPs which fulfill the FRs and then going back to FRs domain to determine the lower-level FRs when needed. This process is continued until the highest-level FRs are satisfied completely.

3 An Idea for Global Cloud Manufacturing Ontology Model

3.1 Defining the First-Level FR and DP

To propose a global manufacturing data integration solution, the chapter uses the axiomatic design framework approach. The overall decomposition approach is shown in Fig. 2. As stated earlier, the high-level FR should be defined. Then using the decomposition approach, we should move to DPs zone to clarify the high-level DP. The chapter then should go back to FR zone to elaborate the high-level FR based on lower-level FRs. Considering the characteristics of global manufacturing

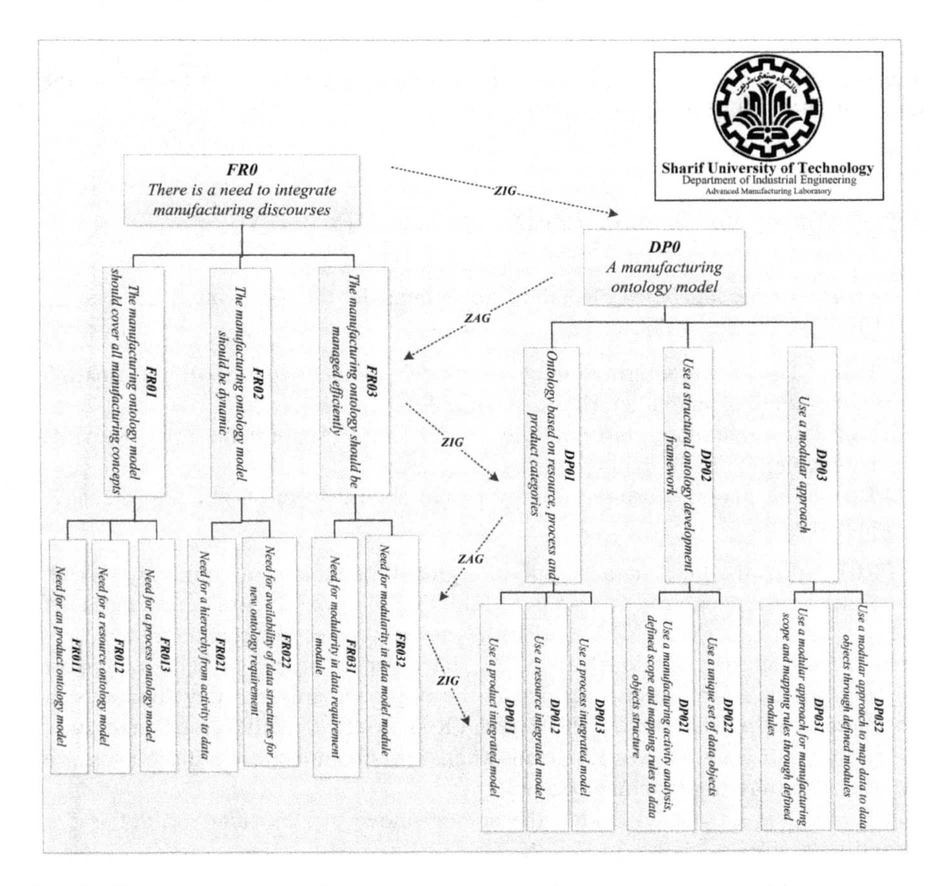

Fig. 2 The axiomatic design zigzag approach for manufacturing ontology model

data integration solution in the aforementioned section, the authors define the first-level FR as:

- FR0 "There is a need to integrate the whole global manufacturing discourses"

The above definition describes the essential need of integrating the different discourses in the global manufacturing paradigm. The solution should be capable of integrating the different manufacturing ontology aspects through whole manufacturing processes. Moving to the DP zone, the authors propose the first-level DP as:

- DP0 "Develop a global manufacturing ontology model"

The first-level DP proposes the main contribution of the chapter. The authors intend to propose a global manufacturing ontology model. The global manufacturing ontology model enables the whole discourses in the global manufacturing operations. As shown in Fig. 2, the authors continue the decomposition processes by moving back to the FR zone. The first-level FR0 should be decomposed to lower-level FRs and then the authors continue the zigzag movement to reach to the required DPs to fulfill the main FR0.

3.2 Defining the Second-Level FRs and DPs

To decompose the first-level FR0, the authors consider the following three second-level FRs as shown in Fig. 2:

- FR01 "The manufacturing ontology model should support all information requirements of manufacturing discourses"
- FR02 "The manufacturing ontology model should support the improvements and developments in manufacturing discourses"
- FR03 "The manufacturing ontology model should have an efficient management structure"

FR01 fulfills the first characteristic of a global manufacturing ontology model. FR01 insures the capability of a proposed solution to cover all data requirements of manufacturing processes. FR02 and FR03 insure the second characteristics of a global manufacturing ontology model. As stated earlier, there is a need to support ongoing improvement procedures due to the improvements and developments in the manufacturing discourses. FR02 and FR03 insure that the global manufacturing integration will be maintained when new manufacturing discourses are introduced or any improvement occurs.

To fulfill the second-level FRs, the authors move to DPs zone and define the second-level DPs as shown in Fig. 2:

- DP01 "The manufacturing ontology model supports the manufacturing discourses using a resource, process and product centric approach"
- DP02 "The manufacturing ontology model uses a development/improvement structure"
- DP03 "The manufacturing ontology model uses a modular approach to manage the model"

The relation among FRs zone and DPs zone in the second-level is shown in Eq. 5. FR01 is fulfilled by DP01. The authors propose an ontology model using a resource-, process-, and product-centric approach. The idea of using a three-categorized approach based on resource, process, and product is supported in recent researches (Lemaignan et al. 2006; Tolio et al. 2010; Valente et al. 2010). FR02 is fulfilled by DP02. To manage the improvements and developments in manufacturing discourses, the authors propose a development/improvement structure. DP02 insures the capability of the global manufacturing ontology to align itself as any improvement or development should be encountered. Finally FR03 is fulfilled by DP02 and DP03. To support an efficient management structure, the authors propose both the development/improvement structure and also using a modular approach which insures the efficient management of manufacturing ontology model (Houshmand and Valilai 2012, 2013; Valilai and Houshmand 2013a, b).

$$\begin{bmatrix} FR01 \\ FR02 \\ FR03 \end{bmatrix} = \begin{bmatrix} X & 0 & 0 \\ X & X & 0 \\ X & X & X \end{bmatrix} \begin{bmatrix} DP01 \\ DP02 \\ DP03 \end{bmatrix} \quad (5)$$

3.3 Defining the Third-Level FRs and DPs

Getting back to FR zone, the authors propose the third-level FRs decomposing the FR01 based on the second-level DPs as shown in Fig. 2:

- FR011 "The manufacturing ontology model should support all information requirements of products in manufacturing domain"
- FR012 "The manufacturing ontology model should support all information requirements of resources in manufacturing domain"
- FR013 "The manufacturing ontology model should support all information requirements of processes in manufacturing domain"

The third-level FRs for FR01 categorize the first characteristics of a global manufacturing ontology model in three information requirement scopes based on the DP01. FR011 states the need for a consistent product information model, while FR012 and FR013 look for a resource and process information model. Fulfillment of FR011, FR012, and FR013 insures the capability of proposed solution for the first characteristic of global manufacturing ontology model (Tolio et al. 2010). The authors propose the third-level FRs decomposing the FR02 based on the second-level DPs as shown in Fig. 2:

- FR021 "The manufacturing ontology model should provide a hierarchical structure to enable manufacturing domain analysis to data requirements"
- FR022 "The manufacturing ontology model should insure an integrated data structures for new data requirements of manufacturing discourses"

As stated in DPs zone, DP02 proposes a development/improvement structure. This structure should insure the capability of global manufacturing ontology model to fulfill the new manufacturing world information requirements. The authors propose the FR021 as a need which enables the manufacturing world analyze for obtaining the needed information requirements. This structure should use a hierarchical structure which analyzes the manufacturing world requirements and leads to the required information structure. The FR022 is proposed to insure that the needed data requirements are included in the global manufacturing ontology model to fulfill the FR021. The authors propose the third-level FRs decomposing the FR03 based on the second-level DPs as shown in Fig. 2:

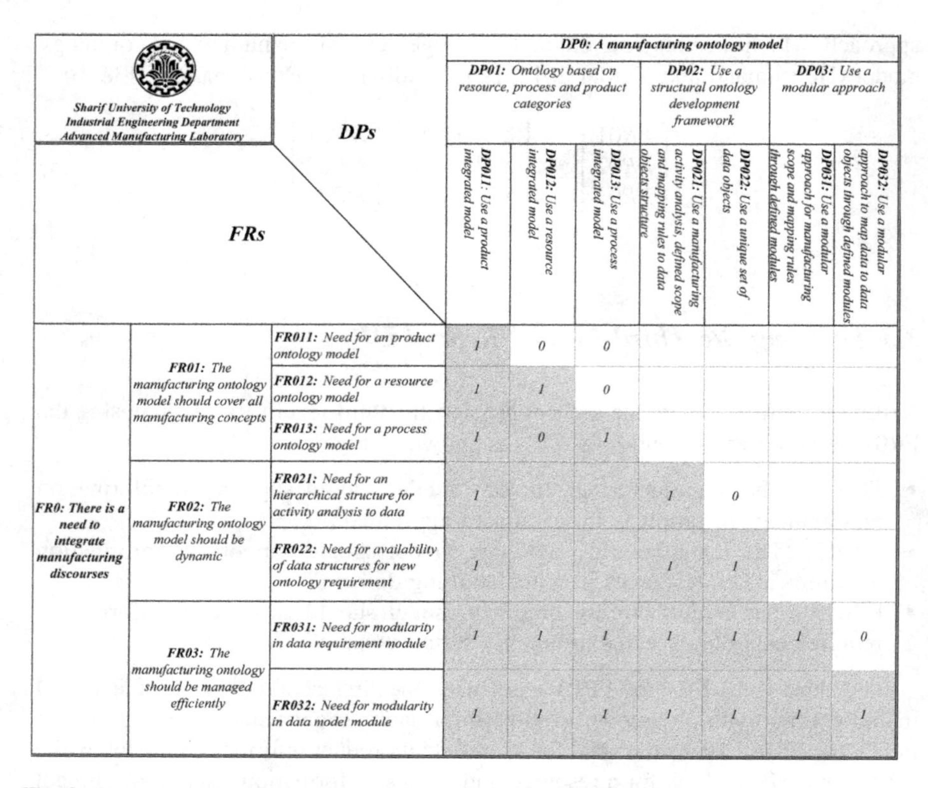

Sharif University of Technology
Industrial Engineering Department
Advanced Manufacturing Laboratory

FRs		DPs	DP0: A manufacturing ontology model						
			DP01: Ontology based on resource, process and product categories			DP02: Use a structural ontology development framework		DP03: Use a modular approach	
			DP011: Use a product integrated model	DP012: Use a resource integrated model	DP013: Use a process integrated model	DP021: Use a manufacturing activity analysis, defined scope and mapping rules to data objects structure	DP022: Use a unique set of data objects	DP031: Use a modular approach for manufacturing scope and mapping rules through defined modules	DP032: Use a modular approach to map data to data objects through defined modules
FR0: There is a need to integrate manufacturing discourses	FR01: The manufacturing ontology model should cover all manufacturing concepts	FR011: Need for an product ontology model	1	0	0				
		FR012: Need for a resource ontology model	1	1	0				
		FR013: Need for a process ontology model	1	0	1				
	FR02: The manufacturing ontology model should be dynamic	FR021: Need for an hierarchical structure for activity analysis to data	1			1	0		
		FR022: Need for availability of data structures for new ontology requirement	1			1	1		
	FR03: The manufacturing ontology should be managed efficiently	FR031: Need for modularity in data requirement module	1	1	1	1	1	1	0
		FR032: Need for modularity in data model module	1	1	1	1	1	1	1

Fig. 3 The proposed manufacturing ontology model FRs and DPs relation

- FR031 " The manufacturing ontology model should support a modular approach to analyze the manufacturing world information requirement"
- FR032 "The manufacturing ontology model should support a modular approach to fulfill the manufacturing information requirement"

As stated in DPs zone, DP03 proposes a modular approach to enable an efficient structure for global manufacturing ontology model. The authors propose two third-level FRs to decompose the second-level FR03. FR031 states the need for modularity when analyzing the manufacturing world information requirement, while the FR032 focuses on modularity in data structures which are designed to fulfill the information requirements.

Moving to DPs zone, the authors define the third-level DPs to fulfill the third-level FRs as shown in Fig. 2. The relation among the third-level FRs and DPs is shown in Fig. 3. As considered, the relation matrix among the FRs and DPs is a low-triangular matrix which obeys the Quasi-coupled design. To fulfill FR0111, FR012, and FR013, the following DPs are introduced in the third-level zone:

- DP011 "Use a product integrated model"
- DP012 "Use a resource integrated model"

- DP013 "Use a process integrated model"

To fulfill the requirements for an information model of manufacturing ontology, the authors propose the application of integrated models in the area of product, resource, and process through DP011, DP012, and DP013. DP011 enables product data integration. DP012 focuses on supporting manufacturing resources involved in flow controls of a shop floor or in a factory and DP013 focuses on supporting manufacturing process data. As discussed earlier, the idea of categorizing the manufacturing ontology model based on a three-dimensional approach of resource, process, and product is supported in recent researches (Lemaignan et al. 2006; Tolio et al. 2010; Valente et al. 2010).

To fulfill the requirements of FR021 and FR022, the authors propose two third-level DPs as shown in Fig. 3:

- DP021 "Use a manufacturing activity analysis, defined scope and mapping rules to data object structures"
- DP022 "Use a unique set of data objects"

To fulfill the third-level FR021, the authors propose DP021. This DP insures a hierarchical structure to enable manufacturing world analysis to data requirements. DP021 proposes the hierarchical structure which starts with manufacturing activity analysis and then leads to formal description of the mapping to the data object structures. Using this hierarchical structure, a manufacturing ontology model can analyze a manufacturing discourse and leads to the data structures for storing and retrieving operations.

The authors also use the DP021 to fulfill the third-level FR022 and moreover propose another third–level DP defined as DP022. This DP provides an integrated and consistent data structure for all manufacturing discourses. DP022 enables FR011, FR012, and FR013 defined at the third-level to have distinct structures for defining the manufacturing discourses activity models but insures their integrity from sets of similar data objects. Using this structure, the improvements and developments in manufacturing discourses can be managed by an integrated data structure, while giving flexibility in analyzing the manufacturing world activity.

To fulfill the requirements of FR031 and FR032 completely, the authors propose two third-level DPs as shown in Fig. 3:

- DP031 "Use a modular approach for manufacturing scope and mapping rules through defined modules"
- DP032 "Use a modular approach to map data to data objects through defined modules"

To fulfill the third-level FR031, the authors apply the DP011 to DP013 and DP021 to DP022. Moreover, the DP031 is introduced to accomplish the FR031 fulfillment. This DP states that the efficient management of a global manufacturing ontology model needs a modular approach. DP031 uses a modular structure in analyzing manufacturing world activities and leading to the formal description of the mapping to consistent data objects.

Besides DP031 to fulfill the third-level FR032, the authors propose the DP032. DP032 focuses on using a modular approach in selecting the applicable constructs from the data objects. This modularity should encompass defined data object modules which are applicable for different analyzed manufacturing world activities.

Considering the overall structure of the approach proposed by the chapter as shown in Fig. 3, the authors proposed DP011, DP012, and DP013 three-level DPs to fulfill the three-level FRs including FR011, FR012, and FR013. This forms the first characteristic of a global manufacturing ontology model. Using the other proposed three-level DPs, the chapter fulfills the second characteristic of a global manufacturing ontology model. The overall relation among the FRs and DPs are satisfying the first axiom of the axiomatic design theory in a Quasi-coupled design form.

4 Implementation of the Structure in the Process Domain

In this section the chapter demonstrates the capabilities of the proposed global manufacturing ontology model by presenting the proper parameter values for DPs in process domain. As stated earlier, PVs in axiomatic design process domain express the required tools to implement the proposed DPs. Considering that the proposed DPs are fulfilling the first axiom of the axiomatic design theory, the chapter uses ISO-based structures as required PVs to fulfill the proposed DPs. This section will show that the proposed PVs relation with DPs insure the first axiom of the axiomatic design theory. This section can be one of the feasible implementation methods of global manufacturing ontology model.

Referring to DPs zone, the authors define the third-level PVs to fulfill the third-level DPs as shown in Fig. 4. The relation among the third-level DPs and PVs is shown in Fig. 5. As considered, the relation matrix among the DPs and PVs is a low-triangular matrix which obeys the Quasi-coupled design. To fulfill the DP0111, DP012, and DP013, the following PVs are introduced in the third-level zone:

- PV011 "Use ISO 10303"
- PV012 "Use ISO 15531"
- PV013 "Use ISO 18629"

To fulfill the requirements of the product-, resource-, and process-integrated model, the authors propose the application of three international standards in the area of product, resource, and process through PV011, PV012, and PV013. PV011 focuses on ISO 10303 (STEP) standard to enable product data integration. PV012 focuses on ISO 15531 (MANDATE) supporting manufacturing resources and manufacturing flow management like flow controls in a shop floor or in a factory and PV013 focuses on ISO 18629 (PSL) standards to support manufacturing process data.

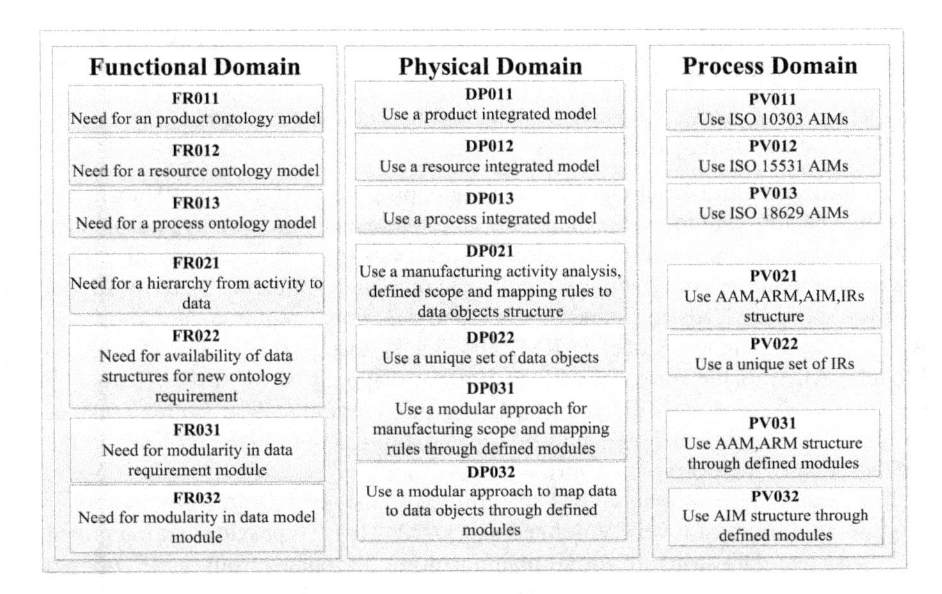

Fig. 4 The proposed PVs for global manufacturing ontology model

Sharif University of Technology ***Industrial Engineering Department*** ***Advanced Manufacturing Laboratory*** ***PVs*** / ***DPs***	***PV011:*** *Use ISO 10303 AIMs*	***PV012:*** *Use ISO 15531 AIMs*	***PV013:*** *Use ISO 18629 AIMs*	***PV021:*** *Use AAM,ARM,AIM structure*	***PV022:*** *Use a unique set of Irs*	***PV031:*** *Use AAM,ARM structure through defined modules*	***PV032:*** *Use AIM structure through defined modules*
DP011 *: Use a product integrated model*	*1*	*0*	*0*				
DP012: *Use a resource integrated model*	*1*	*1*	*0*				
DP013: *Use a process integrated model*	*1*	*0*	*1*				
DP021: *Use a manufacturing activity analysis, defined scope and mapping rules to data objects structure*	*1*	*1*	*1*	*1*	*0*		
DP022: *Use a unique set of data objects*	*1*	*1*	*1*	*0*	*1*		
DP031: *Use a modular approach for manufacturing scope and mapping rules through defined modules*	*1*	*1*	*1*	*0*	*0*	*1*	*0*
DP032: *Use a modular approach to map data to data objects through defined modules*	*1*	*1*	*1*	*0*	*0*	*1*	*1*

Fig. 5 The relation among proposed PVs and DPs for global manufacturing ontology model

To fulfill the requirements of DP021 and DP022, the authors propose two third-level PVs as shown in Fig. 4.:

- PV021 "Use AAM, ARM, AIM and IRs hierarchical structure"
- PV022 "Use a unique set of IRs"

To fulfill the third-level DP021, the authors propose PV021. This PV insures a hierarchical structure to enable manufacturing world analysis to data requirements. PV021 proposes the hierarchical structure of ISO-based data integration which starts with manufacturing activity analysis known as Application Activity Model (AAM) and then leads to defined scopes and functional requirements known as Application Reference Model (ARM) and finally using Application Interpreted Models (AIM) as the formal description of the mapping between the ARMs and the Integrated Resource (IR) models. Using this hierarchical structure, a manufacturing ontology model can analyze a manufacturing discourse and leads to the data structures for storing and retrieving operations.

The authors also use the PV021 to fulfill the third-level DP022 and moreover propose another third-level PV defined as PV022. This PV provides an integrated and consistent data structure for all manufacturing discourses from the ISO-based integrated data structures. PV022 enables the DP011, DP012, and DP013 defined at the third level to have distinct structures for defining the manufacturing discourses activity model through AAM and ARM structures but insures their integrity from sets of similar IRs. Using this structure, a structural ontology development framework is insured and the improvements and developments in manufacturing discourses can be managed by integrated data structures, while giving flexibility in analyzing the manufacturing world activity through AAM and ARM structures.

To fulfill the requirements of DP031 and DP032 completely, the authors propose two third-level PVs as shown in Fig. 3:

- PV031 "Use modular approach of AAM and ARM structure"
- PV032 "Use a modular approach for MIM structure to interpret manufacturing data to IRs"

To fulfill the third-level DP031, the authors apply PV011 to PV013 and PV021 to PV022. Moreover, the PV031 is introduced to accomplish the DP031 fulfillment. This PV states that the efficient management of a global manufacturing ontology model needs a modular approach. This modular approach has been discussed in recent researches as a solution for shortcomings arisen by using the ISO standards like STEP standard (Houshmand and Valilai 2013; Valilai and Houshmand 2013a, b). PV031 uses a modular structure for analyzing manufacturing world activities and leading to the formal description of the mapping between the ARMs and the Integrated Resource (IR) models.

Besides PV031 to fulfill the third-level DP032, the authors propose the PV032. PV032 focuses on using a modular approach in selecting the applicable constructs from the integrated resources as baseline conceptual elements. This modularity is achieved through the baseline constructs with additional constraints and relationships which are specified by entities containing local rules, refined data types, global rules, and specialized textual definitions.

Considering the overall structure of the proposed PVs by the authors as shown in Fig. 4, three ISO standards, as the three-level PVs, are proposed to fulfill the three-level DPs including DP011, DP012, and DP013. This results in forming the first characteristics of a global manufacturing ontology model. Using the other proposed three-level PVs, the authors fulfill the second characteristics of a global manufacturing ontology model. The overall relation among the DPs and PVs are satisfying the first axiom of the axiomatic design theory in a Quasi-coupled design form.

5 Conclusion and Discussions

Today's global enterprises have faced with a growing increase in the competitiveness that forces them to adopt and develop new strategies and methods with relevant challenges in integrating product, process, and enterprise dimensions and lifecycles. Considering the Cloud manufacturing paradigm as a solution for today's global manufacturing enterprises, this chapter discussed the first requirement of the Cloud manufacturing paradigm known as global manufacturing integration. The integration of enterprise-level business systems with manufacturing systems is found to be one of the inevitable drivers for productivity and making businesses more responsive to supply chain demands. However, integrating heterogeneous and autonomous data sources through enterprises' structures which are numerous in number and also in approaches for data management is a significant challenge. In this chapter, the concepts of global manufacturing data integration have been studied comprehensively. Discussing the dominant researches for solutions offered as enabler approaches in global manufacturing data integration, the authors proposed the axiomatic design theory capability to propose an efficient solution for global manufacturing ontology model. Two main characteristics of the solution are described as:

1. Providing a consistent manufacturing lifecycle data integration model.
2. The required structure and procedures to fulfill improvements and developments in manufacturing discourses.

Using the axiomatic design theory, the authors started from the first-level functional domain "There is a need to integrate the whole global manufacturing discourses". Zigzagging among design parameter domain and functional domain, the authors decomposed the first-level functional requirement and proposed a solution in design domain while fulfilling the axiomatic design first axiom. A section is provided to obtain a candidate implementation structure for the proposed global manufacturing ontology model in the process domain. This implementation benefits from international standard organization standards like ISO 10303, ISO 15531, and ISO 18629 to provide a consistent manufacturing lifecycle integrated data structure. Moreover, to form an efficient approach to fulfill the improvements and developments in manufacturing discourses, the authors benefited from the international standard organization efforts for providing modular integration standard frameworks.

Considering the cloud manufacturing paradigm, the authors recommended the proposal of frameworks which use the proposed Cloud manufacturing ontology model for manufacturing data integration services in further researches. The solutions should define an efficient structure which enables the agents engaged in different manufacturing discourses for manufacturing data integration in the form of services. Also as the proposed ontology model insures a consistent information model for improvement and developments in manufacturing discourses, the proposal of collaborative and interoperable platforms especially based on Cloud manufacturing paradigm is suggested. These researches perform the next steps toward achieving the Cloud manufacturing paradigm in global manufacturing.

References

Abdelkafi N, Pero M, Blecker T, Sianesi A (2011) NPD-SCM alignment in mass customization. In: Flavio S, Giovani Fogliatto and da Silveira JC Mass customization engineering and managing global operations. Springer, London, pp 69–85

AbdulKadir A, Xu X, Hammerle E (2011) Virtual machine tools and virtual machining—a technological review. Elsevier Ltd. Robot Comput Integr Manufac 27:494–508

Ahmad MM, Cuenca RP (2012) Critical success factors for ERP implementation in SMEs. Elsevier Ltd. Robotics and Computer-Integrated Manufacturing. http://dx.doi.org/10.1016/j.rcim.2012.04.019

Albano LD, Suh NP (1994) Axiomatic design and concurrent engineering. Butterworth-Heinemann Ltd. Comput Aided Des 26:499–504

Alsafi Y, Vyatkin V (2010) Ontology-based reconfiguration agent for intelligent mechatronic systems in flexible manufacturing. Elsevier Ltd. Robot Comput Integr Manufac 26:381–391

Americon A, Antonio A (2011) Factory Templates for Digital Factories Framework. Elsevier BV Robot Comput Integr Manufac 27:755–771

AMICE. CIMOSA (1993) Open system architecture for CIM, 2nd revised and expanded version. Springer, Berlin

Ang CL (1987) Technical planning of factory data communications systems. Elsevier Science Publishers B.V. Comput Ind 9:93–105

Ang CL (1989) Planning and implementing computer integrated manufacturing. Elsevier Science Publishers B.V. Comput Ind 12:131–140

Bachlaus M et al (2008) Designing an integrated multi-echelon agile supply chain network: a hybrid taguchi-particle swarm optimization approach. J Intell Manufact 19:747–761 Springer

Bae S, Lee JM, Chu CN (2002) Axiomatic design of automotive suspension systems. http://www.sciencedirect.com/science/article/pii/S0007850607614796 CIRP Annal Manufact Technol 51: 115–118

Ball A, Ding L, Patel M (2008) An approach to accessing product data across system and software revisions. Adv Eng Inform 22:222–235

Barbau R et al (2012) OntoSTEP: enriching product model data using ontologies. Comput Aided Des 44:575–590 Elsevier Ltd

Bellatreche L et al (2006) Contribution of ontology-based data modeling to automatic integration of electronic catalogues within engineering databases. Comput Ind 57:711–724 Elsevier B.V

Berio G, Vernadat FB (1999) New developments in enterprise modelling using CIMOSA. Comput Ind 40:99–114 Elsevier B.V

Bloomfield R et al (2012) Interoperability of manufacturing applications using the Core Manufacturing Simulation Data (CMSD) standard information model. Comput Ind Eng 62:1065–1079 Elsevier Ltd

Blornqvist E, Ohgren A (2006) Constructing an enterprise ontology for an automotive supplier. 12th IFAC Symposium on Information Control Problems in Manufacturing. Elsevier IFAC Publications, Saint Etienne, pp 665–670

Bohm A, Kanne C-C (2011) Demaq/Transscale: automated distribution and scalability for declarative applications. Inform Syst 36:565–578 Elsevier B.V

Brown CA (2005) Teaching axiomatic design to engineers theory, applications, and software. SME J Manufact Syst 24:186–195

Carnahan D et al (2005) Integration of production, diagnostics, capability assessment, and maintenance information using ISO 18435. Chicago, Illinois, ISA, 2005. ISA EXPO 2005

Cebi S, Celik M, Kahraman C (2010) Structuring ship design project approval mechanism towards installation of operator–system interfaces via fuzzy axiomatic design principles. Elsevier Inc. Inf Sci 180:886–895

Cebi S, Kahraman C (2010) Extension of axiomatic design principles under fuzzy environment. Expert Syst Appl 37:2682–2689 Elsevier B.V

Cerovsek T (2011) Review and outlook for a 'Building Information Model' (BIM): a multi-standpoint framework for technological development. Adv Eng Inform 25:224–244 Elsevier Ltd

Chang X, Terpenny J (2009) Ontology-based data integration and decision support for product e-design. Roboti Comput Integr Manufact 25:863–870 Elsevier Ltd

Chen K-Z, Feng X-A (2003) Computer-aided design method for the components made of heterogenous materials. Comput Aided Des 35:453–466 (Elsevier Science Ltd)

Chen Y-J, Che Y-M, Chu H-C (2009) Development of a mechanism for ontology-based product lifecycle knowledge integration. Expert Syst Appl 36:2759–2779 Elsevier Ltd

Chituc C-M, Toscano C, Azevedo A (2008) Interoperability in collaborative networks: independent and industry-specific initiatives—The case of the footwear industry. Comput Ind 59:741–757 Elsevier B.V

Curran R et al (2010) A multidisciplinary implementation methodology for knowledge based engineering: KNOMAD. Expert Syst Appl 37:7336–7350 Elsevier Ltd

Cutting-Deeelle AF et al (2006) Information exchange in a CROSS-DISCIPLINARY supply chain: formal Strategy and application. In: 12th IFAC symposium on information control problems in manufacturing, Elsevier IFAC Publications, Saint Etienne, pp 573–578

Cutting-Decelle AF et al (2007) ISO 15531 MANDATE: a product-process-resource based approach for managing modularity in production management. Concurr Eng 15:217-235. DOI: 10.1177/1063293X07079329 (SAGE Publications)

Dangayach GS, Deshmukh SG (2006) An exploratory study of manufacturing strategy practices of machinery manufacturing companies in India. Int J Manag Sci 34:254–273 Elsevier Ltd

Dzeng R-J, Tommelein ID (2004) Product modeling to support case-based construction planning and scheduling. Autom Constr 13:341–360 (Elsevier B.V)

Feng C-M, Wu P-J (2009) A tax savings model for the emerging global manufacturing network. Int J Prod Econ 122:534–546 Elsevier B.V

Feng S, Song E (2003) A manufacturing process information model for design and process planning integration. SME J Manufact Syst 22:1–15

Ferrer I et al (2010) Methodology for capturing and formalizing DFM Knowledge. Robot Comput Integr Manufact 26:420–429 (Elsevier Ltd)

Gazdik I (1996) Zadeh's extension principle in design reliability. Fuzzy Sets Syst 83:169–178 Elsevier Science B.V

Gielingh W (2008) An assessment of the current state of product data technologies. Comput Aided Des 40:750–759

Goncalves-Coelho AM, Mourao Antonio JF (2007) Axiomatic design as support for decision-making in a design for manufacturing context: a case study. Int J Prod Econ 109:81–89 (Elsevier B.V)

Goscinski A, Brock M (2010) Toward dynamic and attribute based publication, discovery and selection for cloud computing. Future Gener Comput Syst 26:947–970 (Elsevier B.V)

Grubic T, Fan I-S 2010) Supply chain ontology: review, analysis and synthesis. Comput Ind 61:776–786 (Elsevier B.V)

Guerra-Zubiaga DA, Young RIM(2006) A manufacturing model to enable knowledge maintenance in decision support systems. J Manufac Syst 25:122–136 (Elsevier)

Hannam R (1997) Computer integrated manufcaturing: concepts to realisation. Edinburgh Gate, Addison-Wesley, London

Heo GY, Lee SK (2007) Design evaluation of emergency core cooling systems using axiomatic design. Nucl Eng Des 237:38–46 (Elsevier B.V)

Hernandez-Matias JC et al (2008) An integrated modelling framework to support manufacturing system diagnosis for continuous improvement. Robot Comput Integr Manufact 24:187–199 (Elsevier Ltd)

Hirani H, Suh NP (2005) Journal bearing design using multiobjective genetic algorithm and axiomatic design approaches. Tribol Int 38:481–491 (Elsevier Ltd)

Houshmand M, Valilai OF (2012) LAYMOD: a layered and modular platform for CAx product data integration based on the modular architecture of the standard for exchange of product data. Int J Comput Integr Manufact 25:473–487. http://dx.doi.org/10.1080/0951192X.2011.646308 (Taylor & Francis)

Houshmand M, Valilai OF (2013) A layered and modular platform to enable distributed CAx collaboration and support product data integration based on STEP standard. Int J Comput Integr Manufact http://dx.doi.org/10.1080/0951192X.2013.766935 (Taylor & Francis)

Hwang R, Katayama H (2009) A multi-decision genetic approach for workload balancing of mixed-model U-shaped assembly line systems. Int J Prod Res 47:3797–3822 (Taylor & Francis)

Janthong N, Brissaud D, Butdee S (2010) Combining axiomatic design and case-based reasoning in an innovative design methodology of mechatronics products. CIRP J Manufact Sci Technol 2:226–239

Jiang,Y, Peng G, Liu W (2010) Research on ontology-based integration of product knowledge for collaborative manufacturing. Int J Adv Manufact Technol 49:1209–1221 (Springer)

Jiao J, Helander MG (2006) Development of an electronic configure-to-order platform for customized product development. Comput Ind 57:231–244

Johansson H, Peter A, Orsborn K (2004) A system for information management in simulation of manufacturing processes. Adv Eng Soft 35:725–733 (Civil-Comp Ltd and Elsevier Ltd)

Koren Y (2010) The global manufacturing revolution; product-process-business integration and reconfigurable systems. Wiley & Sons, Inc, New Jersey

Kosanke K (2005) ISO Standards for interoperability: a comparison. In: First international conference on interoperability of enterprise software and applications, INTEROP-ESA'2005 ,Geneva

Kramer T, Xu X (2009) STEP in a nutshell. In: Xun X, Nee AYC (ed) advanced design and manufacturing based on STEP. Springer, London, p 34

Kretz D et al. (2011) Implementing ISO standard 10303 application protocol 224 for automated process planning. Robot Comput Integr Manufact 27:729–734 (Elsevier Ltd)

Kulak O, Durmusoglu MB, Tufekci S (2005) A complete cellular manufacturing system design methodology based on axiomatic design principles. Comput Ind Eng 48:765–787 (Elsevier Ltd)

Lai Ivan KW, Tam Sidney KT, Chan Michael FS (2012) Knowledge cloud system for network collaboration: a case study in medical service industry in China. Expert Syst Appl 39:12205–12212 (Elsevier Ltd)

Lampathaki F et al (2009) Business to business interoperability: a current review of XML data integration standards. Comput Stan Interfaces 31:1045–1055 (Elsevier B.V)

Lebreton BGM, Van Wassenhove LN, Bloemen RR (2010) Worldwide sourcing planning at Solutia's glass interlayer products division. Int J Prod Res 48:801–819 (Taylor & Francis)

Lee G, Eastman CM, Sacks R (2007) Eliciting information for product modeling using process modeling. Data Knowl Eng 62:292–307

Lemaignan S et al (2006) MASON: A proposal for an ontology of manufacturing domain. Prague, Czech Republic, 15–16 June : Los Alamitos, Calif. : IEEE computer society, ©2006. Proceedings of the IEEE workshop on distributed intelligent systems: collective intelligence and its applications (DIS'06). pp 195–200. http://dx.doi.org/10.1109/DIS.2006.48

Li B-H et al (2010) Cloud manufacturing: a new service-oriented networked manufacturing model. Comput Integr Manufact Syst 16:1–7

Liang W-Y, O'Grady P (1998) Design with objects: an approach to object-oriented design. Comput Aided Des 30:943–956

Lin HK, Harding JA (2007) A manufacturing system engineering ontology model on the semantic web for inter-enterprise collaboration. Comput Ind 58:428–437 (Elsevier BV)

Liu S, Young RIM, Ding L (2011) An integrated decision support system for global manufacturing co-ordination in the automotive industry. Int J Comput Integr Manufact 4:285–301 (Taylor & Francis)

Lu RF, Storch RL (2011) Designing and planning for mass customization in a large scale global production system. In: Flavio SF, da Silveira GJC (eds) Mass customization engineering and managing global operations. Springer, London, pp 23–48

Ma Y-S, Chen G, Thimm G (2009) Fine grain feature associations in collaborative design and manufacturing—a unified approach. In: Lihui W, Nee AYC (eds) Collaborative design and planning for digital manufacturing. Springer, London, pp 71–79

Madni AM, Lin W, Madni CC (2001) IDEONTM: an extensible ontology for designing, integrating and managing collaborative distributed enterprises. Syst Eng 4:35–48 (Wiley Periodicals, Inc)

Maropoulos PG, Ceglarek D (2010) Design verification and validation in product lifecycle. CIRP Annal Manufact Technol 59:740–759 (Elsevier)

Martin RA (2005) www.tinwisle.com/. Accessed 27 Apr 2012 (Online) www.tinwisle.com/iso/RM_SME_SUMMIT05.pdf

Molina A, Velandia M, Galeano N (2007) Virtual enterprise brokerage: a structure-driven strategy to achieve build to order supply chains. Int J Prod Res 49:3853–3880 (Taylor & Francis)

Molina A et al (1995) Modelling manufacturing capability to support concurrent engineering. Concurr Eng 3:29–42. doi: 10.1177/1063293X9500300105 (SAGE)

Naciri S et al (2011) ERP data sharing framework using the generic product model (GPM). Expert Syst Appl 38:1203–1212 (Elsevier Ltd)

Nylund H, Andersson P (2010) Simulation of service-oriented and distributed manufacturing systems. Robot Comput-Integr Manufact 26:622–628

Ogot M (2011) Conceptual design using axiomatic design in a TRIZ framework. Procedia Eng 9:736–744 (Elsevier Ltd)

Oztemel E, Tekez EK (2009) A general framework of a reference model for intelligent integrated manufacturing systems (REMIMS). Eng Appl Artif Intell 22:855–864 (Elsevier Ltd)

Panetto H, Dassisti M, Tursi A (2012) ONTO-PDM: Product-driven ONTOlogy for product data management interoperability within manufacturing process environment. Adv Eng Inf 26:334–348 (Elsevier Ltd)

Panetto H, Molina A (2008) Enterprise integration and interoperability in manufacturing systems: trends and issues. Comput Ind 59:641–646 (Elsevier B.V)

Pappalardo M, Naddeo A (2005) Failure mode analysis using axiomatic design and non-probabilistic information. J Mate Process Technol 164–165:1423–1429 (Elsevier B.V)

Paredes-Moreno A, Martınez-Lopez FJ, Schwartz DG (2010) A methodology for the semi-automatic creation of data-driven detailed business ontologies. Inf Syst 35:758–773 (Elsevier B.V)

Patel Minnie H et al (2006) Air cargo pickup schedule for single delivery location. Comput Ind Eng 51:553–565 (Elsevier Ltd)
Peck J, Nightingale D, Kim S-G (2010) Axiomatic approach for efficient healthcare system design and optimization. CIRP Annal Manufact Technol 2010:469–472
Pierra G et al (1998) Exhgange of component data: The PLib (ISO 13584) model, standard and tools. In: Proceedings of the CALS EUROPE'98 conference, pp 160–176
Pratt Michael J, Anderson Bill D, Rangerc T (2005) Towards the standardized exchange of parameterize feature-based CAD models. Comput Aided Des 37:1251–1265
Ray SR, Jones AT (2006) Manufacturing interoperability. J Intell Manuf 17:681–688
Ridwan F, Xu X (2012) Advanced CNC system with in-process feed-rate optimisation. Robot Comput Integr Manufactu 23(3):423–441. http://dx.doi.org/10.1016/j.rcim.2012.04.008
Rudberg M, Olhager J (2003) Manufacturing networks and supplychains: an operations strategy perspective. Omega 31:29–39 (Elsevier)
Runger G et al (2011) Development of energy-efficient products: Models, methods and IT support. CIRP J Manufact Sci Technol 4:216–224
Santarek K (1998) Organisational problems and issues of CIM systems design. J Mater Process Technol 76:219–226 (Elsevier Science)
Scannapieco M et al (2004) The DaQuinCIS architecture: a platform for exchanging and improving data quality in cooperative information systems. Inf Syst 29:551–582 (Elsevier Ltd)
Shen B et al (2011) Collaborative engineering supporting technology for manufacturing in SOA. Computer integrated manufacturing systems Jisuanji Jicheng Zhizao Xitong/Computer integrated manufacturing systems. CIMS 17:876–881
Shen W et al (2010) Systems integration and collaboration in architecture, engineering, construction, and facilities management: a review. Adv Eng Inf 24:196–207 (Elsevier Ltd)
Shirwaiker RA, Okudan GE (2011) Contributions of TRIZ and axiomatic design to leanness in design: an investigation. Procedia Eng 9:730–735 (Elsevier Ltd)
Steele J, Son Y-J, Wysk RA (2001) Resource modeling for the integration of the manufacturing enterprise. SME J Manufact Syst 1:407–427
Stiassnie E, Shpitalni M (2007) Incorporating lifecycle considerations in axiomatic design. CIRP Annals CIRP 56/1/2007:1–4. doi:10.1016/j.cirp.2007.05.002
Subrahmanian E et al (2005) Product lifecycle management support: a challenge in supporting product design and manufacturing in a networked economy. Int J Prod Lifecycle Manag 1:4–25
Suh NP (1990) The principles of design. Oxford University Press, New York
Suh NP (2001) Axiomatic design: advances and applications. Oxford University Press, New York
Tao F, Hu YF, Zhang L (2010) Theory and practice: optimal resource service allocation in manufacturing grid. Machine Press, Beijing
Thielman J et al (2005) Evaluation and optimization of general atomics' GT-MHR reactor cavity cooling system using an axiomatic design approach. Nucl Eng Des 235:1389–1402 (Elsevier B.V)
Togay C, Dogru AH, Tanik JU (2008) Systematic component-oriented development with axiomatic design. J Syst Softw 81:1803–1815 (Elsevier Inc)
Tolio T et al (2010) SPECIES—Co-evolution of products, processes and production systems. CIRP Annal Manufact Technol 59:672–693. http://dx.doi.org/10.1016/j.cirp.2010.05.008
Tu Y, Dean P (2011) One-of-a-kind production. Springer, London, pp 5–11. DOI 10.1007/978-1-84996-531-6
Tursi A et al (2009) Ontological approach for products-centric information system interoperability in networked manufacturing enterprises. Annual Reviews in Control 33:238–245 (Elsevier Ltd)
Uschold M et al (1998) The enterprise ontology Knowl Eng Rev 13:31–89
Valente A et al (2010) A STEP compliant knowledge based schema to support shop-floor adaptive automation in dynamic manufacturing environments, CIRP Annal Manufact Technol 59:441–444

Valilai OF, Houshmand M (2010) INFELT STEP: An integrated and interoperable platform for collaborative CAD/CAPP/CAM/CNC machining systems based on STEP standard. Int J Comput Integr Manufact 23:1095–1117. http://dx.doi.org/10.1080/0951192X.2010.527373 (Taylor & Francis)

Valilai OF, Houshmand M (2011) LAYMOD; A layered and modular platform for CAx collaboration management and supporting product data integration based on STEP standard. World academy of science, engineering and technology. In: International conference on mechanical, industrial, and manufacturing engineering, vol 78. Amsterdam, The Netherlands, pp 625–633

Valilai OF, Houshmand M (2013a) A collaborative and integrated platform to support distributed manufacturing system using a service-oriented approach based on cloud computing paradigm. J Robot Comput Integr Manufact 29:110–127. http://dx.doi.org/10.1016/j.rcim.2012.07.009 (Elsevier B.V)

Valilai OF, Houshmand M (2013b) A platform for optimization in distributed manufacturing enterprises based on cloud manufacturing paradigm. Int J Comput Integr Manufact (Taylor & Francis). http://doi.org/10.1080/0951192X.2013.874582

Vassiliadis P et al (2001) Arktos: towards the modeling, design, control and execution of ETL processes. Inf Syst 26:537–561 (Elsevier Science Ltd)

Verstichel S et al (2011) Efficient data integration in the railway domain through an ontology-based methodology. Transp Res Part C 16:617–643 (Elsevier Ltd)

Vichare P et al (2009) A unified manufacturing resource Model for representing CNC machining systems. Robot Comput Integr Manufact 25:999–1007 (Elsevier Ltd)

Wang, JX et al (2009) Design and implementation of an agent-based collaborative product design system. Comput Ind 6:520–535

Wang WYC, Chan HK, Pauleen DJ (2010) Aligning business process reengineering in implementing global supply chain systems by the SCOR model. Int J Prod Res 48:5647–5669 (Taylor & Francis)

Whitfield RI et al (2012) A collaborative platform for integrating and optimising computational fluid dynamics analysis requests. Comput Aided Des 44:224–240 (Elsevier Ltd)

Wiendahl H-P et al (2007) Changeable manufacturing—classification, design and operation. Annal CIRP 56:783–810 (Elsevier Ltd)

Wiesner A, Morbach J, Marquardt W (2011) Information integration in chemical process engineering based on semantic technologies. Comput Chem Eng 35:692–708 (Elseiver Ltd)

Wolfert J et al (2010) Organizing information integration in agri-food—a method based on a service-oriented architecture and living lab approach. Comput Electron Agricu 70:389–405 (Elsevier B.V)

Wu B et al (2012) Energy information integration based on EMS in paper mill. Appl Energy 93:488-495 (Elsevier Ltd)

Wu Q, Zhu M, Rao NSV (2009) Integration of sensing and computing in an intelligent decision support system for homeland security defense. Pervasive Mobile Comput 5:182–200 (Elsevier B.V)

Xu X (2012) From cloud computing to cloud manufacturing. Robot Comput Integr Manufact 28:75–86. http://dx.doi.org/10.1016/j.rcim.2011.07.002

Xu XW, Newman ST (2006) Making CNC machine tools more open, interoperable and intelligent—a review of the technologies Comput Ind 57:141–152

Ye Y et al (2008) An ontology-based architecture for implementing semantic integration of supply chain management. Int J Comput Integr Manufact 1:1–18 (Taylor & Francis)

Yi J-W, Park G-J (2005) Development of a design system for EPS cushioning package of a monitor using axiomatic design. Adv Eng Softw 36:273–284 (Elsevier Ltd)

Young RIM et al (2005) Sharing manufacturing information and knowledge in design decision support. In: Bramley A et al (eds) Advances in integrated design and manufacturing in mechanical engineering. Springer, London , pp 173–188

Zauner M, Holzl J, Kopacek P (1993) A modular low-cost CAQ system. Control Eng Pract 1:1063–1068 (Elsevier Ltd)

Zha XF, Du H (2002) A PDES/STEP-based model and system for concurrent integrated design and assembly planning. Comput Aided Des 34:1087–1110 (Elsevier Science Ltd)
Zhang G, Shang J, Li W (2011a) Collaborative production planning of supply chain under price and demand uncertainty. Eur J Oper Res 215:590–603 (Elsevier B.V)
Zhang X et al (2011b) A STEP-compliant process planning system for CNC turning operations. Robot Comput Integr Manufact 27:349–356 (Elsevier Ltd)
Zhao W, Liu JK (2008) OWL/SWRL representation methodology for EXPRESS-driven product information model; Part I. Implement Methodol Comput Ind 59:580–589
Zhao X, Pasupathy TMK, Wilhelm RG (2006) Modeling and representation of geometric tolerances information in integrated measurement processes. Comput Ind 57:319–330 (Elsevier B.V)
Zhao YF, Habeeb S, Xu X (2009) Research into integrated design and manufacturing based on STEP. Int J Adv Manufact Technol 44:606–624
Zhou B-H, Xi LF, Tao L-Y (2007) A framework of order evaluation and negotiation for SMMEs in networked manufacturing environments. Int J Comput Integr Manufact 20:199–210 (Taylor & Francis)
Zhou X et al (2007) A feasible approach to the integration of CAD and CAPP. Comput Aided Des 39:324–338
Zurawski R (2007) Integration technologies for industrial autmated systems. CRC Press; Taylor & Francis Group, New York

Distributed, Collaborative and Automated Cybersecurity Infrastructures for Cloud-Based Design and Manufacturing Systems

J. Lane Thames

Abstract Cloud-Based Design and Manufacturing (CBDM) refers to a product realization model that enables collective open innovation and rapid product development with minimum costs through a social networking and negotiation platform between service providers and consumers. Because of the diversity of cyber resources constituting CBDM systems and because CBDM systems are Internet-enabled, cyber-physical platforms, new types of cybersecurity systems that provide real-time, dynamic, and preemptive protection will be needed. In this chapter, an overview will be provided of emerging global-scale cyber information exchange frameworks that will enable this type of cyber protection. Further, a reference architecture utilizing information obtained from global cyber exchanges for dynamic cyber protection of CBDM systems will be proposed.

1 Introduction

Recently, cloud computing has made its advent to the domain of computer-aided product development. In addition to running computer-aided design (CAD) systems as a service in the cloud, other business-related everything-as-a-service models have started to emerge. One such model relates to manufacturing and aims at utilizing physical resources, for example, 3D printers for additive manufacturing, mills, lathes, and other manufacturing-related equipment, through the cloud. Long-term, computer-aided product development in general (including design, analysis, and simulation, as well as manufacturing) is anticipated to become predominantly cloud based. It is a promising new model to facilitate globally distributed design and manufacture processes that seamlessly integrate both virtual

J. Lane Thames (✉)
Research and Development Department, Tripwire Inc., 30009 Alpharetta, GA, USA
e-mail: lthames@tripwire.com

D. Schaefer (ed.), *Cloud-Based Design and Manufacturing (CBDM)*,
DOI: 10.1007/978-3-319-07398-9_8,

and physical resources. Cloud-Based Design and Manufacturing (CBDM) refers to a product realization model that enables collective open innovation and rapid product development with minimum costs through a social networking and negotiation platform between service providers and consumers within the so-called 'Cloud.'

CBDM systems are, by definition, cyber-enabled systems that utilize Internet communication capabilities along with various forms of information technologies. Unfortunately, cybersecurity continues to be a challenging and complex problem for systems that use Internet communication technologies. Consequently, CBDM systems will require sophisticated cybersecurity infrastructures to protect their highly sensitive and valuable assets.

Because of the diversity of cyber resources constituting CBDM systems and because CBDM systems are Internet-enabled, cyber-physical platforms, new types of cybersecurity systems that provide real-time, dynamic, and preemptive protection will be needed. Traditional cyber defense is no longer applicable on many levels. One particular trend, as evidenced by standards being developed, such as TAXII (Mitre. Taxii: Trusted automated exchange of indicator information. http://taxii.mitre.org), STIX (Mitre. Stix: Structured threat information expression. http://stix.mitre.org), and CyBEX (Rutkowski et al. 2010), is the development of cyber threat information exchanges that can be used for automated and active response to real-time cyber attacks (Lane Thames 2012, 2013). These standards and the resultant technologies that enable real-time cyber defense systems are likely to play a key role for securing Internet-enabled platforms, such as CBDM systems.

The purpose of this chapter is to provide the reader with a general understanding of how CBDM systems can be secured using distributed, collaborative, and automated technologies. To set the stage, the chapter will begin with an overview of CBDM. Next, the chapter with provide a brief overview of common cybersecurity technologies. Then, the chapter will describe a number of emerging initiatives that seek to facilitate the sharing of cybersecurity information. Finally, the chapter will propose a distributed security reference architecture that can utilize shared cybersecurity information to provide more robust, dynamic, and real-time cybersecurity for CBDM systems.

2 Cloud-Based Design and Manufacturing Systems

Most recently, cloud computing has made its advent into the domain of computer-aided product development (Schaefer 2011). As a first step in this direction, companies consider replacing their own Computer-Aided Design (CAD) software licenses with CAD software as a service in the cloud (Schaefer 2011). Running a CAD package on a provider's server(s) through the cloud and paying a small fraction of the original license fee on a pay-as-you-go usage basis is certainly appealing. In addition, time and cost intensive software updates and maintenance issues are out of the picture as well. On the downside, it is obvious that an internal local area network connection allows significantly faster data transfer rates than an Internet connection. In addition, rendering CAD data can be very demanding in terms of computing power and over the Internet one may experience a slight lag in response time. Whether or not the lag is tolerable will depend on the various usage scenarios. While it may be perfectly acceptable in a CAD training environment, rendering delay due to lag may be annoying (and costly) in day-to-day full-scale design operation. In practice, currently most companies choose speed over cost and prefer to run their software locally. One way of minimizing the response lag is to store CAD data and software on the same server. The less data one needs to transfer through the cloud, the better (and faster/cheaper). In light of this, it becomes apparent that providing/renting storage space as a service through the cloud is yet another interesting business model to consider. Pay-as-you-go rates for data storage may be significantly cheaper than purchasing your own hard disk drives and, similar to the SaaS case, hardware maintenance and replacement are no longer an issue. An important issue related to such cloud computing services is that of data security. Not knowing exactly where in the cloud sensitive data is stored and what is going to happen to it in case of a blackout or server crash is a major concern for any company.

In addition to CAD software as a service, other business-related "everything-as-a-service" models have started to emerge (Banaerjee et al. 2011). The same trend will also be seen within the mobile Internet, which will be one particular communication platform that utilizes ubiquitous cloud computing technology for mobile smartphone applications along with mobile devices actually offering cloud services, such as data collection, i.e., smart phone sensors. For example, Morgan Stanley (Morgan Stanley, The mobile Internet report) claimed that the number of mobile Internet users could be 10 times larger than the number of personal computer Internet users.

While computing and information technology is undergoing a seismic paradigm shift from the traditional client/server model to the enhanced cloud computing model, design and manufacturing communities are beginning to consider aspects of cloud computing. A number of companies, including Autodesk and Fujitsu, are attempting to implement this model. For example, Autodesk claims that they are able to provide their customers with greater access to design and engineering documents anywhere and any time (Autodesk. autodesk.com.) Some of the featured services include: (1) Cloud rendering, providing customers with powerful rendering capabilities so as to have better visualization of 3D models; and (2) Software as a service, helping designers to exchange information securely so as to enhance effectiveness and efficiency of team collaboration. Another example is Fujitsu and its engineering cloud, which makes it possible to efficiently consolidate applications and high-volume data formats. Their engineering cloud provides a high-speed thin client environment, server consolidation, and license consolidation (Fujitsu. fujitsu.com.).

One particular and emergent paradigm currently being investigated (Schaefer et al. 2012) is related to manufacturing and aims to explore the potential of extending the cloud computing model to physical resources, such as 3D printers for distributed and Internet-enabled additive manufacturing machines, mills, lathes, and other manufacturing-related resources. Long-term, computer-aided product development in general (including design, analysis, and simulation, as well as manufacturing) is expected to become predominantly cloud based. It is considered a new model to aid future globally distributed design and manufacture processes that seamlessly integrate both virtual resources, such as CAD systems as well as physical resources such as, for example, additive manufacturing machines.

This new model and emergent paradigm is referred to as Cloud-Based Design and Manufacture (CBDM). Previously, my colleagues and (Schaefer et al. 2012) proposed the following definition for cloud-based design and manufacturing:

> CBDM refers to a product development model that enables collective open innovation and rapid product development with minimum costs through social networking and crowd-sourcing platforms coupled with shared service pools of design and manufacturing resources and components.

Figure 1 illustrates the concepts underlying the foundations and principles of CBDM systems aligned with our proposed definition thereof.

Before describing the elements illustrated by Fig. 1, it is noteworthy to explain the use of the term "Cloud". Communication and network engineers have traditionally encapsulated the inherent interconnection complexity of networks with *cloud diagrams*. In essence, a network of any reasonable size is too complex to draw on a diagram. Consequently, cloud diagrams are used to hide the interconnect complexity, while simultaneously revealing the primary details of a particular network diagram. As seen from Fig. 1, the Internet communication 'cloud' forms the basic and required underlay network for any CBDM system in general. As stated previously, CBDM technologies are enabled by Internet-based information and communication technologies. This dependency is represented by illustrating CBDM as an overlay in Fig. 1. Moreover, the figure seeks to illustrate the overall and basic interconnectivity of the primary elements of a CBDM system. For example, the human resources of a CBDM system form their own human-centric network, which is represented by design teams, social networks, and students, just to name a few. Likewise, the cloud resources, which include human, virtual, and physical resources, are illustrated along with their appropriate partitions. One of the primary goals of CBDM is to enable efficient product development and realization processes. Hence, appropriate interconnections are established between this goal and the basic partitions of the diagram. Further, one should observe the needs of the product development and realization process, namely industrial needs and educational needs. These two sectors comprise the basic categories of entities that need the CBDM functionality. Moreover, industrial needs and educational needs are, in general, intricately bound. Industry will use CBDM technology to produce raw goods and services. Obviously, industry depends on educational entities for the following: (1) to educate students on the basic principles and foundations of CBDM systems in order to accomplish their economic goals and (2) to conduct cutting-edge research and development on the underlying details of CBDM systems. Hence, the educational and industrial entities are intricately bound.

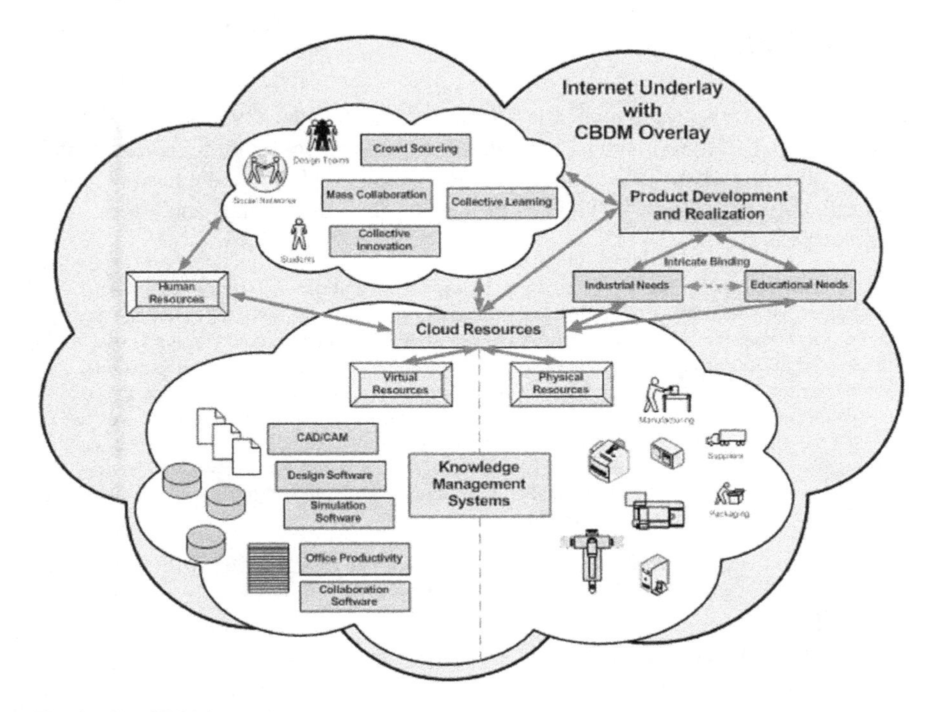

Fig. 1 The CBDM concept

By definition, most if not all of the entities comprising a CBDM system have some type of cyber 'connection.' Even physical entities have cyber interfaces, thereby causing them to be cyber-physical. As such, the entities of a CBDM system will require various types of cybersecurity mechanisms. In the next section, several categories of cybersecurity mechanisms will be presented so that the reader will have an understanding of the types of mechanisms available for securing CBDM systems along with knowledge into the workings thereof.

3 An Overview of Cybersecurity

In this section, the basic concepts underlying firewalls, intrusion detection, and intrusion prevention are introduced. Although other cybersecurity technologies exist, the technologies reviewed in this section are the most relevant for securing CBDM systems, particularly when used in conjunction with cybersecurity information exchanges.

3.1 Terminology

The International Telecommunication Union (ITU) and its Overview of Cybersecurity (ITU-T. X Series: Data networks, open system communications and security. International Telecommunication Union (ITU), Telecommunication Standardization Sector (ITU-T)) provides the following definition of cybersecurity:

> Cybersecurity is the collection of tools, policies, security concepts, security safeguards, guidelines, risk management approaches, actions, training, best practices, assurance and technologies that can be used to protect the cyber environment and organization and user's assets. Organization and user's assets include connected computing devices, personnel, infrastructure, applications, services, telecommunications systems, and the totality of transmitted and/or stored information in the cyber environment. Cybersecurity strives to ensure the attainment and maintenance of the security properties of the organization and user's assets against relevant security risks in the cyber environment. The general security objectives comprise the following: Availability, Integrity (which may include Authenticity and Nonrepudiation), and Confidentiality.

Three key concepts form the core of cybersecurity problems: vulnerabilities, threats, and attacks. A vulnerability is any weakness contained within a computing or networking system that can be exploited. A threat is defined as any process that can potentially violate the security policies of a system. An attack is considered to be any active process that deliberately seeks to violate the security policies of a system.

Organizations seek to protect their computational and networking infrastructures based on defense-in-depth strategies. The objectives of this strategy are to reduce the impact of attacks that successfully exploit vulnerabilities within the system and to reduce the number of threats within the system. The design of systems using defense-in-depth strategies should be founded on the ideas of security architectures and security management infrastructures. These two concepts have been defined by Shirey (2000) as follows. A security architecture is a plan and set of principles that describe (a) the security services that a system is required to provide to meet the needs of its users, (b) the system elements required to implement the services, and (c) the performance levels required in the elements to deal with the threat environment. A security management infrastructure includes system elements and activities that support security policy by monitoring and controlling security services and mechanisms, distributing security information, and reporting security events.

3.2 Firewalls

Firewalls are security mechanisms that control the flow of network traffic, i.e., communication packets, into or out of a communication system (Cheswick et al. 2003). The flow of network traffic through a firewall is governed by *security policies*, which are defined by a collection of *N packet filters* (or filters for short) written in the native language of a particular firewall technology. Sometimes, packet filters are referred to as firewall rules.

Firewalls are dependent on network topology, such that filters are assigned to the direction of traffic flow relative to the orientation of the firewall with respect to the virtual resources it protects. In particular, a firewall provides an interface between *secure resources* and *insecure resources*. Secure resources reside behind an *inside interface* of the firewall, whereas insecure resources reside beyond an *outside interface* of the firewall. Figure 2 illustrates the idea of a firewall and its inside–outside topology-dependent perspective.

Firewall designs are constructed by the *Principle of Least Privilege*, which is expressed as follows:

> All communication traffic attempting to traverse a firewall must be explicitly permitted by a security policy. Otherwise, the traffic must be denied.

In other words, if a packet is processed by a firewall and does not match an explicitly defined filter, then the firewall must deny the packet's access into or out of the communication system. The filtering process provided by firewalls is demonstrated by the arrows of Fig. 2, which represent packet flows entering and leaving the outbound and inbound filter collections. The arrows leaving out of the inbound/outbound filter collections represent packet flows that matched one or more packet filters. Packet denial is represented by more arrows entering a filter collection (inbound or outbound) than ones leaving. The suppressed arrows represent packet flows that are denied by the firewall.

Figure 3 illustrates a simple perimeter firewall system.

As can be seen in Fig. 3, the collection of packet filters assigned to the firewall state that packets from the 'outside' and destined to the inside host named *WWW* with destination port 80, as well as those destined to the inside host named *Email* with destination port 25 are *explicitly* permitted (allowed). Further, the filter written as *allow from inside to any port* provides *unrestricted* access from inside users/resources to any external Internet resource. The last filter, *deny from any to any port any*, is the default filter blocking any packets that do not match the first three filters, which enforces the principle of least privilege.

Table 1 shows a snippet of packet filters from a real-world enterprize-class Cisco PIX firewall. The filters shown in Table 1 have a structure based on the so-called 5-tuple firewall filter, which includes *header field specifications* from the network and transport layers of the TCP/IP protocol stack. In particular, the 5-tuple includes the (*protocol type, source IP address, destination IP address*) from the network layer of the TCP/IP stack and the (*source port, destination port*) from the transport layer of the TCP/IP stack. In general, firewall packet filters can describe any value permissible by the specifications of the associated TCP/IP headers. For example, IPv4 addresses are contained in the source/destination IP headers in the network layer of the TCP/IP stack. Since IPv4 specifies 32 bits for IP addresses, any one or more of the possible 2^{32} IP addresses can be used for the source or destination IP addresses of a packet filter. In most cases, firewall filters specify the

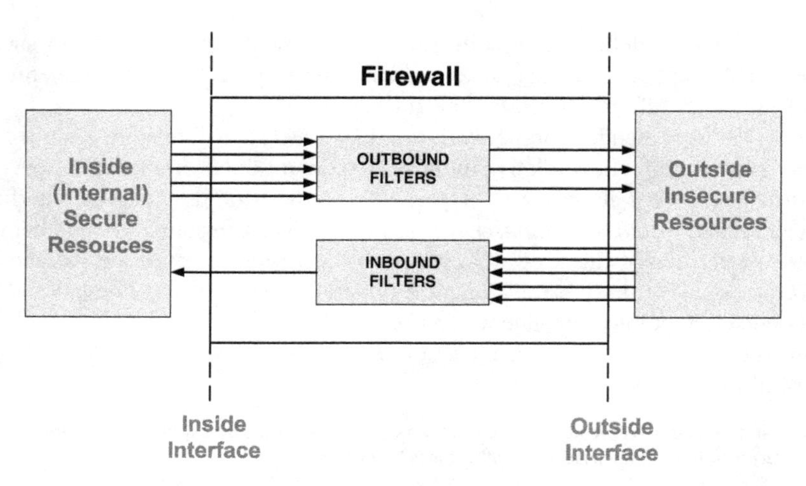

Fig. 2 The inside–outside topology-dependent firewall perspective

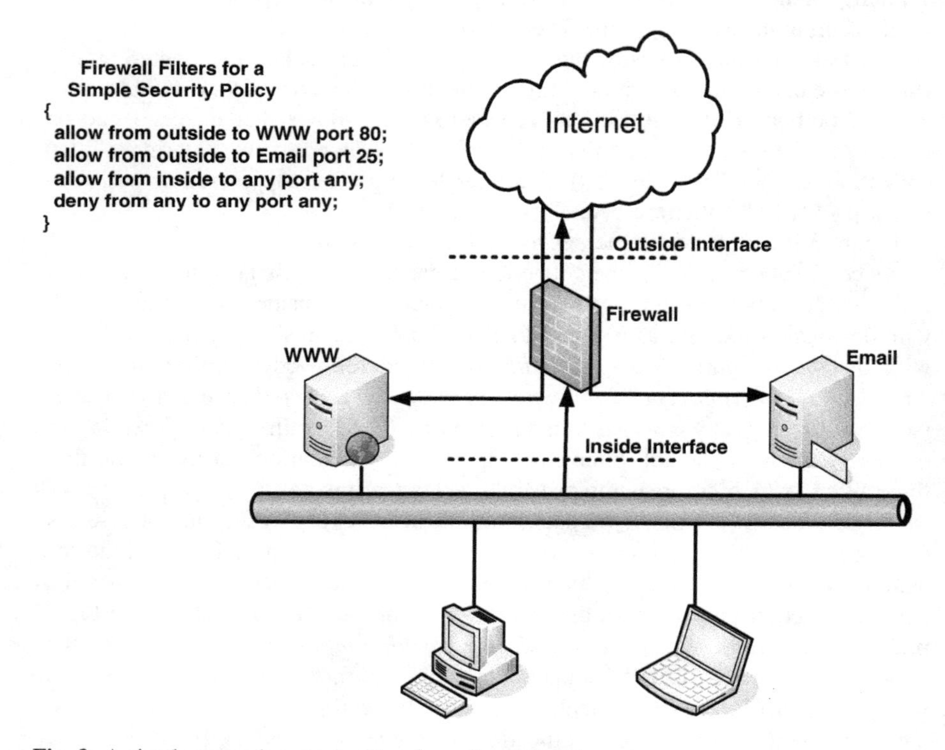

Fig. 3 A simple network system with a firewall and its filter-set

Table 1 A real-world firewall filter-set

Proto	SrcIP	DestIP	SrcPort	DestPort
gre	130.207.134.9	128.61.209.0/26	any	any
tcp	130.207.134.9	128.61.209.0/26	any	eq pptp
gre	130.207.134.9	128.61.209.128/26	any	any
tcp	130.207.134.9	128.61.209.128/26	any	eq pptp
tcp	128.61.252.110	143.215.254.64/29	any	eq https
tcp	128.61.252.110	143.215.254.64/29	any	eq www
tcp	128.61.252.110	143.215.254.64/29	any	range 22 23
tcp	128.61.5.0/24	143.215.254.64/29	any	eq https
tcp	128.61.5.0/24	143.215.254.64/29	any	eq www
tcp	128.61.5.0/24	143.215.254.64/29	any	range 22 23
ip	any	any	any	any
ip	128.61.210.0/24	any	any	any
ip	128.61.211.0/24	any	any	any
ip	128.61.212.0/24	any	any	any
ip	128.61.213.0/24	any	any	any
ip	143.215.254.8/29	any	any	any
ip	143.215.254.56/29	any	any	any
ip	143.215.254.64/29	any	any	any
ip	any	any	any	any

tuples as a collection of *points, prefixes,* or *ranges*. A 'point' is a single numeric value, a prefix is an expression based on the first x-bits of a value followed by wildcards in the remaining $w - x$-bits where w is the number of bits allocated to the particular header field in the TCP/IP stack, and a range is of the form $[L, U]$ where L is the lower endpoint and U is the upper endpoint of a range. Table 2 gives an example of each. Prefixes are specified by a certain number of *most significant bits* and are always inherently tied to the number of bits specified by a particular header field. For example, if the value 01000 * from the second row of Table 2 represents an IP address field, then an IP address where the first 5 bits equal 01000 will match the expression regardless of the remaining $32 - 5 = 27$ bits. In this case, the first 5 bits are the most significant bits of the filter specification 01000 * and the remaining 27 bits are the least significant *don't care* bits with respect to the filter expression. The range specification of the last row of Table 2 specifies any value greater than 1,023. However, since the header fields of the TCP/IP stack are specified by a particular number of w bits, the range specification >1023 implies that the range is interpreted as $[1024, 2^w - 1] = (1023, 2^w - 1]$. For instance, if the range is based on the 16 bits of the TCP port number specification, then the range above is [1,024, 65,535].

Table 2 An example of points, prefixes, and ranges used by packet filters

Point	Prefix	Range
128.61.208.3	128.61.55.0/24	[20, 22]
TCP	01000*	[000, 110]
http = 80	10.20.*.*	>1023

3.3 Intrusion Detection

In general, cyberattack detection mechanisms are processes that collect data from within the cyber environment and utilize classification algorithms to determine the legitimacy of events associated with the data. A cyberattack detection system is a security mechanism that utilizes one or more cyberattack detection algorithms. A cyberattack prevention mechanism seeks to apply countermeasures that prevent adverse impacts caused by cyberattacks. A cyberattack prevention system employs one or more cyberattack detection mechanisms along with the associated techniques required to implement countermeasures.

An intrusion is defined as any unauthorized attempt to access cyber resources. Intrusion detection and intrusion detection systems are particular cases of the more general categories of cyberattack detection and cyberattack detection systems. The same is true for intrusion prevention and intrusion prevention systems. Consequently, intrusion detection/prevention and cyberattack detection/prevention will be used interchangeably throughout this dissertation.

Intrusion detection is concerned with the discovery of intruders attempting to gain unauthorized access to an organization's computing, networking, and information resources (Anderson 1980). Intrusion detection is a process performed by an intrusion detection system (IDS) where events within a cyber environment are analyzed via audit data and sequences of events that indicate violations of an organization's security policies trigger alerts that notify security administrators of the abnormal system activities.

The fundamental premise underlying the design of an IDS is based on the idea of *behavior*. Understanding the behavior of computer users and processes is a central idea underlying the design of detection algorithms for intrusion detection systems. Particularly, the behavior of an intruder (human or agent) is assumed to be significantly different from the behaviors of legitimate users (Stallings 2003). For example, in *most* cases, a legitimate user will not attempt to read to the contents of a system's password database. However, an intruder will *usually* attempt to retrieve the contents of the password database. The terms 'most' and 'usually' are highlighted to illustrate the *statistical* nature that actually underlies the *behavioral premise* of IDS design. But, the behaviors are not always different. In other words, a behavioral *overlap* exists whereby activities of legitimate users versus intruders (equivalently, legitimate cyber activities versus cyberattacks) cannot be perfectly differentiated at all points in time. The preceding arguments imply that an intrusion detection system and its detection algorithm(s) must compensate for their inherent *false negative* and *false positive* detection rates (Axelsson 2000).

An IDS is commonly categorized by two types of algorithmic approaches, which include *anomaly detection* and *signature detection*. Anomaly detection applies statistical analysis to real-time data based on historical archives of data representing legitimate (normal) user activity. Behavior deviating from the statistical profiles of normal activities is considered abnormal (anomalous), and these deviations are used as the basis for detecting intrusions. Signature detection utilizes a collection of rules derived from data and patterns related to known profiles of particular types of intrusive behavior. Activities that generate content with patterns matching one or more signatures within the rule-base are indicative of an intrusion. A special case of signature detection is known as *blacklisting*. A blacklist is a collection of distinct identifiers associated with well-known malicious sources. Blacklists can be composed of values, such as IP addresses, application types, filenames, or transport layer port numbers. The primary distinction between these two intrusion detection approaches is the following. Anomaly detection is concerned with the behavioral (or activity) profiles of legitimate users, whereas signature detection is based on the behavioral or activity patterns (profiles) of intruders.

3.4 Intrusion Prevention

Intrusion prevention can be categorized by two primary security mechanisms related to intrusion activities: (1) proactive mechanisms and (2) reactive mechanisms. *Proactive* intrusion prevention mechanisms enforce security policies with services, such as authentication, authorization, and access control. For example, password systems and firewalls are proactive intrusion prevention systems. These systems implement security policies designed to prevent unauthorized access to cyber resources. *Reactive* intrusion prevention mechanisms function synergistically with intrusion detection systems and provide functionality that aims to stop intrusion occurrences in real-time and/or to recover from any adverse effects caused by a successful system penetration. Intrusion prevention systems capable of both detecting intrusions and reacting with countermeasures to intrusion occurrences are referred to as *active response systems*.

Collectively, intrusion detection and prevention systems are categorized as intrusion response systems (IRS). In Stakhanova et al. (2007), proposed an intrusion response system taxonomy, which is illustrated graphically by Fig. 4. In their taxonomy, an IRS is classified by two top-level characteristics indicating the system's degree of automation and the system's response activity. IRS automation is categorized by notification systems, manual response systems, and automatic response systems, whereas response activity comprises passive and active response. Passive response systems are notification-based systems. Active response systems, as described above, provide real-time countermeasures to

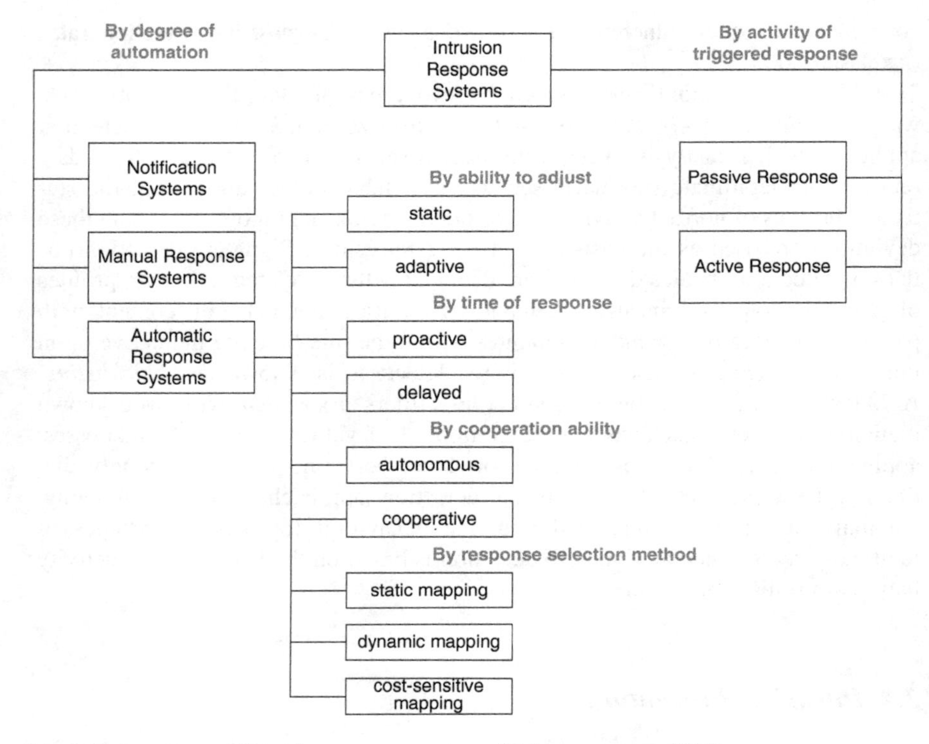

Fig. 4 A taxonomy of intrusion response systems (Stakhanova et al. 2007)

intrusions and attacks. Active response includes countermeasures, such as disabling user accounts, terminating system processes, and traffic filtering. Automatic response systems are further partitioned by its adjustment ability, response time, cooperation ability, and response selection method.

The chapter thus far has presented information overviewing CBDM systems and cybersecurity technologies. In the next section, a discussion will be given of emerging technologies and standards seeking to define standards for sharing cybersecurity event information across organizational boundaries.

4 Global-Scale Cybersecurity Information Exchanges

Information related to cybersecurity events plays an important role for systems that protect cyber assets. One way to increase the robustness of existing cybersecurity systems will be to expand the information horizon from within which we collect security event information.

Fortunately, a number of initiatives based on establishing cybersecurity information exchange technologies are gaining momentum. One such initiative is

TAXII (Trusted Automated eXchange of Indicator Information) (Mitre. Taxii: Trusted automated exchange of indicator information. http://taxii.mitre.org). There are others such as Cybersecurity information EXchange framework (CyBEX) (Rutkowski et al. 2010) and CIF [the Collective Intelligence Framework (CIF The collective intelligence framework. https://code.google.com/p/collective-intelligence-framework/)]. However, technologies associated with TAXII will be the focus herein.

TAXII is an open, community-driven, standardization effort led by the U.S. Department of Homeland Security (DHS) and facilitated by the MITRE Corporation, which is a not-for-profit organization that serves as a community coordinator for TAXII. The basic premise underlying initiatives, such as TAXII is that organizations can benefit by sharing security event information. As such, TAXII's objective is to establish a framework for standardized, trusted, and automated exchanges of cyber threat information.

Organizations can use TAXII services and message exchanges in order to share security event information with one another, so that emerging cyber threats and attacks can be detected and mitigated more quickly than with existing noninformation sharing technologies. For example, if organization ***O*** detects a cyber threat ***T***, it can distribute information that describes ***T*** via TAXII to other organizations. These recipient organizations can then implement preemptive security controls and countermeasures that prevent adverse effects caused by ***T***.

TAXII is not a stand-alone initiative. It is tightly coupled to other initiatives, such as STIX, CybOX, and MAEC. Structured Threat Information eXpression (STIX) (Mitre. Stix: Structured threat information expression. http://stix.mitre.org) is an initiative seeking to define standardized and structured representations of cyber threat information. Cyber Observable eXpression (CybOX) (Mitre. Cybox: Cyber observable expression. http://cybox.mitre.org) is a structured language that describes entities within an organization's cyber operational environment. Malware Attribute Enumeration and Characterization (MAEC) (Mitre. Maec: Malware attribute enumeration and characterization. http://maec.mitre.org) is a structured language for describing malware attributes. STIX, CybOX, and MAEC are, similar to TAXII, led by DHS and facilitated by MITRE.

The inter-relationship of these initiatives is as follows. STIX uses languages, such as (but not limited to) CybOX and MAEC to represent cybersecurity event information, and TAXII serves as the transport mechanism for STIX information.

Recently, Microsoft announced that it intends to support TAXII and STIX. According to an article describing new Microsoft Active Protections Program (MAPP) initiatives (Microsoft. Mapp initiatives. http://blogs.technet.com/b/bluehat/archive/2013/07/29/new-mapp-initiatives.aspx):

> Through this new program, MAPP for Responders, we are working to build new partnerships and community collaborations that will enable strategic knowledge exchange. Microsoft intends to contribute to this effort by sharing threat indicators such as malicious URLs, file hashes, incident data and relevant detection guidance. Employing a "give to get" model, the community will benefit when data they provide is enriched by aggregating it with data from others.

The article (Microsoft. Mapp initiatives. http://blogs.technet.com/b/bluehat/archive/2013/07/29/new-mapp-initiatives.aspx) goes on to say:

> Effective knowledge exchange requires automation and a common format. To accomplish this, we plan to support Mitre's STIX and Trusted Automated eXchange of Indicator Information (TAXII) specifications. As open specifications for the formatting and transport of information, STIX and TAXII are starting to see broad adoption. Regardless of format, we want to serve customers by facilitating the flow of threat intelligence to organizations who can capitalize on it. As such, we will also seek to build transforms for other commonly used formats. This effort is currently in development and we intend to launch a pilot in the near future.

Hence, industry is seeing the potential advantages for developing innovative systems that provide security event information sharing capabilities. Indeed, this is the class of technology that will be needed to provide more robust security for CBDM systems. In the remainder of this section, an overview will be given of the key features of TAXI and STIX.

4.1 Taxii

TAXII is a standardization effort that seeks to enable trusted and automated exchange of cybersecurity event information across organizational perimeters (Mitre. Taxii: Trusted automated exchange of indicator information. http://taxii.mitre.org). In other words, the goal is to provide mechanisms that allow organizations to share security intelligence with one another in a trusted and automated fashion. In this subsection, the important characteristics of TAXII (TAXII. Taxii specification, version 1.0. http://taxii.mitre.org/specifications/version1.0/TAXII_Overview.pdf) are provided.

TAXII is concerned mostly with the 'transport' of security event information. As such, defining an information sharing model is one of its key features. Three primary information sharing models are defined by TAXII: *Hub and Spoke*, *Source and Subscriber*, and *Peer to Peer*.

Hub and Spoke is a centralized information sharing model where the 'hub' acts as the central clearinghouse for all participants (spokes) of the collaboration network. When a participant has information to share, it will encode the security event information and pass it on to the hub. Then, the hub will distribute this new information to all other parties within the network. Source and Subscriber information sharing models allow individual parties to subscribe to other information providers within a collaboration network. Note that hub and spoke communication models are 'broadcast' models, whereby all parties within a collaboration network produce information that is shared with all other parties and consume information produced by all other parties. There is no 'selective' mechanism within the sharing model itself. (Although this does not mean that a party must act on information obtained from another.) However, source and subscriber models are many-to-many communication models, but the production and consumption is based on subscription. If a party wants to receive security event information from a collaborating entity, it must subscribe itself with that entity in order to receive it. One might consider this to be a filtered broadcast information sharing model. Peer to peer information sharing models are similar to hub and spoke. However, new information is distributed between peers instead of a centralized hub. Finally, it is to be noted that variants of these three basic information sharing models are allowed.

The aforementioned information sharing models define the high-level communication architectures that can be used by TAXII implementations. At a lower level, TAXII defines a *Services Specification*, *Protocol Binding Specifications*, and *Message Binding Specifications*. TAXII services specifications define services offered by TAXII implementations. The protocol binding specification defines how messages should be transported and which network protocols should be used for the transport. The message binding specification defines requirements for representing and encoding TAXII messages. For example, a collaboration network might specify that messages be encoded with XML and be transported over HTTPS (secure hypertext transfer protocol).

TAXII categorizes the participants of a collaboration network as *Consumers* and *Producers*. Producers supply the collaboration network with new or updated security event information. Obviously, consumers consume the security event information provided by producers. Any cyber entity within an organization can be both a producer and a consumer. In general, this is the best scenario. However, there are cases where this cannot be achieved. This is especially true for networks, such as CBDM systems because many entities within a CBDM system are so-called cyber-physical entities with limited processing and communication power. Obviously, there are other networked systems with similar limitation properties. In general, we should assume that entities with limited resource are protected by perimeter devices with more powerful resources.

As stated above, TAXII implementations serve as the transport mechanism for trusted and automated exchange of cybersecurity event information. In the next subsection, standards will be reviewed for describing and encoding security information, which produces the content that is actually transported by TAXII.

4.2 Stix

STIX–structured threat information expression–is a standardization effort for defining a language that represents structured cyber threat information (Mitre. Stix: Structured threat information expression. http://stix.mitre.org). A fundamental goal of this standardization effort is to provide a language for representing structured cyber threat information, such as security event information. Moreover, the structured representation should be expressive, flexible, extensible, readable, and automatable.

STIX ties together a number of key architectural components, whereby information related to the architectural components constitute the sources of information to be encoded by a STIX language. Some of these key architectural components include the following: *cyber observables*, *indicators*, *incidents*, *adversary tactics, techniques, and procedures*, *exploit targets*, *courses of action*, *cyberattack campaigns*, and *cyber threat actors*. Cyber observables are 'anything' that can be observed within the cyber infrastructure. For example, it could be samples of network traffic destined to a cloud-based CAD software server or files being uploaded to a cloud-based 3D printer. Indicators are cyber observables along with relevant data related to abnormal (malicious) behavior. Incidents are collections of information related to particular adversarial actions. Adversary tactics, techniques, and procedures refers to information related to cyber activities, such as attack patterns, malware signatures and propagation methods, and infrastructure details, such as command and control systems for botnets. Exploit targets define entities along with their vulnerabilities, weaknesses, and configurations using data, such as common vulnerabilities and exposures (CVE Common vulnerabilities and exposures. http://cve.mitre.org/). Courses of action are information describing, for example, how to protect entities with certain vulnerabilities or how to protect a network from an emerging threat. Cyber attack campaigns include data related to sets of incidents. Finally, cyber threat actors are data that describe, identify, and characterize malicious cyber entities. For example, it could be a black list of malicious IP addresses.

With this comprehensive list of cyber threat information, STIX can, in general, provide a fine-tuned, highly-detailed characterization of an organization's cyber threats and cyber adversaries. When this information is coupled with a transport mechanism, such as TAXII along with a distributed and highly dynamic protective framework, a significant increase in cybersecurity is achievable. As such, the following section will describe one such distributed and dynamic protection framework that can be utilized by CBDM systems.

5 Reference Architecture for Securing CBDM Systems

In this section, a cybersecurity reference architecture for CBDM systems is presented. This reference architecture is referred to as a distributed firewall and active response (DFAR) architecture. DFAR is viewed as a distributed system of nodes within a Trusted Domain of Administration (TDA). The TDA is the lowest level of network granularity with respect to an organization's collection of cyber resources.

The design underlying DFAR assumes that each cyber entity constituting a TDA is a producer and consumer of DFAR information, and this information is used to dynamically protect the overall network. Ergo, each entity within DFAR is a producer and consumer of cybersecurity event information. Moreover, this security event information is sent to or received from other partner organizations in the form of TAXII messages. In this section, the details describing the main

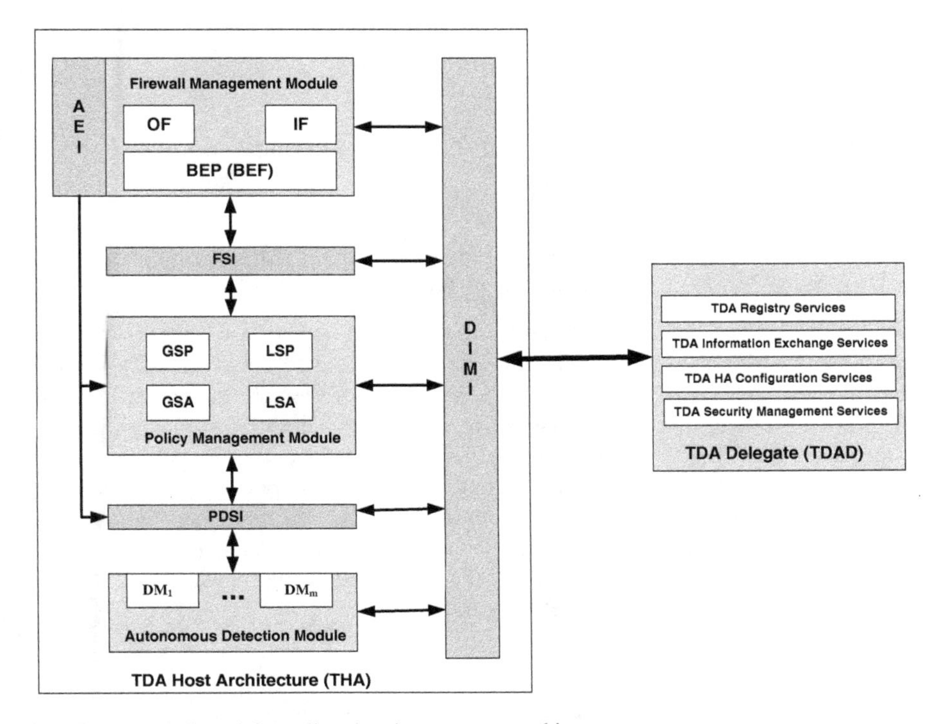

Fig. 5 The distributed firewall and active response architecture

functionalities of DFAR are introduced. Then, the section will tie these pieces together along with information provided in previous sections of this chapter in order to illustrate how CBDM systems can achieve cyber protection in a distributed, collaborative, and automated fashion.

Figure 5 provides a block diagram of the distributed firewall and active response architecture. DFAR is a distributed system of nodes that implement the TDA host architecture (THA). The TDA is viewed as a distributed infrastructure with centralized interfacing and management. In other words, the TDA is a distributed system with centralized control. The TDA Delegate (TDAD) is responsible for centralized interfacing and security management of the overall system along with serving as the primary exchange point of an organization for TAXII messages. DFAR centralized interfacing and security management seeks to solve various challenges that prevent large-scale distributed systems integration and information exchange for systems having large numbers of dynamic cyber resources, i.e., resources within a CBDM implementation.

The TDA host architecture is designed to be modular with communication between THA modules as well as between the end host and the TDAD occurring via communication interfaces. The THA is comprised of three primary modules and four inter/intra process communication interfaces. The primary THA modules are the following:

- Autonomous Detection Module
- Policy Management Module
- Firewall Management Module

The THA communication interfaces include the following:

- Audit Export Interface
- Filter Specification Interface
- Policy Description Specification Interface
- Distributed Infrastructure Management Interface

Two terms are used extensively throughout the description of the distributed firewall and active response architecture and its associated behaviors. These two terms are *host* and *local*, and the precise semantics of these terms are defined as follows with respect to DFAR. A DFAR host is ***any*** cyber entity implementing an instance of the TDA host architecture (as illustrated by Fig. 5 and described below). For example, a host can be a complete physical computing system, such as a desktop computer, a server, or a laptop, or it can be a virtual system such as a software process running within a physical computing system. The term local refers to cyber resources, such as TDA hosts or cyber components protected by TDA hosts that are invariant to network topology.

5.1 Autonomous Detection Module

The autonomous detection module (ADM) is a collection of independent autonomous cyberattack detection modules (DM) that are active on the local system and the collection of rules that define how the ADM must interact with the policy management module. The autonomous detection module (ADM) contains zero or more independent cyberattack detection modules DM_i that are responsible for monitoring audit data collected by the local host and/or collected from collaborating audit data collection systems. These detection modules are responsible for real-time detection of cyberattack events.

The ADM can provide network-based audit event data to the various detection modules via (1) packet capture processes that directly interact with the network interface of the local host, (2) via audit event data collected by the firewall management module and delivered over the policy description specification interface, or (3) via audit event data provided by the TDAD over the distributed infrastructure management interface, i.e., new TAXII messages containing STIX information.

The DM can be as simple as a process monitoring the local host for failed login attempts to as complex as a full-scale IDS deployment. Regardless, the ADM was designed to be modular and extensible. Cyberattacks are dynamic and continuously evolving. As a result, detection modules should be capable of adapting themselves in an autonomic fashion in order to adjust their detection behaviors.

Because cyberattacks are significantly diverse, it is not feasible to have a single, monolithic detection module capable of reliable classification for all possible attack scenarios. As a result, the ADM is designed to be modular such that many different detection modules operate, independently if needed, on a given host. The goal is to enable different types of independent, target-specific, autonomous attack detectors to reside on a single host, thereby providing a more robust defense-in-depth strategy for the local host and the overall TDA.

5.2 Policy Description Specification Interface

The policy description specification interface (PDSI) provides local communication interfacing between the policy management module, the attack detection module, and the audit export interface of the firewall management module. The PDSI delivers new blocking enforcement filters generated by the ADM to the policy management module. Moreover, it receives audit event data from the audit event interface and delivers audit event data to the attack detection module.

5.3 Policy Management Module

The policy management module (PMM) is responsible for management of global and local security policies (GSP, LSP) for the local host. The PMM also manages the archival of global security audits (GSA) received from the TDA via the DIMI along with archival of local security audits (LSA) generated by the firewall management module and the autonomic detection module.

Global security policies and audits provide information representing views from distributed hosts comprising the TDA. This *global perspective* can be used by the host to make, in conjunction with outputs received by the ADM and the AEI, more informed decisions when new cyberattacks are discovered within a CBDM system.

5.4 Filter Specification Interface

The filter specification interface (FSI) provides a communication mechanism that translates new PDSs received from the PMM into the filtering language of the blocking enforcement point (BEP) that implements the firewall functionality for the local TDA host. The FSI creates a blocking enforcement filter (BEF) for the outbound interfaces (outbound filters, OF) and the inbound interfaces (inbound filters, IF). This interface allows for heterogeneous firewall systems, whereby (1) the local host may implement multiple firewall instances if multi-firewall functionality is needed and (2) the TDA is not required to implement any particular type of firewall.

5.5 Firewall Management Module

The firewall management module (FMM) is responsible for dynamically configuring firewall filters when rule structures are received from the FSI. The FMM inserts new inbound filters and outbound filters based on information provided by the FSI. The FMM also collects audit data based on traffic flowing through the outbound/inbound interfaces and delivers the audit data to the PMM and/or the PDSI via the audit export interface (AEI). This audit data is archived via the LSA functionality of the PMM and can be delivered to the ADM detection modules by the PDSI.

5.6 Distributed Infrastructure Management Interface

The distributed infrastructure management interface (DIMI) is the key communication interface that integrates the host members of the TDA into the overall system. It provides direct communication between THA elements and the TDAD, thereby providing collaboration capabilities for the system. The DIMI will be explained further in conjunction with a description of the TDAD.

5.7 TDA Delegate

The TDA delegate (TDAD) is responsible for centralized interfacing and security management of the overall system. It provides generic centralized security services, such as management of encryption mechanisms, security compliance management, and configuration management for TDA hosts. It also serves as the primary exchange point for sending and receiving TAXII messages.

The TDAD is also responsible for information exchange services for the overall system. The goal is to distribute information based on newly discovered cyberattacks to each member of the TDA. However, this goal can be challenging for large systems containing many mobile hosts and other dynamic cyber resources. Reachability issues in terms of maintaining the IP addresses of dynamic nodes as well as dynamic nodes located behind perimeter firewalls belonging to foreign networks roamed by mobile nodes cause challenges with seamless information distribution.

In order to solve these problems, the TDAD provides a virtual hub-and-spoke communication mechanism based on connection-oriented command and control (C2) protocols. As such, the TDAD serves as a centralized interfacing system for each TDA member, whereby information exchange describing the sources of new cyberattacks is distributed to TDA members via the TDA delegate and its C2 protocol. When a TDA member detects a new cyberattack, the PDS is delivered to the TDAD via the DIMI. The TDAD, in turn, initiates a command via its C2 channels bound to the remaining hosts within the system. The TDAD reverse-connects to each member's PDSI with a connection-oriented C2 protocol. Once each member receives the PDS from the TDAD via its DIMI, the PDS is processed by its PMM and then delivered to its firewall management module.

5.8 Putting it All Together

As of now, a number of technologies have been introduced. The end goal of this chapter is to show designers of CBDM systems how they might implement secure CBDM systems. Traditionally, cybersecurity is an 'afterthought' of most cyber designs. However, it is the purpose of this chapter to provide enough background information along with feasible architectures and frameworks, so that designers can build security into CBDM systems from inception.

So, how do the technologies described thus far in this chapter come together to provide robust cybersecurity for CBDM systems. All that remains, actually, is a 'big picture' discussion that shows how the pieces fit together. Hence, the purpose of this section is to do just that.

Recall from Sect. 2 of this chapter, the following definition of CBDM systems:

> CBDM refers to a product development model that enables collective open innovation and rapid product development with minimum costs through social networking and crowd-sourcing platforms coupled with shared service pools of design and manufacturing resources and components.

By definition, most if not all of the entities comprising a CBDM systems have some type of cyber 'connection.' Even physical entities have cyber interfaces, thereby causing them to be cyber-physical. Each one of these entities can, in general, be actors within cybersecurity information exchange technologies, such as TAXII and STIX as well as TDA hosts within the distributed firewall architecture presented in the previous section. Moreover, TAXII/STIX can serve as a global-scale information exchange framework for the distributed firewall architecture. As a result, this chapter suggests that CBDM systems can be dynamically protected by distributed firewall architectures and, furthermore, that these systems can share cybersecurity event information (collected and processed by CBDM elements acting as THAs in DFAR) on a global-scale using technologies defined by STIX and TAXII.

Figure 6 illustrates how the technologies described in this chapter come together to implement a distributed, collaborative, and automated cybersecurity infrastructure for CBDM systems.

As the figure reveals, CBDM systems comprised of elements, such as CAD/CAM, design, or simulation software can be protected by one or more implementations of the DFAR TDA host architecture (THA). Depending on the type of system, the THA might protect an entire cloud of CBDM software services, or it could be implemented as a traditional firewall mechanism. Similarly, traditional physical manufacturing elements, such as CNC machines, 3D printers, measurement equipment, and milling machines (to name just a few) are envisioned to be

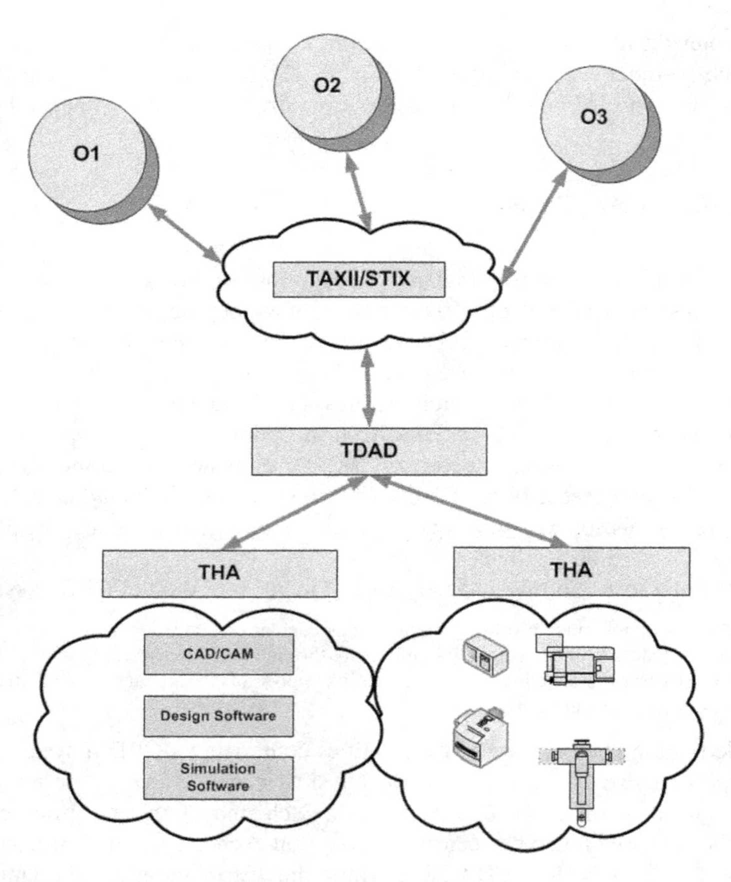

Fig. 6 CBDM distributed and collaborative cybersecurity

cyber-physical systems in CBDM, whereby each is somehow interfaced with an Internet communication device/node. These systems can also be interfaced with a corresponding THA. The figure goes on to illustrate that these aforementioned CBDM systems (from Fig. 6) is managed within DFAR by the TDAD. Moreover, the TDAD serves as the entity that sends and receives cybersecurity event information to and from other partner organizations via the TAXII and STIX information sharing standards and corresponding technologies.

6 Closure

Cloud-Based Design and Manufacturing (CBDM) refers to a product realization model that enables collective open innovation and rapid product development with minimum costs through a social networking and negotiation platform between service providers and consumers. Because of the diversity of cyber resources constituting CBDM systems and because CBDM systems are Internet-enabled, cyber-physical platforms, new types of cybersecurity systems that provide real-time, dynamic, and preemptive protection will be needed.

The goal of this chapter was to provide an overview of the main technologies required to secure emerging CBDM systems. The chapter started with an overview of CBDM. Then, the chapter presented a number of cybersecurity technologies that are considered necessary for securing CBDM systems. Next, the chapter discussed TAXII and STIX, which are two emerging initiatives aimed at standardizing technologies that enable the exchange of cyber-security event information across organizational boundaries. Then, the chapter proposed a distributed firewall architecture that can be seamlessly coupled with cybersecurity information sharing technologies, such as TAXII and STIX and showed how these coupled architectures combined as an overall reference architecture leading to a distributed, collaborative, and automated cybersecurity infrastructure for robust cyber protection of CBDM systems. Since cybersecurity is an 'afterthought' of most traditional cyber designs, a major purpose of this chapter was to provide enough background information along with feasible architectures and frameworks, so that designers can build security into CBDM systems from inception.

References

Anderson J (1980) Computer security threat monitoring and surveillance. James P. Anderson Co., Fort Washington

Axelsson S (2000) The base-rate fallacy and the difficulty of intrusion detection. ACM Trans Inf Syst Secur 3(3):186–205

Banaerjee P, Bash C, Friedrich R, Goldsack P, Huberman B, Manley J, Patel C, Ranganathan P, Veitch A (2011) Everything as a service: powering the new information economy. IEEE Comput 44(3):36–43

Cheswick WR, Bellovin SM, Rubin AD (2003) Firewalls and internet security, 2nd edn. Addison-Wesley, Boston

Rutkowski A, Kadobayashi Y, Furey I, Rajnovid D, Martin R, Takahashi T (2010) Cybex the cybersecurity information exchange framework (X.1500). ACM SIGCOMM Comput Commun Rev 40(5):59–64

Schaefer D (2011) Take e-cad to the cloud? In: Control Design, pp 44–45, Nov 2011

Schaefer D, Thames JL, Wellman R, Wu D, Yim S, Rosen D (2012) Distributed collaborative design and manufacture in the cloud: motivation, infrastructure, and education. In: ASEE 2012 Annual Conference & Exposition, June 2012

Shirey (2000) Rfc 2828: internet security glossary. IETF Network Working Group, May 2000

Stakhanova N, Basu S, Wong J (2007) A taxonomy of intrusion response systems. Int J Inf Comput Secur 1(1/2):169–184

Stallings W (2003) Network security essentials: applications and standards, 2nd edn. Prentice Hall, Upper Saddle River

Lane Thames J (2012) Advancing cyber security with a semantic path merger packet classification algorithm. PhD Thesis, Georgia Institute of Technology

Lane Thames J (2013) Advancing cyber security: a semantic computing approach. Lambert Academic Publishing

Teaching Creativity in Design Through Project-Based Learning in a Collaborative Distributed Educational Setting

Teruaki Ito, Tetsuo Ichikawa, Nevan C. Hanumara and Alexander H. Slocum

Abstract This chapter presents the challenge of teaching creativity in design through project-based learning (PBL) in a collaborated distributed educational setting. First, PBL engineering class examples regarding computer-aided design for a toy modeling and original design/modeling for a remote controlled robot are presented, as a starting point of this challenge, from two different institutions, or the University of Tokushima in Japan and Massachusetts Institute of Technology in the USA. After reviewing these classes, several critical elements are identified for the success of these classes. Considering these elements, PBL provides not only an effective approach for teaching creativity in education in a university setting, but also could be applied more generally in a global setting. The second part of this chapter presents the challenge of teaching creativity in a global project using the web-based design and manufacturing of a dental milling machine, followed by a dental headrest project by the process of expectation management. Reviewing the critical roles of conventional learning management systems in these PBL classes and the current trends of cloud computing, this chapter shows the potential of cloud-based design and manufacturing to support creativity in design education.

T. Ito (✉)
Institute of Technology and Science, The University of Tokushima,
2-1 Minami-Josanjima, Tokushima 770-8506, Japan
e-mail: tito@tokushima-u.ac.jp

T. Ichikawa
Institute of Health Bioscience, The University of Tokushima,
3-18-15, Kuramoto-cho, Tokushima 770-8504, Japan
e-mail: ichi@tokushima-u.ac.jp

N. C. Hanumara · A. H. Slocum
Department of Mechanical Engineering, Massachusetts Institute of Technology,
77 Massachusetts Avenue, Cambridge, MA 02139, USA
e-mail: hanumara@mit.edu

A. H. Slocum
e-mail: slocum@mit.edu

D. Schaefer (ed.), *Cloud-Based Design and Manufacturing (CBDM)*,
DOI: 10.1007/978-3-319-07398-9_9, © Springer International Publishing Switzerland 2014

Keywords PBL · Design education · Collaboration education · LMS

1 Introduction

Creativity education is a challenge for engineering classes. Just as there are many different types of people and learning styles, there are various approaches to teaching creativity in engineering education. Among those, project-based learning (PBL) has been reviewed as one of the most effective approaches to creativity education (Simon 1996). Under these circumstances, learning management system (LMS) has been playing a critical role to support PBL classes so far (Hanson and Robson 2003). Recently, the IT sectors have significantly benefited from cloud computing (Armbrust et al. 2010) through, for example, (i) on-demand self-services, (ii) ubiquitous network access, (iii) rapid elasticity, (iv) pay-per-use, or (v) location-independent resource pooling (Linthicum 2009). Considering the current trend of paradigm shift of cloud computing (Liu et al. 2012; Wu et al. 2012), it is expected that the conventional web-based LMS would shift toward cloud-based systems in the near future.

This chapter starts with two successful examples of PBL classes from two different institutions to review and to show the critical elements for the success. One is Computer-aided design class (CAD-EX) from the University of Tokushima (UT) in Japan and the other one is 2.75 Design and Manufacturing class (2.75) from Massachusetts Institute of Technology (MIT) in USA. Section 3 covers CAD-EX, whereas Sect. 4 covers 2.75. Both classes are designed for sophomore students of Mechanical Engineering department. Reviewing the critical factors observed in these classes, five critical elements for the success of these classes are presented in Sect. 5. These elements include healthy and effective social activities of the class, clear goal setting in the class, appropriate problem setting in the class, continuous improvement on the class by the teacher, and creativity catalyst function of the teacher.

Effectiveness of PBL approach toward creativity education in these examples is based on a local setting, but it could also be applied to a PBL on a global setting. Therefore, Sect. 5 also presents the challenging idea of global collaboration over the computer network (Ito 2011) using the example of Dental Milling Machine (DMM) project.

Dental milling machine project is a global project because a problem setting was proposed by UT and the project was performed at MIT. However, the activities of DMM team members were basically at MIT local, whereas offshoring of fabrication and project review were conducted at UT. It is true that digital engineering makes it possible to perform global collaboration not only in design, but also in manufacturing. DMM project showed how this could be carried out in the global research/teaching project for design and manufacturing of the DMM. As a result of DMM project, it was suggested that digital engineering is not merely for big corporations, but also could be the basis for research projects upon which more and more universities collaborate as part of their teaching process.

Another international collaboration project or the Dental Headrest (DHR) project is presented in Sect. 6 in order to challenge the true international collaboration project. Since the participants of PBL are not always sure what would happen next during the project, especially in global settings, expectation management (Boehm et al. 1999) was taken care of for the success of DHR project.

Learning management system played a critical role in these PBL projects and supported the collaboration in a very effective manner for both local and global projects. Considering the current trend of cloud-based computing in various areas, LMS systems are also shifting from conventional web-based to cloud-based systems. Reviewing pros/cons of LMS in PBL classes, Sect. 7 discusses how to make use of cloud-based LMS (CBLMS), or design and manufacturing systems (CBDM).

2 Project-Based Education Toward Creativity Education

2.1 Creativity Education and Methods for Teaching Creativity

Creativity in education has been regarded as a very important goal not only for students, but also for faculty. To teach creativity, the faculty members themselves need to be creative in their delivery of the material. The faculty members should always actively propose new ideas to the class and use feedback from the students to improve the quality of the class. Staff cannot stagnate if they are to teach others to create.

Just as there are many different types of people and learning styles, there are various approaches to teaching creativity in engineering education. Like a balanced diet, there is no best method, but rather different styles reinforce each other, and continual evolution is probably the key to robustness. Here are typical approaches of teaching creativity.

2.1.1 Hands-On Education

The Education Consortium for Product Development Leadership in the twenty-first century (PD21) developed a design challenge as the first experience in a new Master's degree program in product development (Frey et al. 2000). Students in the program rate a design challenge as a very good introduction to the program and agree that the exercise provides materials for discussion of system architecture and organizational process. The hands-on approach highly integrated with the studio approach (Burroughs 1995; DeLoughyry 1995) has been reported with good success (Starrett and Morcos 2001). As an innovative approach to providing physical demonstrations in the engineering classroom, a haptic paddle was introduced and achieved great success. (Richard et al. 1997; Okamura et al. 2002).

2.1.2 Teamwork-Based Learning

A report issued by the MIT Commission on Industrial Productivity suggested that both industry and academia begin focusing on teams and suggested that academic team projects have two evaluations: (i) a team evaluation and (ii) an individual evaluation (Dertouzos et al. 1989). Building courses on the technical information and concepts developed in introductory required courses, and adding reality to the design process by using engineering teams, students' motivation within laboratory courses increased significantly (Byrd and Hudgins 1995).

2.1.3 Project-Based Learning

Design is widely considered to be the central or distinguished activity of engineering (Simon 1996). It is also said that the purpose of engineering education is to graduate engineers who can design effective solutions to meet social needs (Sheppard et al. 1998). "Appropriate design" courses appear to improve retention, student satisfaction, diversity, and learning (Dym et al. 2005).

2.1.4 World Wide Web-Based Lectures

The World Wide Web (WWW) presents an ever expanding opportunity to present and disseminate curricula interactively over a wide geographical area. The effectiveness of teaching approaches with Web-based tools has been reported and concluded that the WWW may be an enabling technology to provide new opportunities for students to play a more active role in the acquisition of knowledge (Wallace and Mutooni 1997; Wallace and Weiner 1998). WWW-based teaching is pertinent to both local and distance education, and literature studying the efficiency of the WWW in distance education is supportive of the premise that WWW-based lecture can be very effective. Perhaps the greatest feature is that students can easily replay a lecture or part thereof and review that which was not clear the first time.

2.2 PBL Classes for Creativity Education

Among various types of approaches toward creativity education (Belcheir 2000), this chapter focuses on PBL-based education, where cloud-based and/or network-based environment plays a critical role. Using some PBL classes where the authors supervised, the following sections show how creativity education was successfully implemented in these classes.

The first case study of PBL is the computer-aided design class called Computer-aided design exercise (CAD-EX) at UT in Japan. CAD-EX teaches students the basic knowledge about digital engineering, and guides students to work in small teams to create digital engineering models. The details of the class will be given in Sect. 3.

The second case study of PBL is the introductory design and manufacturing class called 2.007 (formally 2.70) at MIT in the USA. 2.007 teaches students fundamental principles of design, such as St. Venant's Principle and Reciprocity, and guides them apply these principles to the practical exercise of designing and building remote controlled machines that compete in a final contest (http://pergatory.mit.edu/2.007). Students also learn from each other by a peer review evaluation process (PREP). The details of the class will be given in Sect. 4.

These two PBL classes were designed and offered to students in each institution, respectively, under the traditional settings of LMS environment. Using questionnaire data obtained from the students in these classes, Sect. 5 analyzes the educational effectiveness of these classes. Even though these classes evolved separately based on each history and different requirements, several significant common ideas were recognized in terms of creative education. Section 5 also discusses pros/cons of traditional settings of LMS based on the review of Sects. 3 and 4.

3 Creativity Education Through Toy Model Engineering PBL

3.1 Course Overview

Creativity enhancement education (Ito and Oyama 2005) has been emphasized in the design education curriculum in the Department of Mechanical Engineering at UT. The CAD exercise class, called CAD-EX (Ito et al. 2004) plays a major role in this curriculum. Up until 2002, CAD-EX taught students how to design/implement drawing software using C++ as well as how to use CAD application software. Students were offered the chance to understand the internal procedure of drawing during implementation of drawing software application. With references to commercial software, students learn how to follow basic design procedures including defining requirements, creating basic designs, detailed designs, and design evaluation. As a result, it was recognized that students understand CAD software much better than just by teaching CAD application using the software's tutorials.

Thanks to joint efforts with Center for Advanced Information Technology at UT, solid modeling software was introduced to CAD-EX in 2003. Revising the curriculum of CAD-EX, teamwork-based education was introduced with the goal of future collaboration with industrial partners. It has been reported that educational effects were recognized in the revised version of CAD-EX (Ito 2005).

Toward further improvement of education in CAD-EX, a web-based education system, or Web-based LMS was developed and introduced to the course. The web-based LMS provides administration information and teaching materials to students, and collects feedback and reports from the students. The system also provides an environment for both teaching staff and students to share data and information. As a result, a paperless environment was built that does not have the time lag associated with turning in, collecting, grading, and giving back assignments on paper.

3.2 Overview of CAD-EX

CAD-EX started in 1999, based on its predecessor class, which covered computational drawing, for the primary purpose of teaching sophomore students how to use 2D CAD software. In addition, CAD-EX students learned how to implement basic graphical programming so that they should understand CAD software not as a black box. In 2003, CAD-EX started using 3D CAD software, or solid modeling software.

Currently, CAD-EX schedules 3 h of class every week for a total of 15 weeks during the spring semester. A total of 120 sophomore students are grouped into 20 teams with six students each, and they work on a team-based assignment. The assignment given to each team is to select a plastic model with a minimum number of parts, which the team is to reverse engineer and build a solid model (Fig. 1). Being allotted his/her own role in each team, students work together to achieve the goal according to the course schedule.

- (Week 1 and 2) Team assignment and member role: A total of 120 sophomore students are divided into 20 teams of six students each. Each team discusses and decides upon a team leader and other roles including responsibilities for modeling, animation, manual, presentation, and report preparation. The target plastic model, which each team selects, must satisfy several requirements, including minimum number of parts, availability on the market, and mechanical function. Each team is also required to submit a project schedule for the remaining weeks.
- (Week 3) Lectures on solid modeling: Students attend lectures to receive an overview of solid modeling software as well as the computer room environment. Simple modeling exercises are given to students to acquire basic modeling skills including design intent. Instruction manuals are available to students from the web-server.
- (Weeks 4–8) First half of exercise: Students start working on each project according to the course schedule. Each team has meetings with teaching staff to show work progress and discuss their respective projects. Teaching staff guides each team toward the goal of accurately completing their project on time.
- (Week 9) Intermediate presentation: Each team gives an intermediate presentation to show what they have done during the first half of the course and to show their revised schedule for the second half based on a review of materials

already submitted in Week 8. Shortly after each presentation, peer-review type comments are fed back to teams.
- (Weeks 10–14) Second half of exercise: Students keep working according to the revised project schedule. In the meantime, those issues, which found in the intermediate presentation, will be discussed separately with each team during this period. Week 14 is the due date for deliverables including the final report, presentation materials, solid model data, and animation data. Completion of submission is required to receive permission (called "getting a ticket") to make a final presentation.
- (Week 15) Final presentation: Nominated teams make final presentation based on the presentation materials submitted by Week 14. Animation of the solid model and its assembly/disassembly animation are shown in the presentation as well. Peer review evaluation is given to each presentation and feedback is given back shortly after each presentation. At the end of class, overall class assessment is conducted using course evaluation sheet prepared for CAD-EX.
- (Week 16) Final report: Each team is required to present a hardcopy of the final report, which was to have been digitally uploaded to the server in Week 14.

3.3 Outcomes of CAD-EX

At the end of final presentation in Week 15, a class assessment was conducted using an evaluation sheet in which all of the students gave feedback. The survey questionnaire composes five categories of questions, which include (1) independent attitude to problem solving, (2) project planning ability, (3) teamwork ability, (4) creative thinking, and (5) adaptability to basic information technology. Since the detailed report has been published, this section covers the summary of each category.

3.3.1 Self-Motivation

Attendance of students was extremely high in the past 5 years. For the task difficulty question, 50 % of students answered that they chose a difficult task, which shows a positive attitude toward the class work. Reviewing the overall deliverables in 2004, 17 teams did not show any desire to seek the minimum requirement of 30 parts, but instead took a rather difficult design object to solid model. More than 50 % of the students commented that they wanted other people to see what they have built.

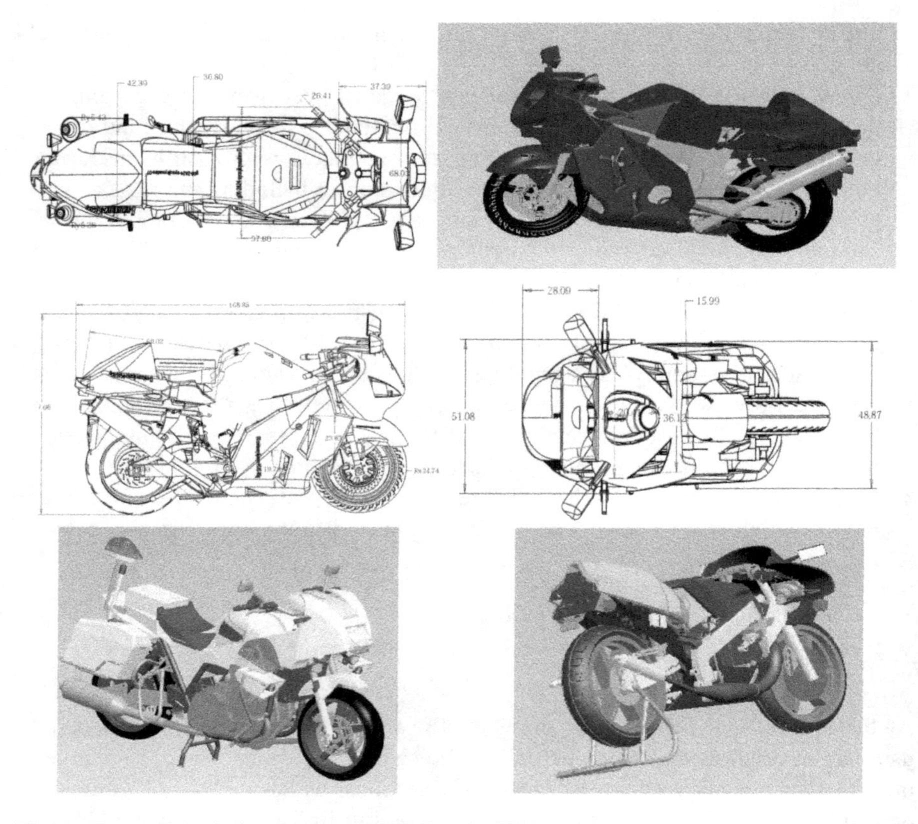

Fig. 1 Toy models designed by CAD-EX class in 2013

3.3.2 Project Planning

Even though a high level of performance was assumed by most of the teams at the initial start of the course, 70 % of students answered affirmative for achievement and 80 % did well for theme selection. At the intermediate presentation, most of the teams were behind schedule. However, all of the teams completed their project by Week 15. A total of 102 students answered the question: "How long did you need to become accustomed to the solid modeling software?", and the average time was 7.9 h, which corresponds to around 3 weeks. Therefore, speed of modeling was accelerated after 3 weeks, which explains the results quite well. Seventy percent of students worked after class hours to meet their schedule. The average time spent outside of class hours was 20 h, ranging from 3 to 90 h. This shows that each student was willing to work outside of class hours to develop the skills needed to contribute to the team.

3.3.3 Teamwork Ability

Ninety percent of students answered they liked cooperative teamwork with 5–6 students in a team. Seventy-eight percent of students were satisfied with their role in each team, which shows that the teamwork exercise worked well. Ten percent of students proposed a different ideal number of team members, ranging from 10 to 1. Only a few students would prefer to work by themselves instead of on a team.

3.3.4 Creative Thinking

It was the first time for all of the students to use solid modeling software, and 85 % of students were initially concerned about solid modeling. After solid modeling, 60 % of students showed a strong desire to build a physical model based on their solid model. Fifty percent of students said they would like to design industrial products as digital design objects, whereas the rest preferred plastic models as they did this year.

3.3.5 Web Tools

The CAD-EX website was built and used in 2004. Most of the students liked the use of the web-based system for communication, material distribution, and report submission. Some students were not good at even basic computer use. However, most of the students did not show any resistance to use computers not only as IT tools, but also for digital engineering.

3.4 Discussions on CAD-EX

Based on the evaluation questionnaire, and the presentations at the intermediate/final sessions, this section discusses the educational effects of CAD-EX.

3.4.1 Educational Effects in Self-Motivation

Our experience shows that the students in 2003 taking CAD-EX had a tendency to set up their theme as easy as possible within the required constraints. For example, if the minimum number of parts is 30, they would try to find a plastic model, of which number of parts is as close as 30. However, the 2004 class was different. With the constraints of 30 parts or more, the number of parts in a plastic model in each team was a maximum of 148, and the average was 80 parts. Most of the team did not care for the less number of parts, but rather cared to select a plastic model that interested them. Each team set a high goal and class attendance was very good. Each student contributed well to the team to which he/she belonged.

According to the questionnaire, more than 70 % of students spent their time working on the project out of normal class hours. It was recognized that students did not want to take an easy path, but rather they had a strong wish to take responsibility for their role and contribute to their team. As for the reason why the students were much more motivated in the 2004 class, several things were thought to be important. The year 2004 was the second year to use the 3D cad software along with the plastic models. Based on the experiences in the previous year, the CAD-EX class was presented in a better way due to the efforts of the teaching staff. These efforts include organization of the class schedule, orientation of the students toward higher motivation, and effective information sharing using web technologies.

3.4.2 Educational Effects in Project Planning and Its Execution

The CAD-EX class gave an opportunity to students to learn how to work in a team for project planning, plan execution, and self-review.

The assignment of project planning was given to students 2 weeks prior to the start of solid modeling. The students started working on project planning even without prior knowledge, and they learned many things from difficulties they encountered. In fact, most of the teams were not able to follow the schedule at the time of the intermediate presentation. The students did not yet have appropriate solid modeling skills and they regarded the project planning as an easy task. As a result, the students understood that their planning did not work well, and that they needed to reformulate their plan.

Many teams thus made big changes in their plans after the review of their intermediate presentations. With solid modeling skills enhanced by class lectures, self-criticism of their utopian schedules, and a review of the overall schedule, each team steadily carried out their projects according to their updated more realistic schedules. Learning from failure, the students rethought their schedules to enable them to achieve their goals.

Note that some teams did work cooperatively together even in the first half, and carried out their tasks according to their original schedule. Good teamwork of these teams was due to all members not only working together at the time of presentation, but also during class hours and through information sharing made possible by the web-based class information system. As a result, the students motivated each other, which helped lead to the improvement of teamwork in the second half of the class. It was recognized that these interactions among students were also regarded as an additional positive educational effect.

The computer room was available for all class hours and also extra hours as long as other classes did not require it. The CAD-EX class advises students to carefully manage time-budgets, so that they can effectively use the computer resources. However, time-budgeting may not be so easy for students who are accustomed to working on their own PCs anytime when they want. The CAD-EX students are required to work as a team and to take responsibility in their own role in the team. Team members were required to work together within the time-budget to achieve the common goal.

3.4.3 Teamwork-Based Learning for Cooperative Work

Pairs of students within the teams was the basic structure for the students in the CAD-EX class before 2002. However, a workload imbalance was often observed in many pairs. From our review, it was determined that the role of students in each pair was not so clear. As a result, it often happened that only one student of the pair did most of the work and the other student observed. Because of differences in modeling skill and in modeling progress, the paired students often faced difficulty in working together. Depending on each other was an easier choice than working together. Cooperation with other pairs was not so easy either, because design objects were different from one another in order to avoid copying. When a student skipped class without any sense of guilt and avoided taking responsibility, his/her partner student had no choice but to take care of the assignment. In such uncommon cases, the positive effect of teamwork-based education was not recognized by a few.

Reviewing these past experiences, the curriculum of CAD-EX class was revised to focus on teamwork-based education. First, the role of the students in each team must be clearly defined and assigned at the beginning of the semester, so that each student should clearly understand his/her responsibility on the team. Each student needs to finish his/her modeling so that the team could work on assembly of parts in the final stage of modeling, otherwise the team could not finish the project. The workload for each student is rather heavy, which means that each student has to take care of his/her modeling responsibility, because no other student is in a position to take care of another member's task. Of course, students quite often teach and help each other in the team activities, but almost no student asked another to take care of his/her own responsibility.

Once any of the students in the team acquired basic modeling skills, mutual cooperation among the team started, which accelerated other students to learn the skill efficiently. Even if a student had some reason to skip the class, most of these students took responsibility in finishing his/her assigned task and reported it to the team members so that any inconvenience was avoided as much as possible. It was recognized that students acquired the importance of consciousness in finishing assigned tasks by the due time.

3.4.4 Different Characteristics of the Two Sections

The 120 students were divided into two sections, or A/B in a random manner when they registered to the department in the freshman year, which has nothing to do with entrance examination scores. Both A and B sections take CAD-EX at the same class hours. However, an apparently different atmosphere had been recognized between these two sections after 1 year of college life.

Teamwork education in CAD-EX class gave a chance to students to work cooperatively and collaboratively over these two sections, which could not only enhance improving the modeling skill of each student, but also encourage them to learn how to contribute to the team. As a result, the students learned what they should do to achieve the common goal of the team, which also contributed to create a better learning atmosphere.

3.4.5 Web Server to the Class

The computer environment for the CAD-EX students is supported by Center for Advanced Information Technology at UT. In addition to that, the teaching staff of CAD-EX have developed the web-based system for CAD-EX and introduced it to the class in 2004. Students were comfortable using the web services. Inter-team or intra-team information sharing is required for teamwork learning, and the web-based system introduced in the CAD-EX class worked quite well for this purposes. It may be one of the reasons which explain the better results of the 2004 class compared to that of the 2003 class.

However, the maintenance of web-based system was a heavier burden on the teaching staff than expected. The system needs to be running continuously without problems. The system needed to be updated according to the class schedule, and the teaching staff had to spend lots of time maintaining the system, whereas they should spend most of their time in class. An open issue is how to economically maintain, update, and troubleshoot the web-based system.

4 Creativity Education Through Original Model Design and Its Building PBL

4.1 Course Overview

Perhaps the mother of all robot contests, 2.007 at MIT is a 13 week course that has been replicated worldwide in engineering schools and on television (Wright 2005). Course 2.007 begins in February, when each one of 130 students in the class is given a kit of nearly 100 items ranging from electric motors to structural elements (wooden slats, aluminum sheets, steel rods, plastic tubing, and rubber bands) to

Fig. 2 Machine examples in 2.007

springs, gears, and bearings (2.007 web). Machines designed and built for 2.007 must fit into a box roughly the size of a picnic hamper. A critical element is a battery powered electric drill that students find most useful for building their machines; in addition, the drill battery is the power source for their machines. On contest night, each machine has 45 s to complete certain tasks—gathering plastic bottles, ping-pong balls, or hockey pucks, moving glass marbles, or playing tug of war—while a competing machine does the same tasks within the playing area.

Course 2.007 teaches a creative deterministic design process based on the scientific method. Students have 3 h of lecture for the first 6 weeks of class and 4 h of lab for the entire 13 week course. Students are given a detailed set of course notes at the beginning of the class along with the class schedule. Students are responsible for doing the reading before they come to class, so in class the lecture can take the format where the professor uses material from the reading to actively design a portion of a machine according to the class schedule. This way the students see the teacher using the knowledge they are supposed to get from the reading to do design that the students are also supposed to do (Fig. 2).

Lectures thus aim to teach the creation, engineering, and manufacturing of a remote controlled mechanical machine (Ma and Slocum 2006) to compete in a major design contest at the end of the semester. Students learn to identify a problem and create, develop, and select best strategies, and concepts using the fundamental principles, appropriate analysis, and experimentation. The students then divide their best concept into modules and after developing the most critical modules (MCMs) first in descending orders of criticality, proceed to system integration, testing, and debugging. Project and risk management are introduced as tools to keep the development process under control in order to deliver a robust working machine on time and on budget. The physics and application of machine elements to enable students to create and engineer their machines are learned by reading and in lecture by example. Throughout the course, engineers' professional responsibilities are stressed. At the beginning of the course students are assumed to be competent at parametric solid modeling, spreadsheets, or MATLAB software, and basic machine shop skills, which they have learned in the previous required course 2.670. Educational, reference, and design assistance materials are provided on-line.

4.2 Course Format

The general operating hypothesis of 2.007 is that a big design course can be run very efficiently using the very process that it seeks to teach:

- Deterministic design (everything happens for a reason, and we should be able to think about what will happen beforehand, but do not be afraid to experiment).
- Schedules are important.
- Student (customer) feedback can be very useful if thoughtful changes are made every year in response.
- Teamwork is of course very important, but there has to be an overall manager (Captain of the ship) who makes decisions once opinions are aired, else staff meetings can become long and drawn out.
- KISS: change a little bit each year.
- Seek industry participation, feedback, and sponsorship.

2.007 is composed of lectures and lab sections, all of which are clearly focused on the final contest. The lecture topics T in Table 1 stress this design process as well as fundamental principles of mechanical design.

The design schedule breaks the course into roughly three, similarly to that practiced by many companies: Creative design, detailed design, build and test. Weekly mile-pebbles and periodic milestones help keep the students on schedule. Each week the instructors give the students points toward their final grade:

- Week 1: Mile-pebble 1: (a) Get Kit and Locker and Play with kit elements and the table. (b) Complete Solid Model of Table
- Week 2: Mile-pebble 2: (a) Assemble motor gearboxes, (b) 3 Strategies on FRDPARRC, (c) Preliminary analysis: scoring, power budget…
- Week 3: Milestone 1: (a) Sketch models and start simple car, (b) Best Strategy Selected and FRDPARRC
- Week 4: Mile-pebble 3: (a) Appropriate analysis, (b) Bench level experiments, (c) 3 Concepts on FRDPARRC sheets

Table 1 Class schedule overview of 2.007

Week	Contents
1	T1[a] passionate design; T2[a] creating ideas, T3[a] fundamental principles
2	T4[a] linkage, T5[a] power transmission and elements 1
3	T6[a] screws and gears, T7[a] force, torque and power sources
4	T8[a] structures
5	T9[a] structural interfaces, T10[a] Bearings
6	T11[a] Control system, T12[a] manufacturing
7	T13[a] Ethics and professionalism
8	Spring break
9	No lecture; build, test and debug
10	No lecture; build, test and debug
11	No lecture; build, test and debug
12	No lecture; build, test and debug
13	No lecture; build, test and debug
14	Contest

[a] T1–T13 corresponds the textbook "Fundamental principles of mechanical design"

- Week 5: Milestone 2: (a) Complete simple car, (b) Best Concept FRDPARRC and solid model, (c) Modules defined
- Week 6: Mile-pebble 4: Work on MCM
- Week 7: Milestone 3: Demonstrate MCM
- Week 8: Spring Break: RELAX (UNLESS behind, Lab is open all week)
- Week 9: Mile-pebble 5: (a) Begin details on other modules, (b) Show schedule for rest of semester
- Week 10: Milestone 4: (a) All Engineering Complete, (b) Begin working on other modules
- Week 11: Mile-pebble 6: (a) Build, test, and integrate modules
- Week 12: Mile-pebble 7: (a) Build, test, and integrate modules
- Week 13: Milestone 5: (a) To-Do list, risks, countermeasures, and final plan, (b) Demonstrate good working machine
- Week 14: Mile-pebble 8: (a) In-class seeding contest, Get T-Shirt!, (b) Pack and "ship" machine
- Week 15: Milestone 6: (a) Final Contest!, (b) Reflection

The 130 students for 2.007 work in a lab that is divided into two parts: a machine shop with six lathers and six milling machines as well as sheet metal equipment and large bandsaws; and a 6 m × 20 m assembly area with benches and six sets of drill presses belt sanders and small bandsaws. There is also a soldering bench and a fasteners cabinet. Table 2 shows the supporting staffs assigned for teaching 2.007 in 2005.

Table 2 Overview of the number of supporting staff of 2.007 in 2005

Week	Role	Number of staff
Head instructor	Instructor in charge	1
Administrator	Administration for overall course	1
Section instructor	Faculty and TA students	1[a]
Web manager	Web administration	1[a]
Kit manager	Robot kit management	1[a]
Shop staff	Machine shop staff	5
Technical staff	Lecture room and kit preparation	3[a]
TA	Section leader, assistants	5
UA	Advisor from former 2.007ers	19
Supporting staff	video recording, contest, security	[b]
Sponsors	Donation of parts and funding	21

[a] Including TA
[b] Hired by the class

4.3 Contest in 2005

The contest table for year N is newly designed every year by students and staff who completed the course in year $N - 1$. The 2005 table (MIT 2.007 2005) was inspired by the new undergraduate dormitory at MIT called Simmons Hall, and designed in the summer of 2004 as shown in Fig. 3.

Each student's machine must begin the competition within an 18 inch by 18 inch square. The starting zone extends to 26 inches above the table surface. Robots can be located anywhere within this volume. However, they still must fit into the sizing box (16 × 16 × 24) when in the starting zone. The starting zone is demarcated by a 1 inch line. The score is the sum of the total points scored by placing colored foam blocks into the various scoring bins as shown in Fig. 4. A student can only score with the foam blocks of the color of their side of the table, but they can place a block in any scoring bin (Fig. 5).

4.4 Peer Review Evaluation Process

One of the key elements in 2.007 is the PREP (Graham et al. 2007). Each student in 2.007 designs and builds his/her own machine, but he/she begins to learn about teamwork through the process of peer review. Indeed, an important aspect of any design process is to empower individuals with the ability to create individually and then have their ideas heard. Each individual of a design team first thinks alone, and writes down his/her ideas in whatever format is required by the particular design method used. Next, team members acting as peers individually review each other's ideas and provide written comments. PREP is an important catalyst of making any design process more efficiently by reducing the amount of discussion needed to converge on an idea. PREP plays a very important role in teaching students how to design the robots, and how to act as professional engineers.

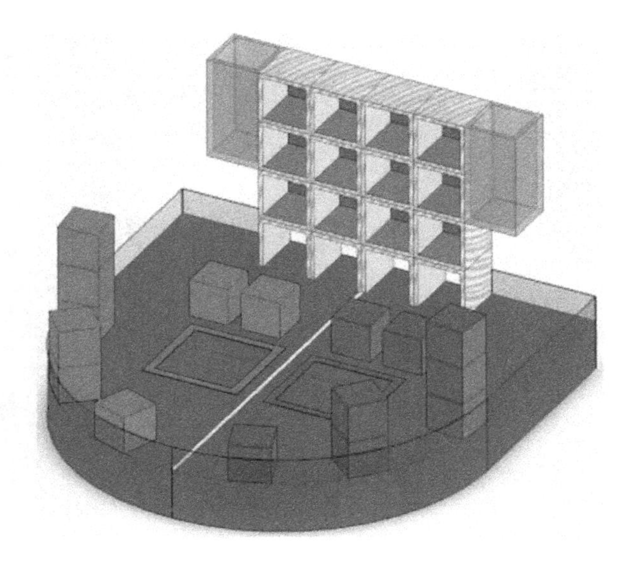

Fig. 3 Contest table for 2.007 in 2005

Fig. 4 Point distribution in contest table

4.5 FRDPARRC Table

Everything happens for a reason, and students need to apply resources and focus in order to discover and understand the issues that would otherwise lead to uncertainty. Minimizing uncertainty, and hence risk, makes a design more deterministic. Good design practice indicates that there is a minimum set of issues to be addressed: The functions a design must perform are the Functional Requirements (FR). The means

Fig. 5 The final contest of 2.007

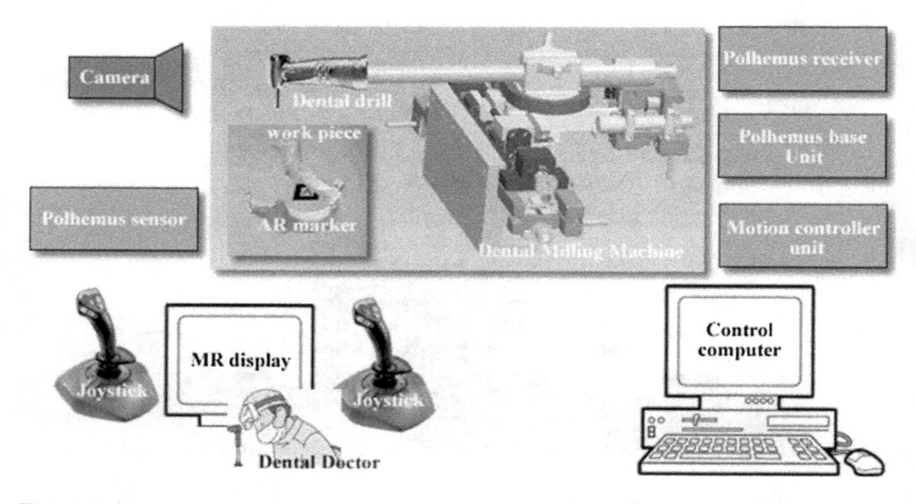

Fig. 6 Dental milling machine developed in DMM project

by which the functions can be accomplished are the Design Parameters (DPs). Analysis (A) is often required to develop the FRs and DPs. References (R) are often consulted. Risks (R) associated with each FR and DP must be identified. For each risk, countermeasures (C) must be thought of ahead of time.

A useful way to represent these basic issues is to make a table with each of the design information elements as column headings. Such a table is called a FRDPARRC table. In 2.007, students are taught to use the FRDPARRC during the design process.

5 Expansion of Creativity Education from Local PBL to Global PBL

There are a number of open research issues associated with teaching creativity to students in design related classes. Sections 3 and 4 showed some of the potential answers to these issues. Section 5.1 discusses how these issues are addressed from these potential answers.

CAD-EX and 2.007 were offered to local students based on local PBL settings. Based on the discussion earlier, Sects. 5.2–5.6 discusses how to expand the creativity education toward global collaboration using a case study of DMM project (Fig. 6). DMM project was conducted between UT and MIT to grapple with the expansion of creativity education in the global environment setting.

5.1 Research Questions and Potential Answers on Creativity Education

Dental milling machine was initiated from the observation of the fundamental requirements on creativity education based on the two case studies presented in Sects. 3 and 4. Therefore, before presenting the case study of DMM project, this section reviews these fundamental requirements and presents some potential answers to these open-issue questions, which the authors have obtained from these two classes.

Sections 3 and 4 offered two different formats of PBL, where teachers can use to offer an engineering design class to enhance the creativity of students. CAD-EX and 2.007 were conducted to implement creative engineering education in two separate organizations under the supervision of the authors. Authors discussed and compared these two classes to find out a clue to creativity education. As the results of discussion, it was agreed that the environments for teaching classes are very different in each organization, but several key elements were shown to be mutually important in terms of offering a creative engineering class. These key elements include healthy social activities, clear goals, appropriate problems, continuous improvement, and creativity of teaching staff.

Resources differ between schools. There are some classes that are well funded by corporate sponsors. The number of teaching staff in the class also varies. However, the most important element is the positive creative attitude of the teaching staff. Like the students if they have a sense of ownership and passion, they can accomplish amazing things. If the faculty is creative, it should always be possible to design and offer a class, which enhances the creativity of students. The remaining part of this section discusses the fundamental requirement mentioned earlier.

5.1.1 Does PBL Offer Healthy Social Activities?

From the students' perspective, taking classes does not always mean that a class is attractive, because often it is mandatory. If a class is mandatory, but not attractive to students, they will go through the motions of learning, but retain little. Giving students ownership and responsibility for a goal that will then be exhibited in a competition seems to be an effective way to make a class attractive.

Another way the 2.007 class invites ownership and stimulates passion is by inviting many people to help teach ranging from the students in the previous years to design experts from industry. For example, former students support the class as undergraduate assistants (UAs), and design experts from industry act as instructors for lab sections. The shop technical instructors (machinists who assist the students as they make all their own parts) also help students by offering design for manufacturing advice. The final contest is the special event where all of the staffs involved in the 2.007 can join and enjoy. The parents of the 2.007 students, the corporate partners, the former students, and many other guests are all invited to the final contest, which becomes a big celebration. The 2.007 students learn not only the basic robot design and building techniques, but also enjoy so many things through the healthy social activities from peer review to working together in the shop to helping each other at contest time.

The CAD-EX class, on the other hand, is offered by a small number of teaching staff without any corporate partners. However, the idea of providing the healthy social activities has also applied to the class. The team-working to build the target solid models and the presentation of the achievement by the team work are recognized as the most interesting part of the class according to the survey results from the students. Since the survey results show that the students enjoy the interaction with other students, the key elements to make the class success is how to use these interactive activities in the class.

The basic idea recognized in these two classes in common is to regard the class as an environment, which offers the healthy social activities.

5.1.2 Does PBL Offer a Clear Goal?

A clear goal for the class provides good motivation to students. The goal should not be too high or too low. The goal should not be too complicated or too simple. The goal should be something that students view as challenging but that they can accomplish.

The first goal of 2.007 is to complete a working machine. Since there is only one "winner" statistically, most would "lose" if winning the contest was THE goal. The contest is the thus primarily the basis for fun, healthy, and friendly competition where each student can show off their machine to the crowd. Even though some students see the goal is to win the contest, most see the true goal as to attend the contest and enjoy the event, and this goal can be pursued and achieved by all of the students. As a result, all of the students attend the 2.007 class in a self-motivated mode and enjoy the whole class. This goal does not only work for the students, but also for the supporting members, corporate sponsors, and parents of the students.

The goal of the CAD-EX students is to show off their modeling skills by giving a good presentation in the final stage. Since the deliverable of the class is solid models, it is not possible to make a physical competition as in 2.007. However, the solid models presented by each team are peer reviewed by fellow students and are given scores to select the best solid models. The idea behind the presentation is the same. The students were very motivated to be able to give the best presentation with the best possible deliverables, and in the process, show off their skills.

Ownership, passion, a chance to display what you have accomplished for all your peers to see appears to be the key to a healthy happy productive learning environment.

5.1.3 Does PBL Offer Appropriate Problems?

Project-based learning classes are popular, and when well designed, they blend theory and practice. Engineering science plays a very important role in many engineering designs. However, if students are faced with real problems that they have to solve in a short time and they are not comfortable with the engineering science, they will sometimes use trial and error. His circumventing the intended learning experience can be prevented with good mentoring, and if appropriate problems are given to the students in the first place.

2.007 is a PBL class that appears to have reached a good balance between theory and practice. By requiring FRDPARRC sheets, students' designs rise to the level of their current engineering ability and then some. The joy students experience when they realize that the engineering science they have been learning is actually useful is wonderful to behold, and keeps the staff interested in teaching the course.

As for the CAD-EX class, it is an inverse design exercise. The physical model to design is in front of the students. The goal of the students is to create a corresponding solid model assembly to mirror the physical model as precisely as possible. The CAD-EX students eagerly worked on creating the solid models to make them as real as possible, and it helped them realize the feeling of deep satisfaction that an engineer obtains when they create a new exciting design.

5.1.4 Does PBL Keep Continuous Improvement?

The two classes presented in Sect. 3/Sect. 4 have their origin in very different places. However, both classes have one thing in common, to offer a creative teaching and learning environment. One of the ways this is obtained is through continuous improvement by changing the challenge each year and by listening to students as if they were customers. Teaching staff update the assignments every year, and update the contents of the classes,which is the most difficult part of the class management. Continuous improvement cannot be achieved without significant effort of the teaching staff and volunteers in both courses.

5.1.5 Do PBL Teachers Function as Creativity Catalysts?

What types of class and teaching methods are most effective in enhancing the creativity of the students is an open issue that may never be fully addressed. Much more discussion will be required to find appropriate answers. An added variable is that education has to be carried out in different schools each with different resources. 2.007 might not work well at UT, whereas the CAD-EX might not work well at MIT unless in addition to physical resources, teaching staffs can also be replicated. It is the teaching staff in both classes who truly make the classes creative. Creative teachers are the most important resource available for teaching creativity. There is no magic kit one can buy from a catalog.

5.2 A Case Study of Dental Milling Machine Project for Creativity Education

The benefits of new and emerging technologies on internet/intranet-based systems, or web-based design and manufacturing systems (WBDM) have become increasingly recognizable in recent years in many areas, including both industry and academia. In addition, they are also significant in training and education, especially in PBL, where the goals of projects are achieved by teamwork with various view points in design and manufacturing.

This section presents a case study of DMM project conducted between MIT and UT, based on the 2.75 Precision Machine Design (PMD) course project at MIT. DMM project was proposed by UT, which was picked up from the term project for the MIT graduate course 2.75, and was expanded to become a global research project with UT. The student team of MIT worked on the project under the collaboration with UT staff, and fabrication/evaluation were performed both at MIT/UT. The lessons learned from DMM project are important for future academic team and web-based design and manufacturing projects.

Here is the brief overview of DMM project. In the medical field, there is a trend toward fully or partially automating various medical procedures. Automation leads to more precise movements and shorter operating times. One potential application for automation is tooth crown preparation. Students at Harvard School of Dental Medicine are currently being trained with a haptic device, which allows them to prepare a virtual tooth. A natural extension of this technology is a computer driven milling machine, which performs crown preparations on actual patients. Figure 6 shows the overview of DMM designed and manufactured in DMM project.

An important feature is that the device utilizes haptic input so that the dentist can "feel" what they are doing, which is key in accurate use of a position controlled cutting device. The geometry of teeth and crown preparation features required that the milling machine have three active axes and four passive axes. The three active axes have a minimum workspace of 2 cm^3, which is the maximum size of one tooth. The two passive axes allow for the dental burr to be oriented along the long axis of the tooth. The other two axes position the active axes' workspace over the tooth that is being prepared. Due to limited space inside the mouth, the bulk of this milling machine was located outside of the mouth and the dental drill was mounted on a cantilever into the mouth. In order to create a fixed reference between the machine and the patient's teeth, a rubber-based impression material mold was used as an attachment and registration feature against the patient's other teeth. To prevent patient fatigue due to the weight of the device, an overhead spring-based weight compensation mechanism was developed. The milling machine was controlled by digital input, which could be generated by a haptic stylus. The final goal was that the dentist would receive video feedback as well as force feedback from the haptic stylus.

5.3 iCampus: A Web-Based System to Support DMM Project

For conducting the DMM project, web-based tools of iCampus (iCampus, http://icampus.mit.edu) and PREP played a critical role. iCampus is a web-based program provided by MIT. PREP is a design methodology evolved by Slocum et al. (2005) under the environment of iCampus. Since PREP tool was covered in Sect. 4.4, this section mainly focuses on iCampus.

Initiated in 1999, iCampus is a research collaboration between Microsoft Research and MIT whose goal is to create and demonstrate technologies with the potential for substantial change throughout the university curriculum (Graham et al. 2007). iCampus-sponsored innovations have had broad and significant impact throughout MIT, and they are continuing to evolve through worldwide multi-institutional collaborations. iCampus projects are selected from MIT faculty responses to annually issued requests for proposals. In addition, iCampus has awarded over $1.5 M for projects conceived, initiated, and run entirely by students. More than 400 faculty and research staff, and 300 students, have participated in iCampus-sponsored projects. Most MIT undergraduates have taken subjects whose evolution was at least in part sponsored by iCampus—over 100 subjects in all.

Areas of innovation have included: using Web Services to enable a new educational information technology framework of software and services shared among universities worldwide; transforming the classroom experience by replacing traditional passive lectures with active learning experiences supported by information technology; and educational applications of emerging technologies such as speech recognition and pen-based computing.

Characteristic features of iCampus projects include Document library, Event list, Task monitoring, Meeting workplaces, Picture libraries, Discussion boards, and Email alerts. Details of each function are found at the iCampus website (iCampus).

5.4 *Global Collaboration for Design and Manufacturing*

During the 2.75 project, the group discussion took place in several environments, including the official/unofficial meetings, e-mail, and file-sharing with SharePoint PREP tool developed as part of iCampus. Especially, each part and its assembly were designed and shared among the group in SharePoint, which enables not only the collaboration among the team, but also global collaboration for manufacturing at UT. This section shows how the web-based design and manufacturing was conducted at UT site based on the designed data prepared at MIT site.

5.4.1 Collaboration with Different CAD Tools

The 2.75 team at MIT uses several design software tools, including SolidWorks, Pro/Engineer, and Unigraphics. The design of the machine in this project was mainly made by SolidWorks. Figure 7 shows an example of a solid model for motor-and-mount assembly. In the meantime, the design software tools used by the UT team were Pro/Engineer and CATIA. The first issue to be solved was how to take care of the CAD data built in the different CAD software. SolidWorks exports solid model files to the standard format of Pro/Engineer. This would be one of the solutions if the exported data will not be edited for redesign. Another possible approach would be to convert them to one of the standard formats such as STEP or IGES, so that any CAD tool could load the data. In this project, both approaches were adopted to share the data between the remote members.

5.4.2 Rapid Prototyping (Fused Deposition Rapid Prototype)

Prior to the manufacturing, two types of rapid prototyping, or Fused Deposition Rapid Prototype (FDRP) and Subtracting Rapid Prototype (SRP), were conducted to confirm the design. The STL files required to make the rapid prototype parts were converted from the design data file described in the previous section. Figure 8 shows an example of FDRP model, which is based on the Solid Model data of Fig. 7.

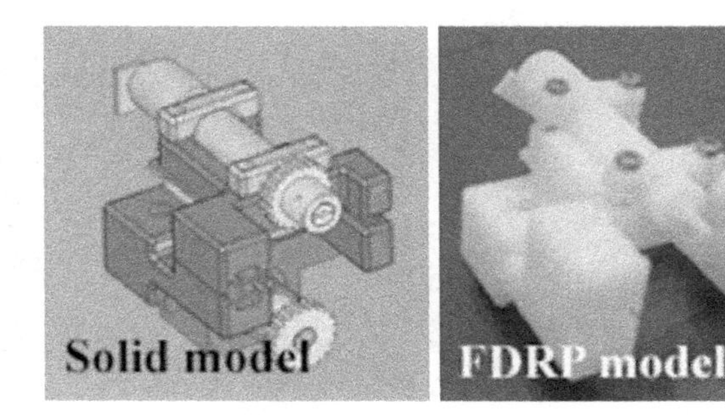

Fig. 7 A motor-and-mount assembly solid model and its rapid prototype

The procedure for this rapid prototyping includes the following four steps.

F1. The STL data for the model is loaded to the RP software.
F2. Configuration of the data is designed and converted to the forming data to be processed by the RP machine.
F3. The RP machine produces the rapid prototype model with supporting material.
F4. The supporting material is removed by dissolving with alkaline solution.

5.4.3 Rapid Prototyping (Subtractive Rapid Prototyping)

Subtractive rapid prototype modeling using a three axis surface machining machine is also used in this project. The procedure of this rapid prototyping includes the following four steps.

- S1. The STL data for the model is loaded to the CAM software for processing. The CAM tool is 3D software that allows you to create tool paths by directly importing surfaces and polygon models in a variety of CAD file formats used by popular CAD systems.
- S2. Configuring the sizes and types of the end-mill and work materials, the appropriate cutting path is designed based on the model and its feasibility is confirmed using the simulation module of the software.
- S3. After the cutting path is designed properly, the milling process is given to the work materials using a three axis surface machining machine.

This rapid prototyping method is not feasible for a part as shown in Fig. 6. However, the SRP type of rapid prototyping was conducted to some of the parts as well. The upper-left portion of Fig. 8 shows an example of SRP rapid prototype for the machine based design, as opposed to the lower-left portion of the FDRP type.

5.4.4 Manufacturing with Milling Machine (NC Manufacturing)

As for manufacturing, the two dimensional drawings were prepared based on the imported CAD model data. Machining process was conducted to manufacture the designed parts. In some cases, however, the machine processing codes were

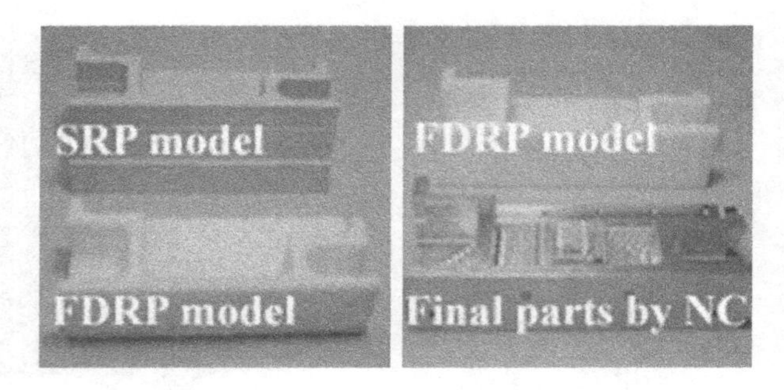

Fig. 8 Mounting plate made by rapid prototype and NC milling machine

created from the 3D CAD model data, and it was processed by NC milling machine to produce the parts. The lower-right portion of Fig. 8 shows the final parts made in the NC milling machine procedure.

5.4.5 Special jig for Manufacturing

The physical shape of some of the machine parts was too complicated to make it based only on the CAD drawings. In that case, special jigs were designed and prepared to enable the manufacturing in a precise manner. Figure 9 shows the examples of special jigs for this purpose. An important design learning experience for students is thinking about the special jigs that may be required to manufacture the parts, and how the parts can be designed to facilitate their holding in a jig. Finally, each machine part was manufactured and assembled based on the design. Figure 10 shows an example of the manufactured assembly corresponding to the model in Fig. 7.

5.5 Tips for the Success of Global Collaboration PBL

Web-based approaches are playing a very critical role not only in industries, but also in academia. The benefit of the technology covers not only the local collaboration activities, but also design and production in the global scale. WBDM made it possible for DMM project to carry out on a global scale. This section covers the results and discussion based on DMM project, and presents as tips for the success of PBL under a global collaboration.

5.5.1 Breakdown the Barriers of Personality, Characteristics, Language, and Culture by PREP

"People with shy or non-assertive personalities can have their ideas evaluated on an equal footing with other, more aggressive members of the team, and team members

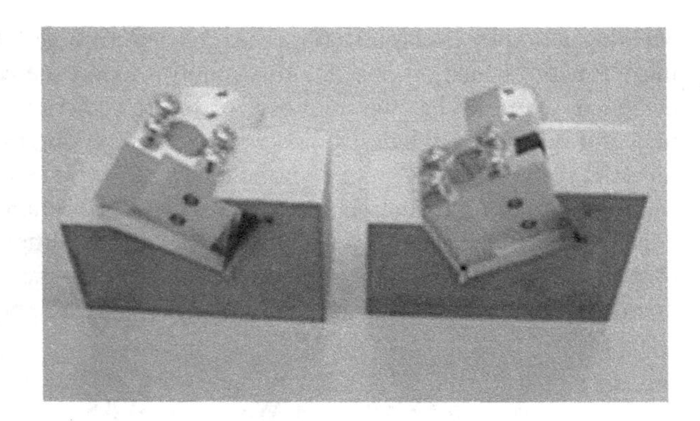

Fig. 9 Mounting plate made by rapid prototype and NC milling machine

Fig. 10 Final product made by NC milling machine

who are subordinate can feel free to express an idea. These social equalizations benefit the team as much as the individual, since the goal is to achieve the very best design—which may have originated with anyone on the team." Considering the ever increasing need for global collaboration, obstacles for collaboration are not only different in design approaches, but also languages and cultures. Even though the native language of the members of UT team was not English, all the members had equal workplaces, Picture libraries, Discussion boards, and Email alerts. All of these functions made it possible to strongly support the collaboration activities of design in teams. What the members of the teams could share were not only the idea and information, but also concepts, design production, and test and assembly ideas. However, SharePoint is not mandatory to achieve this objective. Integrating each individual tool such as file server system, communication board, or regular email system, the same kind of function could be implemented to achieve this objective. The critical thing would be to achieve the way in which the team members could share them in a convenient and seamless manner.

5.5.2 Use WBS to Utilize Idea, Data, Information, and Design

Since the PREP in iCampus was based on the SharePoint software, it provided a very powerful collaboration tool. For example, Document library, Event list, Task,

Meeting workplaces, Picture libraries, Discussion boards and Email alerts. All of these functions made it possible to strongly support the collaboration activities of design in teams. What the members of the teams could share were not only the idea and information, but also concepts, design production, and test and assembly ideas. However, SharePoint is not mandatory to achieve this objective. Integrating each individual tool, such as file server system, communication board, or regular email system, the same kind of function could be implemented to achieve this objective. The critical thing would be to achieve the way in which the team members could share them in a convenient and seamless manner.

5.5.3 Produce in a Concurrent Way with WBS

It would be possible to make the product on a remote site if the design is given in a digital way or in a conventional way. Section 5.4 showed how the production could be carried out in the DMM project. The minimum requirement for its manufacturing would be the mechanical drawings of the parts. However, without the collaboration along the way, the chance of parts not fitting together would have increased greatly.

With the CAD data for each part and its assembly, rapid prototyping could be possible as presented in Sect. 5.4. With the CAD model, FEM analysis with CAE tools would also be possible. If the native CAD data is shared, redesign of the model would also be possible. Therefore, WBDM is a powerful basis to pursue concurrent engineering in a project.

5.5.4 Keep Face-to-Face Interaction Effectively

According to the experience in the research project presented in this chapter, the importance of face-to-face interaction among team members was recognized. Even if web-based tools are beneficial, the most critical things in design proved to be face-to-face discussions. This chapter concludes that the web-based tools will support design activities, but it will never replace the need for face-to-face discussions. At worst by web-based meetings, at best by live meetings.

Note that too-many-meetings can also hinder progress. Indeed, face-to-face discussion is not always necessary. It is sometimes merely a time consuming task and even nonproductive for most of the people who work in a very tight schedule on any part of the world. Therefore, WBS such as the PREP tool (and Sharepoint or its equivalent) provides a productive alternative.

5.5.5 Make Systems Interoperable as Much as Possible

Standard file format or file conversion enables us to share data among team members even if the CAD tools are not the same. However, the use of the same

CAD tool makes it much easier to work on redesign, editing, and regeneration in an alternative output. Even though most of the rapid prototyping and manufacturing were based on the standard file format, the same software (SolidWorks) was adopted and used in the UT team as well in the end. Ideally, just like all cars use a common fuel and anyone can drive them, one day, all CAD systems will use the same basic file format, which is also editable, and they will differentiate themselves based on their interface.

5.6 What is the Next Step for Global PBL?

How does a designer systematically imagine and think about what a customer wants to his/her design? An answer to this question is to follow a deterministic design approach (Slocum 1992), which is taught in the 2.75 PDM class, and has proven to be effective as evidenced by the success of DMM project. This emphasizes a systematic, clear decision-making process that continually seeks to minimize risk, as opposed to shoot from the hip design, while extracting the maximum from creative designers even a global project like DMM.

Digital engineering makes it possible to perform global collaboration not only in design, but also in manufacturing. The DMM project showed how this could be carried out in a global research/teaching project for design and manufacturing of a DMM. As a result of DMM project, it has been suggested that digital engineering is not merely for big corporations, but also might be the basis for research projects upon which more and more universities could collaborate as part of their teaching process.

As for another finding of DMM project, WBDM played a critical role to support the digital engineering. The importance of WBDM was mentioned in the local PBL projects offered by CAD-EX and 2.007. It was also proved that WBDM supported a global collaboration for DHR project, which was one of the key technologies to grapple with the expansion of PBL from local to global. The next section presents how DMM project was taken over to further collaboration for global PBL project.

6 PBL Under Global Collaboration Toward Creativity Education

6.1 From DMM Project to DHR Project

The DMM project, or the first joint project between UT and MIT, was conducted in 2005, and centered on the challenge of designing a robotic dental mill to precisely shape teeth to receive crowns (Ito and Slocum 2007, 2008). DMM project was proposed to the PMD class as a part of the Dental Engineering joint research at UT and a novel prototype was successfully built within the 14 week class as presented in Sect. 5. Critical to the success of the project was the colocation of a UT faculty member who mentored the project closely at MIT during the whole semester. In addition, a local dentist lent his enthusiastic support.

When an international collaboration project was again suggested after a few years of interval, the authors decided to collaborate in the space of dental technology. However, this time it was not possible to locate a UT faculty member at MIT. In addition, the MIT surgeon who originally helped with the milling machine project already left MIT and could not help with the project. Thus emerged the proposal to conduct a global collaboration experiment where the project sponsors would be in Tokushima, Japan and the design team in Cambridge, USA.

Since the project proposal for PMD class is very competitive, with the final decision being made by the students, the discussion focused on getting not only practical topics, but also topics in which students would be interested. As a result, two proposals were discussed with the MIT course staff: a dental wire bending machine and a multiple degree of freedom (DOF) DHR.

Wire bending for dental treatment was manually performed by a skillful dental technician; therefore, it was proposed to design a table-top, computer controlled, dental wire bending machine. However, even though the idea of dental wire bending machine looked new, industrial 3D wire bending machines already exist and simply creating a scaled-down version would not be seen as new and exciting by students. The other reason is feasibility.

The other proposal for the creation of a new DHR was also reviewed carefully by both MIT and UT course staff. A brief review of prior DHR technology indicated that there was room for a new direction. The following comments from Dr. Ed Seldin, retired from MIT's campus health service, convinced the team of the feasibility of DHR project:

> There are some good headrests out there and simpler designs seem to trump more complex ones in the field. There might be a high-end market for a headrest with the capacity to store the spatial coordinates for the positioning of patients of record—perhaps a menu of positions based on the procedure being undertaken and the dental location. The problem with most headrests is a lack of effective dampening, with jerky movement and loss of head support during some adjustments of position, especially with anesthetized or sedated patients. I have had occasion to envision a hydraulically actuated headrest coupled with a haptic interface/joy stick controller and digital memory.

The project was presented to the class in September 2011 and selected by a team who decided to call themselves "SmoothMotion," which gives an indication of their most important FR.

First, Sect. 6.2 presents a brief overview of the PMD class, the deterministic design methodology, and the two types of supporting tools, namely PREP and FRDPARRC table (Graham 2007). Expectation management was also a critical issue to be considered in the DHR project in a global collaboration. Then Sect. 6.3 covers the specific exemplary DHR project, Sect. 6.4 shows how the expectation management was conducted for DHR project, and Sect. 6.5 presents behind-the-scene commentary and the results of this global collaboration.

6.2 MIT PMD Class Overview

MIT PMD class delivers both a project-based engineering experience to students and solutions to real clients while focusing on best engineering practices and fundamental design principles. The goal is to emulate a fast-paced, professional R&D environment, where there is no room for excuses (DeLoughyry 1995). The hypothesis in designing this course is that with advanced students, who are well on their way to becoming practicing engineers, collaborative projects aimed at solving real challenges can effectively advance both learning and research for students, clinicians, and instructors (Fig. 11).

To this end, since 2004 Boston area clinician-investigators have been paired with small teams of graduate and senior undergraduate students to develop proof-of-concept prototype medical devices for unaddressed clinical challenges. As show in Fig. 11, each summer project proposals are solicited from the local medical community and a subset selected, based upon the potential to create novel hardware, for "pitching" to the class in the first week. Students make the ultimate selections and self-form into 3–6 person project teams. Then, partnering with their clinicians, teams begin a 14 week, industry modeled design process, which is roughly broken into three parts:

1. Detailed problem understanding and investigation of prior art, followed by brainstorming of basic strategies for achieving the desired FRs.
2. Selection of best strategy and development of concept embodiments, selection of the best, modularization of design and identification of the MCM.
3. Detailed design, fabrication of parts, system assembly, testing, debugging, and documentation and presentation.

A useful (and fun) tool implemented in the class is the FRDPARRC table, which partners with the deterministic design process. Each FR leads to specific DPs, such as required torques, backed up by analysis (A) and referenced (R) information. However, the students are asked to investigate the possible risks (R) of failure and for each develop countermeasures (C).

Lectures on advanced mechanical design are delivered in conjunction with the design process during the first two thirds of the semester; the final third is focused entirely on completing and testing the prototype device. Teams receive a $3–4,000 purchasing and fabrication budget, which is overseen by their team mentor.

Course instructors meet weekly with each team and serve as project managers/ design consultants/mentors, demanding focus and passion. Documentation is emphasized, and it is critical that team member and mentors maintain a notebook in which they record their contributions. A key element is the PREP, which specifies that team members should first brainstorm/design individually and then, as a group, review each others' ideas. PREP is an important catalyst for making any design process more efficient by reducing the amount of discussion needed to converge on an idea.

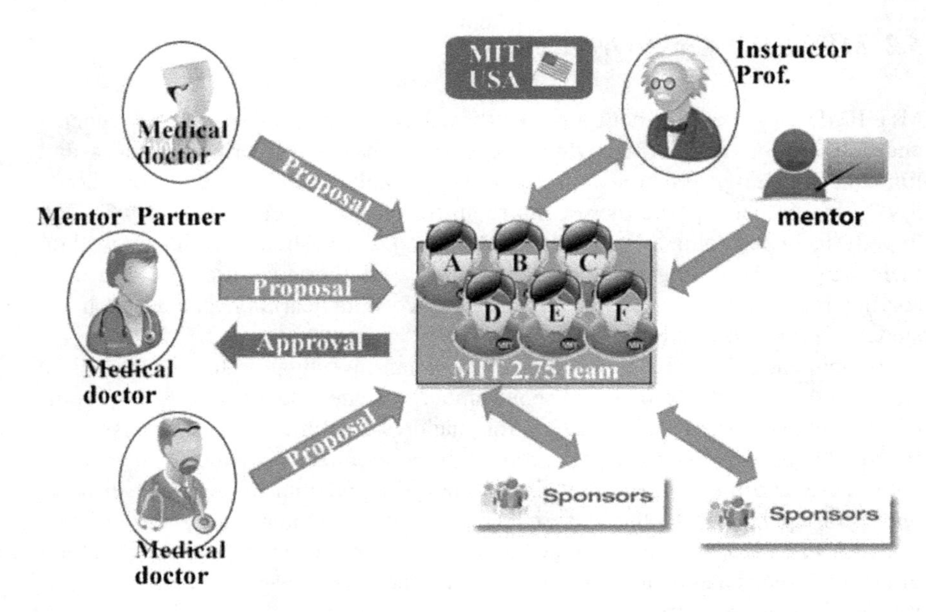

Fig. 11 Framework of PDM for regular 2.75 class project

An online Wiki is used as a collaboration and archiving tool. Students are encouraged to upload scanned notebook sketches, written test results, photos, and videos as the design process progresses. This not only ensures intra team communication, but also allows mentors to participate, regardless of geographic location (Wallace and Mutooni 1997, 1998). Three in-class design review presentations allow for the entire class to participate in the collaboration process (Bourne 2005).

Teams complete the semester by demonstrating their working prototype to an academic, clinical, and industry audience and documenting their project with a publication quality research paper, which they are then encouraged submitting to a conference or a journal. Recently, the course model has been successfully extended to encompass other industry sponsored challenges.

Projects initiated in the course have generated pending patents, peer-reviewed publications, senior and graduate theses, and several funded start-up ventures. Both course alumni and their TAs often go on to work as mechanical designers in the medical device industry, while research conducted in the course has helped clinicians advance their careers (Fig. 12).

6.3 Dental Headrest Project

Section 6.3 presents the DHR project under the collaboration between UT and MIT, including its background, project overview, global collaboration, and final prototype designed/developed by the project.

	Discovery
1	Opening
2	**Clinician presentations** Form teams
3	Define problem Review prior art
4	Brainstorm *Strategies* Bench level experiment
5	**Present 3 *Strategies***

	Design Engineering
6	Select *Strategy* Identify FRs
7	*Brainstorm 3 Concepts* Bench-level prototype Select *Concept*
8	Begin solid model **Present 3 *Concepts***
9	Identify *most critical module* (MCM) & supporting modules

	Building & Testing
10	Fabricate MCM Demo. MCM
11	Fab. other modules **Present final design**
12	Complete fabrication Integrate modules
13	Complete prototype Test! Debug. Test!
14	**Present Prototype** Document

Fig. 12 Course design process schedule

6.3.1 Educational Framework of DHR Project Under Expectation Management

DHR project was conducted between MIT and UT, based on the 2.75 PMD course project at MIT as shown in the working framework of Fig. 13. As opposed to the conventional PMD class project as shown in Fig. 11, the clinician or the customer of the project was allocated in Japan over the globe. In addition to the local mentor at MIT, a global mentor was also appointed at UT to take care of the global collaboration. As for the SmoothMotion team of 2.75, the design and manufacturing process is exactly the same as that of the typical process in PDM class project under the collaboration settings with Boston area clinicians.

Clear communication within the design team and with the client is essential in today's global workplace. Therefore, this poses challenges of time zones, languages, and cultures to DHR project. As opposed to the conventional setting for local partners, mutual understanding and communication between the team and their global partner at UT were not identical. Therefore, the supporting staffs of both UT/MIT took extra-care especially on expectation management in the project so that expectation should be always considered in design/manufacturing processes. Using an example of design project, DHR project, this section reviews how the management of expectations were controlled as part of an international, multidisciplinary global collaboration between UT and MIT.

6.3.2 Overview of DHR Project

Correct patient head position is important for patient and clinician comfort and, ultimately, greatly affects the outcomes of dental treatments. The headrests of dental chairs should be designed to securely support the patient's head position while allowing frequent adjustments by the clinician. However, typical headrests in current dental chairs only permit adjustment about three DOF, which is not

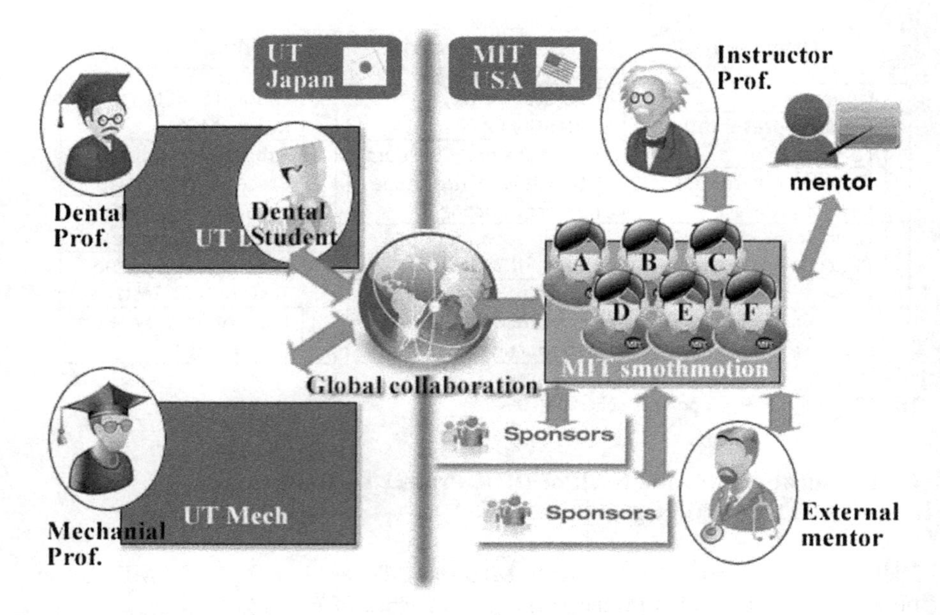

Fig. 13 Educational framework of DHR under global collaboration

enough to satisfy the actual requirements of both patients and doctors, when considering ergonomics issues as well as clinical practices. Figure 14 shows the typical motions, which are extension and tilt. The headrest can be extended along the neck axis by pressing the button and locked in position by releasing the button when the desired adjustment is reached. A lever at the headrest neck is used to release and lock position adjustment. Operation on headrest position adjustment requires both hands of the operator (Morita, http://www.morita.com/).

Furthermore, positioning mechanisms often have too much friction in the joints, so motion resolution can be limited and motion jerky. Clinicians desire to adjust the head position in an intuitive manner with an easy lock and release, as well as provide much more comfort to the patients. The main FRs initially identified were:

- Smooth positioning: The headrest should move easily from position to position with no apparent singularities.
- Quick and Easy position lock/release: A single action should release the mechanism to enable its repositioning.
- Stable and precise positioning: Once the position is fixed, it should be locked firmly and securely.
- Head position sensing: Position sensing might be required in order to implement a computer controlled dental treatment support system. This was not a mandatory requirement.

These customer requirements were provided to the MIT team by UT team and careful discussions were made to develop a shared understanding among the

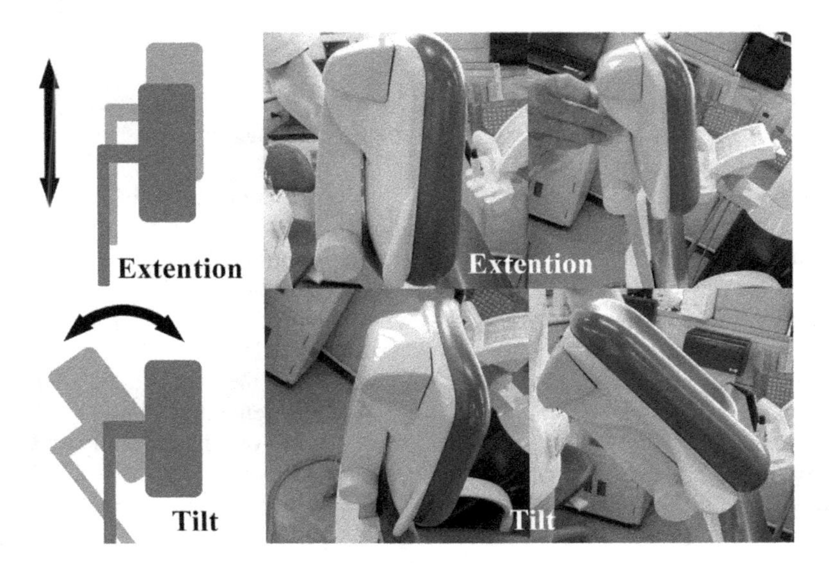

Fig. 14 Typical headrest motion

members of both teams. The resulting mission statement of DHR project was to develop a dental chair headrest with intuitive, continuous, and smooth adjustability, which facilitates head positioning that is comfortable for both patient and dentist.

6.3.3 Global Collaboration on DHR Project

The DHR project was conducted as an international collaboration between UT and MIT. The MIT student team was composed of six undergraduate students from Department of Mechanical Engineering at MIT under the supervision of the MIT course staff. The UT team was composed of Graduate students of Department of Dentistry under the supervision of UT staff from Department of Dentistry as well as Mechanical Engineering. Additional support was provided by Dr. Grace Collura, head of MIT campus dental service, who provided medical user input to the MIT team.

Even though DHR project was an international project, the project schedule and procedure followed that of all the other class projects.

The only difference from other 2.75 projects was that SmoothMotion team and UT clinicians were located on opposite sides of the globe. Therefore, weekly regular project meeting between MIT/UT teams were held using video conference and email, and data/information was shared via the secure MIT Wiki site. This occurred in addition to the regular weekly meetings between the MIT student team and staff and in class presentations.

6.3.4 Final Prototype of DHR Project

The project was conducted based on the deterministic design methodology, which is taught in 2.75 class, and the final prototype was designed/built at the end of the semester, of which photo is shown in Fig. 15.

The core idea of the proposed design is based on jamming media technology with headrest ergonomics to provide a headrest that is intuitive, continuous, and smooth, facilitating comfortable head positioning for both patient and dentist. The headrest uses a vacuum regulation system to achieve rigidity in an otherwise flowing media, which allows six degrees of freedom to the headrest. The headrest consists of two main modules, or an extension module, and a bag joint. The extension module provides one DOF in extension along the neck's axis. The bag-joint, which is continuously adjustable, provides the remaining five degrees of freedom, including three rotations and two translations. An ergonomically friendly handle allows one handed control of the system, which a dual button release ensures a firm grip on the handle before the vacuum on the system is released and structure loses rigidity.

According to the demonstration of video clips, the UT team confirmed that the customer requirements given by UT were mostly fulfilled in the prototype. Two kinds of design updates, or extension mechanism and single hand operation, were also implemented in the final prototype as discussed in the final meeting. However, direct feedbacks of bench-level experiments, which show the stiffness longevity and instantaneous force behavior could not be confirmed in the video clips. Therefore, the UT team concluded that attendance at the final meeting was the only solution to confirm ergonomic function.

The final class presentation was held in December, 2011 at MIT campus, hosting all of the team members, clinicians, teaching staffs, and invited guests. The presentation covered the five final presentations from the energy-focused topics and nine final presentations from the medical-focused topics.

6.4 Expectation Management in Each Stage of Deterministic Design Process

Managing the expectations of both clients and designers is an essential part of every product design and development project, necessary in order to set attainable goals, while retaining space for passion and creativity. The "best designed" product will be deemed a failure if it does not satisfy the customer's true needs, which may not always be clearly expressed. Likewise, as the designer shares his experience and expectations this can broaden the solution space and catalyze the client to consider new possibilities. In truth, a well-executed process can make the combined dreams of both the customers and designers come true. This section covers how expectation management was conducted in DHR project.

Fig. 15 The final prototype

6.4.1 Project Team Formation to Manage Expectations in Problem Understanding

A presentation covering the overview of projects, explaining their problems and significance was presented to students by clinicians in the beginning of 2.75 class. Considering the clinicians' presentations, the students select the projects on which they would want to work and form teams. This procedure provides two advantages. Each team selects a favorite project, which helps to increase the motivation of the student team. In addition, clinicians take great care to present their product desires, so that their problem will be selected by one of the teams. As a result, the matching of student team and a clinician(s) increases the motivation on both sides, and leads to the better teamwork activity (Byrd and Hudgins 1995; Dertouzos et al. 1989; Simon 1996) throughout the semester.

This matching structure applied to the DHR project as well and worked effectively to share common understanding, considering the expectations on each side toward the common goal. One of the characteristic features of the DHR project was that it was an international collaboration project across the globe between the two different cultures, institutions, and languages. Therefore, both teams tried to understand each other's expectations, both technical and professional. This expectation and understanding worked effectively toward the success of DHR project.

6.4.2 Schedule Expectation for the Global DHR Project

At the time of project proposal presentation, no UT staff could be in residence at MIT, so UT staff prepared a presentation and MIT staff made the presentation in the class. After the project was selected, MIT and UT teams worked together.

A regular weekly meeting was held using a video conference system (Skype) between the campus of MIT and the two separate campuses of UT, which means that a three site-connection was used. The time difference between UT and MIT is

13 h during summer and 14 h during winter. When the project started, a regular meeting was scheduled at 8:30JST or 19:30EST on a weekly basis. Even though MIT team members attended the meeting, it was hard for UT team members to regularly attend the meeting because of the conflict of other local meetings and clinical schedules.

Even though the time difference is one of the difficulties to overcome in global collaboration, MIT/UT teams generally managed the difference very well. However, one episode happened in November. Generally, both MIT/UT team members were on-line and waited for the starting time 5–10 min prior to the meeting time in every week. However, one day, in November, the UT team did not find the MIT members on-line even after 15 min passed without any notice or e-mail messages, which had never happened before. There was no reply to an inquiry message sent by UT team to MIT team. If an attendee does not appear in the meeting at the time of appointment, the meeting may be automatically canceled in most of the cases. However, UT team had a feeling that there must be a reason why the MIT team did not show up because the meeting was expected. Thirty minutes later, a message delivered from MIT team to UT team, explaining the time change of 1 h delay due to daylight savings time, which is an uncommon custom for the UT team in Japan. Expectation in the DHR project time schedule worked appropriately to comply with the delay of meeting time 1 h later.

6.4.3 Problem Understanding and Functional Requirement Identification

The project desires were shared among the team composed of students and clinicians, but it was hard to share details over the video conference, and this was where the Wiki helped greatly. The MIT/UT team shared, for example, pictures of existing headrests in the UT hospital, brochures of dental chair products, and video clips showing a simulated scene of UT hospital. Even though the clinicians at UT did not use the definition of FRs, what was required in the clinic was clearly presented to MIT team. The video clip presentation proved to be the most effective method to show the requirements, even though its preparation took much time and effort.

After sharing the basic idea of problem in the project, further web-based meetings made the FRs clear to the SmoothMotion team, so it could work on its design.

SmoothMotion team expectation: The team used as a benchmark a current product on the market made by Sirona (Sirona, http://www.sirona.com/), which is very flexible in adjusting. Their goal was to design a more flexible and innovative product.

UT expectation: Research topic proposal was based on clinical experiences and the UT team was not aware of the Sirona product at the time of proposal. When UT reviewed the benchmark product presented by MIT team, it was confirmed that the adjustable movement of Sirona is good but does not fully satisfy the requirements.

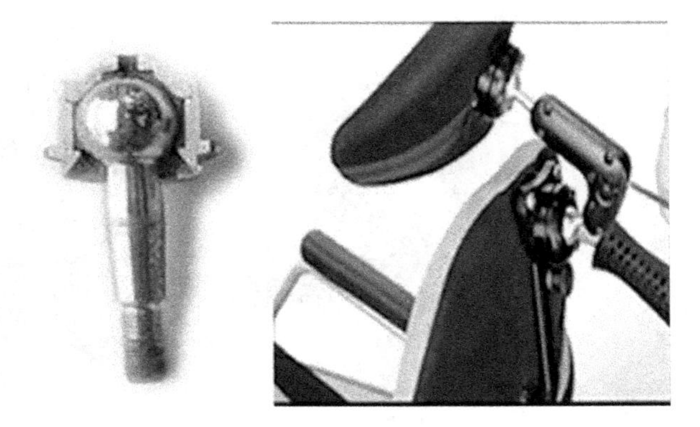

Fig. 16 Proposed concept: ball-joint

For both sides, the expectation was that it should be possible to design an innovative new DHR.

6.4.4 Management of Expectation in Design Strategies and Concepts

In deterministic design, the methods can be applied at each step along the way. SmoothMotion team followed the step of deterministic design ranging from coarse to fine, creating possible strategies, concepts, module, and component. As a result, the team proposed three concepts, which are ball joint, hexapod, and jamming.

The first concept, based on ball joints (Drutchas 1991), is similar to the bone structure of an arm with shoulder and elbow joints replaced with ball and pin joints, respectively. The serial mechanism achieves the required range of motion through a combination of rotations allowed by the joints. Figure 16 shows the example image of this concept.

The second concept, a hexapod (Newport http://www.newport.com), supports the headrest by six links working in parallel, allowing triangulation of three points in space while deterministically defining a plane. Figure 17 shows the example image of this concept.

The third concept, based on media jamming (Brown et al. 2010), uses granular media subject to a critical compressive stress to form a pseudo-solid. It is possible to rapidly switch a cleverly designed support from free-forming and freely adjustable to rigid, pseudo-solid by just applying an appropriate vacuum. Figure 18 shows the example image of this concept applying to robotic arm gripper.

SmoothMotion team built three concept prototypes based on these initial concepts, and posted them to the Wiki for the UT team to review prior to an on-line design reviews. Power point slides and descriptive explanations were first used to share the idea among the MIT/UT teams. However, it was very hard to share the

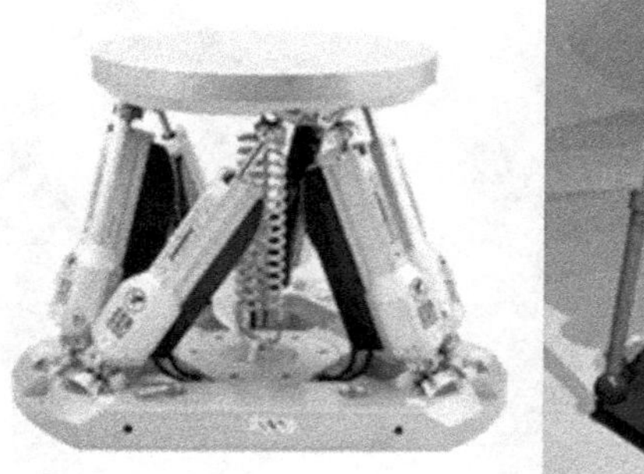

Fig. 17 Proposed concept: hexapod

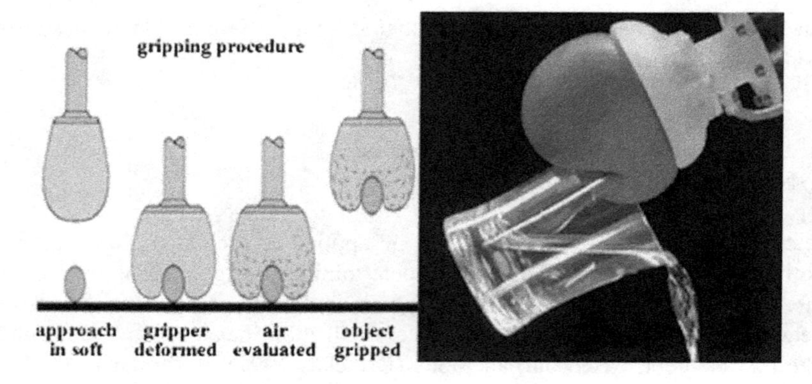

Fig. 18 Proposed concept: media jamming

concept ideas with slides. Therefore, just as UT team did in problem presentation, video clips of physical sketch models were prepared by the MIT team and shared with UT team. Since the motion of the headrest is critical to understand the proposed concepts, video clips worked effectively. In this way, design reviews of the concepts were undertaken by MIT/UT team members. Questions posed by the UT team included:

(Q1) According to the video, hexapod and ball-joint types are too bulky and it may become a problem in clinical operation. The size of Jamming type looks like the most suitable one judging from the video clips.

The sketch model prototypes recorded in the video clips were made to show the mechanism idea. Therefore, the SmoothMotion team did not pay too much attention to appearance and size. However, the clinicians of UT team saw the mechanism and also focused on the expected final imaginary product from the video clip.

This led to a perception of unmet needs. The expectation at the concept model stage of SmoothMotion was somehow different from that of UT, which thought a misunderstand occurring. However, the comments from the UT team provided what is required, or expected by the customer to the SmoothMotion team, which then worked successfully to share their idea of customer expectations in the DHR project team. Back-and-forth discussions include:

(Q2) We cannot see the locking mechanism of Jamming type. Please explain to us about that.

UT Expectation: Locking mechanism is critical and expected to be implemented.

(Q3) Is it possible to install a lock button on the back of the headrest, as is similar to the existing headrest?

UT Expectation: Usability of simple operation is expected.

(Q4) Would it be possible to make downsizing of the connecting part between the chair and headrest? The ideal size would be similar to the existing headrest (this is related to 1)

UT Expectation: Compact design is expected.

(Q5) The level of compressor noise sounds larger that the current level. Would it be possible to decrease the noise level and keep it much quiet?

UT Expectation: Comfortability for both doctors and patients are expected.

(Q6) Please show us the actual adjustable area in each prototype.

UT Expectation: FR fulfillment is expected.

(Q7) As far as we can see from the videos, jamming type is the most favorable one.

UT Expectation: Innovative idea and compact design are expected.

6.4.5 Management of Expectation in Prototype System

Based on the best concept selected, SmoothMotion built a working prototype and presented it to UT team by way of video clip demonstration. It was confirmed that the prototype basically met the FRs given by UT team. However, the following two points were found be inadequate: extension mechanism and single hand operation.

Since the extension mechanism of 100 mm translation along the neck axis was a required specification, the implementation of this mechanism was requested. Even though the final presentation was scheduled in 2 weeks ahead, SmoothMotion team decided to work on the implementation of that function. Reviewing the feasible candidate design ideas considering the project time limit, an additional linear sliding mechanism was designed to make the extension and added to the prototype.

The prototype on the video clip also showed two-hand operation for position adjustment because of the multiple buttons to control the vacuum lock of the media. After additional engineering work was completed on the prototype, a single hand operation was realized.

The expectation of the UT team was well understood by the SmoothMotion team, not only through the design to manufacturing stage, but also in the rework on the final prototype. It was recognized that the SmoothMotion team tried to meet the expectations of the UT team and the UT team believed that the SmoothMotion team could comply with its expectations. At first, the UT team was not aware of the deterministic design process used, the SmoothMotion team followed the deterministic design process, which helped to achieve the goals of the project, and thus helped the UT team to also learn the deterministic design process.

6.4.6 Expectation of Off-Line Meeting to Find an Unsolvable Solutions On-Line

Mechanism and function were mostly understood by using video clips along with figures, photos, written descriptions, and oral explanations. However, the following points were not well understood by the UT team.

According to the bench-level experiment, the headrest becomes very stiff after vacuum is applied and could support the patient head appropriately. However, how long does it keep the same stiffness? Does the stiffness decrease over time? When an impulse force is applied to the headrest, which could occur during clinical treatment, what would happen?

As for the single hand operation to adjust the headrest position, which is one of the critical requirements, the operation looked possible. However, how easy it is to make adjustment? How long does it take the vacuum to make the headrest stiff after adjustment?

Both the MIT and UT teams expected to carry out the DHR project online as far as they could. On the other hand, they also expected to have an off-line meeting to directly share the project achievements. In the event of difficult understanding on-line, one of the solutions is to hold an off-line meeting.

6.4.7 Expectation to Meet in the Final Presentation

The 2.75 final presentations were held at MIT in December 2011, where the teams demonstrated prototypes to the class, instructors and invited visitors who signed non-disclosure forms, so IP rights could be maintained before patents were applied for if required. In the session, teams (including clinicians) respond to questions from the audience, and audience feedback is formally provided to teams.

In order to clarify the questions mentioned in the previous section, a UT staff attended the final presentation. Since this was the first face-to-face meeting, both the team and UT were quite excited to share ideas, and review experiences and results. Even if audio–video communication and information/data sharing is available over the network, face-to-face meetings cannot be replaced by any other media, which was recognized in the final presentation. In the end it is hard to share a teambuilding meal or toast via the internet!

6.5 Behind the Scene of Global Collaboration

The goal of the DHR project was to design an innovative mechanism for the positioning of a dental chair headrest so as to satisfy both the needs of a patient for comfort and a clinician for flexibility and access. The DHR project resulted in an innovative design of headrest adjusting mechanism that was implemented in a prototype. Moreover, the students, faculty and clinicians benefited from the experience of innovative design collaboration in a multidisciplinary, global team.

The project was conducted based on managed expectations and deterministic design, starting from design strategy, followed by concept design, module design, and engineering, prototype building and test, and design reviews. This section showed how expectation management was well organized and was used successfully during the development of the DHR project based on the deterministic design, which has led the project to success in terms of product development as well as PBL-based education (Sheppard et al. 1998).

As for a comment from behind the scene of global collaboration in DHR project, one of the things to point out here is that WBS played a critical role for communication, data exchange, information sharing, and many more. Without the use of WBS, not only the digital engineering, but also communication, deterministic design process, expectation management, etc. were not performed so smoothly. However, WBS used in DHR project was basically a traditional environment. The next section will review the pros/cons of traditional WBS and discuss how the delivery of a joint PBL such as DHR would/could be changed if one utilizes Web-bases system vs. Cloud-based system. I will also discuss on what potential challenges are to overcome, in terms of technology from a student's and a teacher perspective.

7 From Web-Based LMS to Cloud-Based LMS

7.1 Review of Web-Based LMS in CAD-EX and 2.007

Ranging from the toy model engineering PLB in Sect. 3 to the global collaboration PBL in Sect. 6, several types of web-based systems have been used effectively.

The first system to review in this section is the web-based LMS system for CAD-EX class. This LMS has been supported by various class activities in CAD-EX class for many years, for example, information distribution regarding the class work, schedule, assignment materials, reference papers, and on-line manual. Figure 19 shows the collections of snapshot for CAD-EX user interfaces indicating typical functions.

Fortunately, the usage of the web-based CAD-EX LMS increased every year, so that the maintenance work of the system became a burden to the teaching staff. The system needed to be running continuously in a stable condition. If something

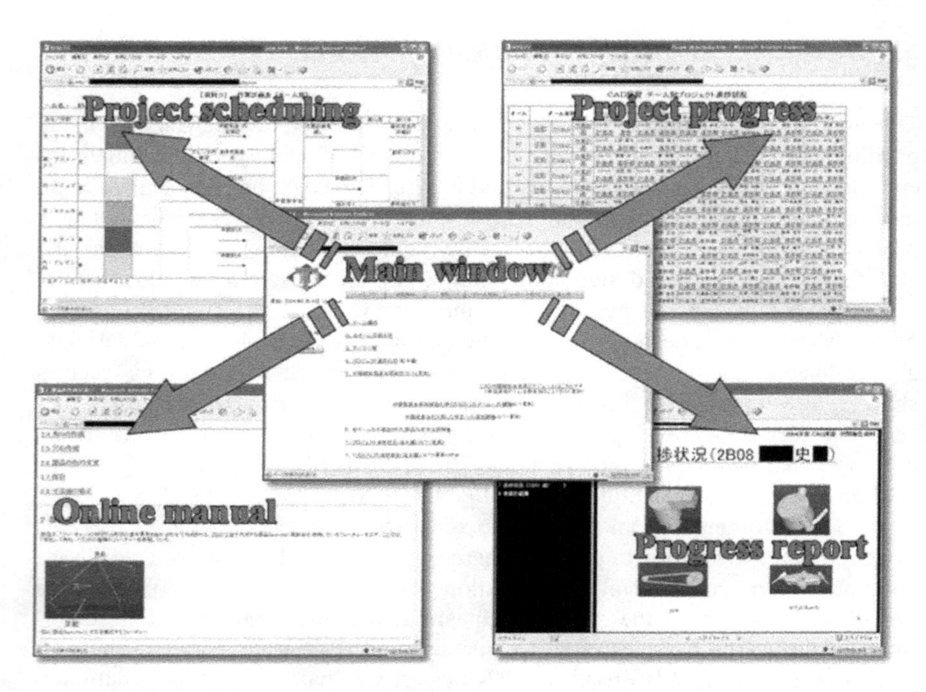

Fig. 19 Snapshots of CAD-EX web-based LMS

happened, system maintenance was expected in a timely manner. The system needed to be updated according to the class schedule, and the teaching staff had to spend many hours in maintaining the system, whereas they were supposed to use most of their time in class activities. An open issue is how to economically, stably, and securely maintain, update, and troubleshoot the web-based system.

The second system to review here is the web-based 2.007 LMS system. Thanks to the tremendous efforts by the teaching staff for many years, the 2.007 LMS worked as a portal to access to various information/data for the class, including course syllabus, calendar, lecture notes, citations, and course materials including sample data, template, and software. Figure 20 shows a snapshot of main windows of 2.007 LMS. The TA staff of 2.007 spent many hours to maintain the web all the time, which was almost the similar situation to that of CAD-EX LMS.

Both of these LMS websites were developed locally and maintained by the teaching staff of each class. For the first few years after the introduction, the usage of the system was not so high and its maintenance was not a big issue. These systems were proved to be effective in both institutions and getting to play a critical role gradually. As a result, system maintenance became one of the biggest issues.

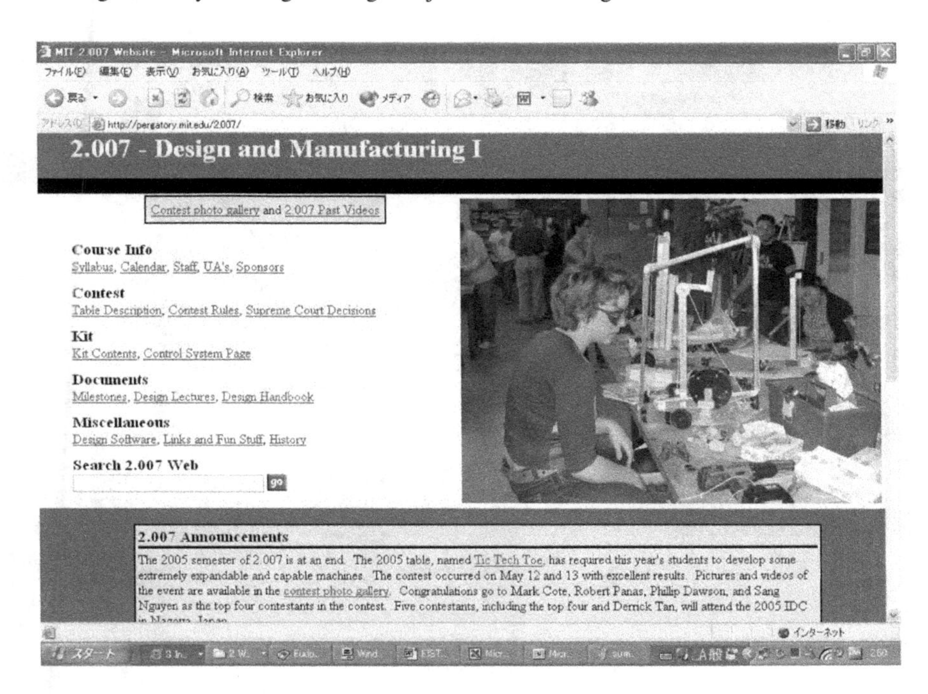

Fig. 20 Snapshot of 2.007 web LMS

7.2 Review of Web-Based Design and Manufacturing Systems in DMM

Section 5.3 presented how iCampus, or the web-based LMS, was applied to DMM project. Characteristic features of iCampus projects including document library, event list, task monitoring, meeting workplaces, picture libraries, discussion boards, and email alerts were all functioned quite well in the project.

iCampus system strongly supported the DMM project because all of the team members of DMM project were on the use of iCampus at MIT area or its neighbors. Even though it was a global project, most of the activities were conducted at MIT by those team members. Since one of the faculty staff of UT colocated at MIT during the whole project, the staff used the iCampus as well. However, other UT staff in Japan did not have an access capability to iCampus. Therefore, the UT staff at MIT collaborated with UT staff in Japan over a web-based system prepared by UT. Figure 21 shows the framework of DMM computer settings.

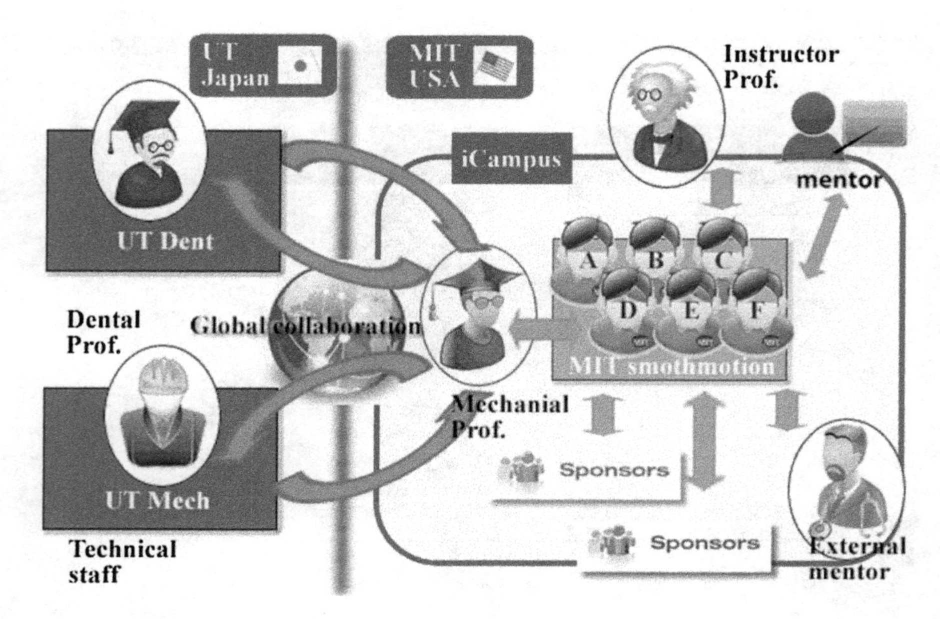

Fig. 21 DMM framework

7.3 Cloud-Based LMS at UT

Cloud computing is the current big trend, which sounds nebulous, but may be ambiguous from the perspective of education. While detailed discussion about cloud computing could be found in other references, a basic breakdown of cloud computing would include software as a service, utility computing, web services in the cloud, platform as a service, managed service provider, service commerce platform, and internet/intranet integration.

Cloud computing might also be more accurately described as "sky computing," with many isolated clouds of services, which IT customers must plug into individually. Under "sky computing", the idea of loosely coupled services running on an agile, scalable infrastructure should eventually make every enterprise a node in the cloud.

Under these circumstances, education support system at UT, or Tokushima University Network System (TUNES), has also been shifting toward cloud computing. Figure 22 shows the overview of the updated TUNES network system at UT. As a result of system renovation, TUNES was rebuilt as a hybrid cloud system including private/public cloud computing system and on-premises system. Private cloud is the network backbone supporting campus network system including i-Collabo (i-Collabo, http://www.nec.co.jp/educate/products/i-collabo/). Public cloud supports the mail service and portal services. UT students are automatically registered in i-Collabo when they register their classes in each semester. This section focuses on i-Collabo.LMS of UT and the current LMS system for CAD-EX

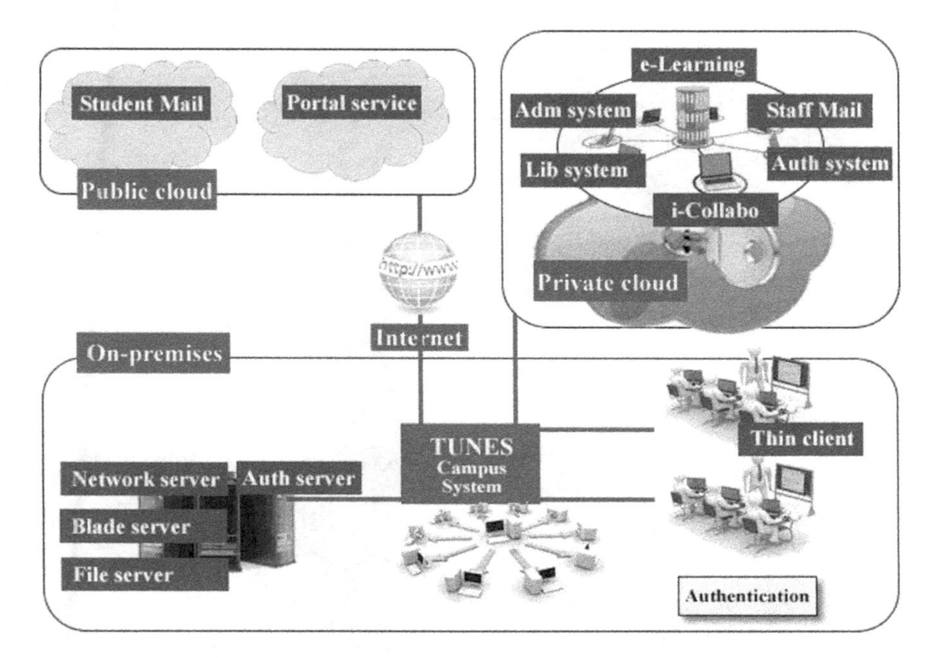

Fig. 22 Overview of TUNES at UT

is presented. Original i-Collabo.LMS was developed by NEC and customized version of i-Collabo.LMS has been introduced to UT in 2011 and applied to CAD-EX class.

Figure 23 shows the brief outline of i-Collabo.LMS, which is integrated with TUNES and runs in the private cloud environment with highly secured accessibility for all users of TUNES. i-Collabo supports UT students all over the campus since the introduction in 2011.

7.4 Advantages and Drawbacks of i-Collabo for CAD-EX

CAD-EX class applied i-Collabo.LMS in the class of 2012 and 2013, where the web-based LMS system used to be supporting the class since 2004 until 2011. When we compare the two LMS systems, the following pros/cons were observed in each comparison item as shown in Table 3. Advantages of i-Collabo were recognized, whereas some issues were addressed.

In terms of teacher perspective, the teaching staffs became free from the maintenance of LMS web server. Therefore, the teaching staffs could focus more on the class. Data collection and assignment management are quite effective in i-Collabo.LMS so that each student were taken care of individually. Thanks to the various build-in functions, cgi-like programming is not necessary. However, the

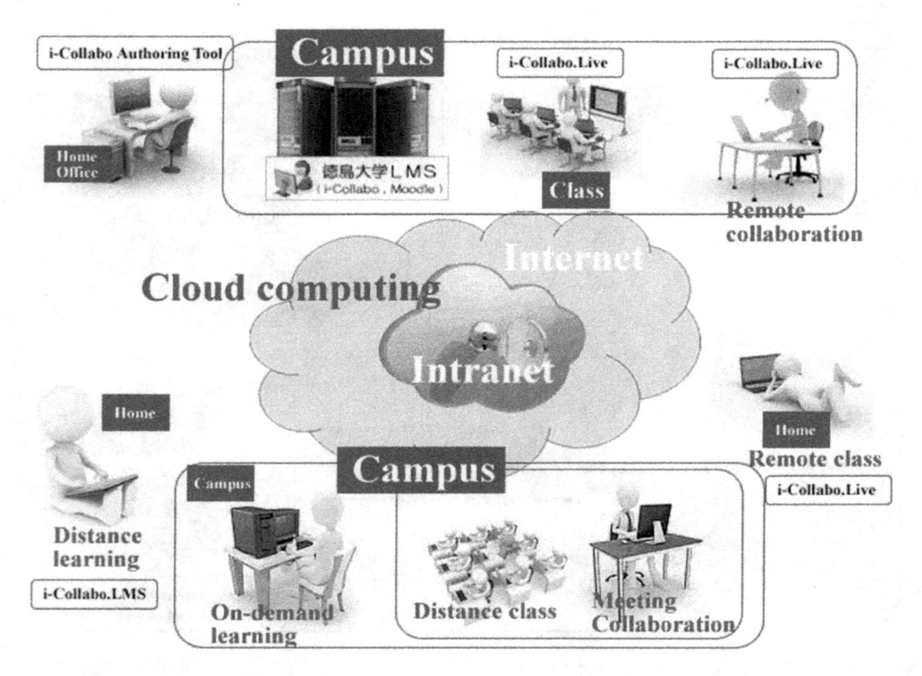

Fig. 23 Overview of i-Collabo.LMS at UT

Table 3 Usability comparison of web-based LMS and i-Collabo.LMS

	Comparison item	Web-based	i-collabo
1	Maintenance	[a]Hard	[b]Easy
2	Customization	[b]Flexible	[a]Version up
3	Lecture preparation	[b]Flexible	[c]Rigid
4	Information distribution	[b]Easy	[b]Easy
5	Data collection	[a]Network folder	[b]Built-in
6	Data organization	[b]Flexible	[a]Rigid
7	Teacher/student interaction	[c]BBS cgi	[b]Built-in
8	Assignment management	[c]cgi	[b]Built-in
9	Security	[b]Access control	[b]By account
10	Accessibility	[a]On campus	[b]On/off campus

[a] Bad
[b] Good
[c] Medium

limit of data size of i-Collabo was one of the critical issues in CAD-EX, because the maximum data size supported by i-Collabo is far less than needed. As a temporally solution, a secured network share folder was prepared separately from i-Collabo.LMS in order to upload/download the large size of CAD data.

In terms of student perspective, i-Collabo provides a general LMS for all of the class of UT and CAD-EX is one of them. i-Collabo environment for each student is automatically prepared when the student registers the class. This is a good advantage. Moreover, Virtual Desktop Service is provided to all of TUNES users. Therefore, it is possible for students to use TUNES system under the same environment at off-campus as well as at on-campus.

In terms of system management perspective, customization was quite easy for the web-based LMS systems, because the teaching staff not only took care of the web service software, but also the web server machine itself. Various new ideas were proposed and discussed in the CAD-EX class management, and some of those ideas were under review through the web service by implementing cgi scripts. After the introduction of i-Collabo.LMS, the system is stable but not flexible. As a result, the class management of CAD-EX was limited within the range of i-Collabo.LMS.

7.5 Discussions Toward Cloud-Based LMS/DM System

i-Collabo.LMS is an effective LMS system to support engineering class in general. By sharing the data and information through i-Collabo.LMS, not only the conventional local class but also e-learning, or global education could be effectively supported. However, i-Collabo.LMS is not designed for PBL support. In order to apply it to PBL class, something is missing in CBLMS systems in general. This section points out several issues to be considered to make it applicable to PBL class.

PBL class requires engineering data, such as solid model data built by CAD software. i-Collabo.LMS is not suitable for sharing those data, not merely because of the volume size, but also the structure of the data. When these engineering data is shared by team members of PBL class, data handling is completely different from general document files, which could be shared in a common shared folder. Engineering data holds data structure, for example, an assembly file composed of several different types of part files attached with drafting files. The files are always updated by the team members. Therefore, in order to take care of these engineering data, PDM types of function may be required. This PDM should be very user friendly in such a way that PBL team members can use it quite easily.

As for sharing ideas, FAQ function was implemented in CAD-EX web system. Questions and answers between students and teachers, or discussions among students were available over the network, and its effectiveness was proved as mentioned in Sect. 3. This type of communication function could be used for PLB class to share information. However, sharing ideas is another issue to be considered. Since we had experiences in 2.007 class under iCampus project, a cloud-based iCampus type of system may be expected for PBL class.

For supporting deterministic design, how to share design strategy and concepts may be a big issue. While a designed prototype could be shared in the form of solid model of CAD software, the strategy and concepts, which are formed before building the prototype, are very hard to share over the network. Face-to-face discussion is strongly required to prepare a strategy to build a prototype through concept formation. As for DHR project, the team members held a regular meeting using a video communication system with video clip presentation. CBDM system may need an innovative function to support this kind of sharing for global PBL projects.

As for remote fabrication, public cloud-based design/manufacturing system may be required, because the manufacturing staff and facility may be out of campus. As is often the case with 2.75 projects where the team members place an order of required parts for prototype building, they may want to place an order of remote fabrication through the cloud-based CBDM system. Before actually placing an order, a simulation-based presentation regarding the remote fabrication may be required to make sure that the order should be suitable to the prototyping.

As for team forming and design evaluation, a face-to-face meeting was a very critical event in DHR project. Even if a video communication system technology has been advanced, maybe with virtual reality environment, nothing could take over the real experience of meeting face-to-face and evaluating the real prototype, which was the common comments among the global team and staff of DHR project. Attending an international conference virtually, without actually visiting the conference venue, would lose something important, which could be obtained only by visiting the actual venue. Even if many parts of collaboration could be maintained over the network for global collaboration project, a face-to-face meeting, whether for team forming or final evaluation, cannot be avoided to achieve the real goal of PBL.

8 Concluding Remarks

PBL has been reviewed as one of the most effective approaches to teach creativity in design education. In fact, the two PBL examples of CAD-EX and 2.007 presented in this chapter shows the effectiveness of PBL approach in engineering education classes. The critical factors in these examples were also applicable in global PBL project as described in the DMM project. When the project level is further global such as DHR project, expectation management is another critical approach to be considered in PBL projects.

Reviewing the several PBL projects in this chapter, it is confirmed that LMS has been playing a critical role in these projects. Considering the current trend of cloud-based computing in various areas, it is likely that the usage of CBLMS systems would gradually increase to support PBL classes by way of various benefits, such as on-demand self-service, ubiquitous network access, rapid elasticity, or location-independent resource pooling. This chapter showed the current situation of CBLMS system for CAD-EX, which has been shifted to cloud-based system.

The major target of creativity education in engineering would be how to teach students to be creative in design, activities, attitude, manufacturing, etc. CBLMS systems would play a key role as a supporting tool of PBL in achieving the target of creativity education.

Acknowledgments The author would like to acknowledge Prof. David Gossard and Prof. Ernesto Blanco of MIT for discussion regarding design class. The author would like to thank Daisuke Yonekura and Hiroyuki Ukida, and Center for Advanced Information Technology at UT for supporting CAD-EX.

The authors would like to acknowledge 2.75 Team for DMM project with Ryan Griffin, Melissa Read, Gerald Rothenhofer, and Josh Young. The authors would like to acknowledge the remote manufacturing support of Tokushima University Machine Shop with Tetsuya Sato and Junji Tamatani.

The authors would like to acknowledge Dr. Takaharu Goto of UT and the members of Ichikawa Lab for supporting DHR project. The authors would like to acknowledge the SmoothMotion team members including: Zachary D. Nelson, Wesley D. McDougal, Sammy M. Khalifa, Andrew T. Carlson, David C. Parell, and John W. Romanishin. The authors would like to thank Drs. Ed Seldin and Grace M. Collura, Chief of MIT Dental Service, for their advice and local mentorship.

The authors would like to thank Prof. Shuichi Fukuda from Stanford University, for his comments and advice to the projects. The authors would like to thank Center for Integration of Medicine and Innovative Technology for its support and J. Morita Corporation for donation of experimental materials.

References

Armbrust M, Fox A, Griffith R, Joseph AD, Katz R, Konwinski A, Lee D, Patterson, Rabkin A, Stoica I, Zaharia M (2010) Above the clouds: a view of cloud computing. Commun ACM, 53(4), pp 50–58.

Belcheir MJ (2000) The National Survey of Student Engagement: results from Boise State freshmen and seniors. Research report 2000-04, Boise State University, pp 1–15

Boehm BW, Abi-Antoun M, Port D, Kwan J, Lynch A (1999) Requirements engineering, expectations management, and the two cultures. In: 4th IEEE international symposium on requirements engineering (RE '99), 7–11 June 1999, Limerick, Ireland, pp 14–22

Bourne J, Harris D, Mayadas F (2005) Online engineering education: learning anywhere, anytime. J Eng Educ 9(1), pp 131–146

Brown E, Rodenberg N, Amend J, Mozeika A, Steltz E, Zakin MR, Lipson H, Jaeger HM (2010) Universal robotic gripper based on the jamming of granular material. In: Proceedings of the National Academy of Sciences (PNAS), Nov 2, vol 107, no 44, pp 18809–18814

Burroughs R (1995) Technology and new ways of learning, ASEE Prism, January, pp 20–23

Byrd JS, Hudgins JL (1995) Teaming in the design laboratory. J Eng Educ 85(4):335–341

DeLoughyry TJ (1995) Studio' classrooms. The chronicles of higher education, March, 31, pp A19–A21

Dertouzos ML et al (1989) Made in America. MIT Press, Cambridge

Drutchas G (1991) Ball joint. U.S. patent 4986689, Filed Sept 11, 1989, and issued Jan 22, 1991

Dym CL, Agogino AM, Eris O, Frey DD, Leifer L (2005) Engineering design thinking, teaching, and learning. J Eng Educ 95(1):103–120

Frey DD, Smith M, Bellinger S (2000) Using hands-on design challenges in a product development master's degree program. J Eng Educ 90(4):487–493

Graham M, Slocum A, Sanchez RM (2007) Teaching high school students and college freshman product development by Deterministic Design with PREP. ASME J Mech Design (Spec Iss Design Eng Educ) 129:677–681

Hanson P, Robson R (2003) An evaluation framework for course management technology [electronic version]. Educause Centre for Applied Research, 14(research bulleting). http://www.educause.edu/ir/library/pdf/ERB0314.pdf. Accessed Oct 12, 2006

iCampus [online] http://icampus.mit.edu/

i-Collabo [online] http://www.nec.co.jp/educate/products/i-collabo/

Ito T (2005) A web-based approach to teamwork-based learning in 3D CAD exercise class. In: 3rd international conference on education and information system: technologies and applications, Orland, FL, USA, pp 98–30

Ito T (2011) A challenge of global collaboration towards creative engineering education. Design engineering workshop 2011, vol 11, no 11, pp 109–111, Tosu, Nov

Ito T, Oyama A (2005) Studies on the effect of feedback-based evaluation method in fundamental creative engineering class. J Jpn Soc Eng Educ 53(1):41–46 (in Japanese)

Ito T, Slocum AH (2007) Teaching collaborative manufacturing: experience and observation. Int J Internet Manuf Serv 1(1):75–85

Ito T, Slocum AH (2008) Teaching creative engineering: education in Japan and the USA. In: Proceedings of the ASME 2008 international design engineering technical conferences & computers and information in engineering conference, vol DETC2008, no 49324, pp 1–10, New York, USA, Aug

Ito T, Yonekura D, Ukida H (2004) A new approach to towards team working education in 3D CAD exercise class. J Jpn Soc Eng Educ 52(4):62–65 (in Japanese)

Linthicum D (2009) Cloud computing and SOA, convergence in your enterprise: a step-by-step guide. Addison-Wesley Professional, Indianapolis

Liu F, Tong J, Mao J, Bohn R, Messina J, Badger L and Leaf D (2012) NIST cloud computing reference architecture: recommendations of the national institute of standards and technology, (Special Publication 500–292). CreateSpace Independent Publishing Platform, USA. (ISBN:14781680219781478168027)

Ma H, Slocum AH (2006) A flexible-input, desired-output (FIDO) motor controller for engineering design classes. IEEE Trans Educ 49(1):113–121

MIT 2.007 (2005) 2.007 classes and contest. [online] http://pergatory.mit.edu/2.007

Morita [online] http://www.morita.com

Newport hexapod products online catalog [online] http://www.newport.com/hexapod

Okamura AM, Richard C, Cutkosky MR (2002) Feeling is believing: using a force-feedback joystick to teach dynamic systems. J Eng Educ 92(3):345–349

Richard C, Okamura AM, Cutkosky MR (1997) Getting a feel for dynamics: using haptic interface kits for teaching dynamics and controls. In: ASME IMECHE 6th annual symposium on haptic interface, Dallas, TX, Nov 15-21

Sheppard S, Johnson M, Leifer L (1998) A model for peer and student involvement in formative course assessment. J Eng Educ 88(4):349–354

Simon HA (1996) The science of the artificial, 3rd ed. MIT Press, Cambridge

Sirona Dental System [online] http://www.sirona.com

Slocum A (1992) Precision machine design. Society of Manufacturing Engineers. (ISBN-10: 0872634922)

Slocum D, Long P, Slocum A (2005), Teaching the next generation of precision engineers. In: Proceedings of the 5th EUSPEN international conference, Montpelier, France, S5P1

Starrett SK, Morcos MM (2001) Hands-on, minds-on electric power education. J Eng Educ 91(1):93–99

Wallace DR, Mutooni P (1997) A comparative evaluation of world wide web-based and classroom teaching. J Eng Educ 87(3):211–219

Wallace DR, Weiner ST (1998) How might classroom time be used given www-based lectures? J Eng Educ 88(3):237–248

Wright SH (2005) Robot contest puts design into action. MIT Tech Talk 49(26):4

Wu D, Thames JL, Rosen DW, Schaefer D (2012) Towards a cloud-based design and manufacturing paradigm: looking backward, looking forward. In: Proceedings of IDETC/CIE2012, DETC2012-70780, August 12–15, Chicago, Il, USA

Printed in the USA
CPSIA information can be obtained
at www.ICGtesting.com
LVHW060514070823
754480LV00006B/388